Synthetic Polymers

Synthetic Polymers

TECHNOLOGY, PROPERTIES, APPLICATIONS

Dorel Feldman

Concordia University, Montreal, Canada

and

Alla Barbalata

Université de Moncton, Edmundston, Canada

CHAPMAN & HALL

London · Weinheim · New York · Tokyo · Melbourne · Madras

Published by Chapman & Hall, 2–6 Boundary Row, London SE1 8HN, UK

Chapman & Hall, 2–6 Boundary Row, London SE1 8HN, UK

Chapman & Hall GmbH, Pappelallee 3, 69469 Weinheim, Germany

Chapman & Hall USA, 115 Fifth Avenue, New York, NY 10003, USA

Chapman & Hall Japan, ITP-Japan, Kyowa Building, 3F, 2-2-1 Hirakawacho, Chiyoda-ku, Tokyo 102, Japan

Chapman & Hall Australia, 102 Dodds Street, South Melbourne, Victoria 3205, Australia

Chapman & Hall India, R. Seshadri, 32 Second Main Road, CIT East, Madras 600 035, India

First edition 1996

Typeset in 10/12pt Times by Academic & Technical Typesetting, Bristol
Printed in Great Britain by St Edmundsbury Press Limited,
Bury St Edmonds, Suffolk

ISBN 0 412 71040 4

A catalogue record for this book is available from the British Library

Library of Congress Catalog Card Number: 96-83490

(∞) Printed on permanent acid-free text paper, manufactured in accordance with ANSI/NISO Z39.48-1992 and ANSI/NISO Z39.48-1984 (Permanence of Paper).

Contents

Preface

In order to produce polymers it is, of course, necessary to find ways in which small molecules, known as monomers, may join together in large numbers. This process is known as polymerization; there are a number of techniques in which the polymerization can be realized, but these can all be classed under the following two main types: chain reaction polymerization and step reaction polymerization. The simplest examples of chain reaction polymerization are the formation of polyethylene from ethylene and poly(vinyl chloride) from the corresponding monomer, the vinyl chloride. Polyamides, polyesters, polyurethanes, epoxy polymers and many others are examples of polymers obtained through step reaction processes.

The versatility of polymerization resides not only in the different types of reactants (monomers) which can be polymerized, but also in the variations allowed by copolymerization, stereospecific polymerization, ring opening polymerization, etc.

It is well known that the main technologies for polymer formation may be carried out with monomer alone (bulk or mass), in an organic solvent (solution), as an emulsion in water (emulsion), as droplets, with each droplet comprising an individual bulk polymerization suspended in water (suspension) and in a gas phase fluidized bed. All these techniques are commercially applied.

This book is designed as an introduction to the technology of polymer synthesis, with a survey of their properties and applications. It is intended for undergraduate and graduate students of chemical engineering, chemistry, engineering, material science and biochemistry. It will be useful to those engineers and professionals interested in how the basic materials for plastics, elastomers, synthetic fibers, adhesives, coatings, sealants, etc. are produced. Specifically, the book aims to broaden the perspective of specialists in different technical areas concerning the technology of polymer synthesis as well as their properties and applications. It is the authors' hope that this book will stimulate an interest in polymer technology and will provide the ground work for more comprehensive reading.

The book, organized in two parts, emphasizes the most recent technologies of the synthesis of well known commercial polymers. It is an outgrowth of the polymer course taught by the authors as well as research studies on different polymers and polymer technology aspects and their industrial experience.

The first part covers the technologies based on chain addition polymerization to obtain polymers such as polyolefins (polyethylene, polypropylene, polyisobutylene and butyl rubber), vinyl polymers (polystyrene, poly(vinyl chloride), poly(vinyl acetate), fluoropolymers, polyacrylics) and diene polymers (polyisoprene, polychoroprene, polybutadiene).

The second part deals with the commercial polymers produced through different types of step growth polymerization processes. Thus, phenolics and aminoplastics are obtained by polycondensation processes; polyesters and polyamides are mainly obtained by polycondensation and by ring opening polymerization, and polyurethanes and epoxies are prepared by step addition polymerization.

In the same family of polymers the different techniques which lead to polymers for plastics, fibers, elastomers or other macromolecular products, are discussed. Each technological flow chart is presented in a simple way with the intention to show the reader the principal operations in a certain technology and/or the most important equipment used.

Besides the principal technologies used for the synthesis of the above mentioned polymers, their main properties and applications are also discussed. It is also necessary to mention that it was not the authors' intention to cover the specialty polymers.

The authors wish to thank the following authors and publishers who allowed the use of data and illustrations from their publications:

Drs C.M. Blow, J.A. Brydson, C. Hepburn and J. Mijovic
American Chemical Society
Applied Science Publishers Ltd
ASM International
Butterworth-Heinemann Ltd
Carl Hanser Verlag
Chapman & Hall
McGraw-Hill Inc.
M. Dekker Inc.
Elsevier Science Publ. Ltd
Van Nostrand Publ. Co.
Taylor & Francis Inc.
The Plastics and Rubber Institute
Vulkan Verlag GMBH
J. Wiley & Sons Inc.

Acknowledgements are also addressed to Dr P. Hastings, Senior Editor at Chapman & Hall, and his team for their cooperation during our work. Special thanks go to our families for encouragement and support.

D. Feldman (Montreal)
A. Barbalata (Edmundston)
July 1995

Abbreviations

AA	acetoacetanilide
ABS	acrylonitrile–butadiene–styrene terpolymer
AIBN	azo-bis-isobutyronitrile
AN	acrylonitrile
B	butadiene
BAN	butadiene–acrylonitrile copolymer
BDA	benzyldimethylamine
BDO	1,4-butandiol
BMC	bulk molding compounds
BP	benzoyl peroxide
BPA	bis-phenol A or 2,2′-(4-hydroxyphenyl) propane
CFC-11	trichlorofluoromethane
CGE	*p*-cresyl glycidyl ether
Cp	cyclopentadienyl
C_p	specific heat
CPE	chlorinated polyethylene
CR	controlled rheology
DABCO	triethylenediamine
DAIP	diallyl isophthalate
DBPIB	α, α'-di(*tert*-butylperoxy)-*m*/*p*-diisopropyl benzene
DBPH	2,5-dimethyl-2,5-di(*tert*-butylperoxy)hexane
DETDA	3,5-diethyltoluene diamine (2,4- and 2,6-isomers)
DGEBA	diglycidyl ether of bis-phenol A
DGEEG	diglycidyl ether of ethyleneglycol
DMAMP	dimethylaminomethylphenol
DMA	dynamic mechanic analysis
DMF	dimethylformamide
DMS	dynamic mechanical spectroscopy
DMSO	dimethylsulfoxide
DSC	differential scanning calorimetry
EAA	ethylene–acrylic acid copolymer

EC	ethylene carbonate
ECTFE	poly(chlorotrifluoroethylene-co-ethylene)
ECPE	extended chain polyethylene fibers
EDTA	ethylenediamine tetra-ethylacrylate copolymer
EEA	ethylene–ethyl acrylate copolymer
EG	ethylene glycol
EMA	ethylene–methylacrylate copolymer
EPDM	ethylene–propylene–diene terpolymer
EPR	ethylene–propylene rubber
Et	ethyl
ETFE	poly(tetrafluoroethylene-co-perfluoropropyl vinyl ether)
EVAc	ethylene–vinylacetate copolymer
FEP	poly(tetrafluoroethylene-co-hexafluoropropylene)
Flu	Fluorenyl
FTIR	Fourier transform infrared spectroscopy
HA	4-hydroxyacetanilide
HDI	1,6-hexamethylene diisocyanate
HDPE	high density polyethylene
HEN	hydrocarbon epoxy novolac
HER	1,3-di(2-hydroxyethyl) resorcinol
HIPS	high impact polystryrene
H_{12}MDI	4,4′-dicyclohexylmethane diisocyanate
HMTA	hexamethylenetetramine
HMWPE	high molecular weight polyethylene
HQEE	1,4-di(2-hydroxyethyl) hydroquinone
HR	high resiliency
H_6XDI	1,3-bis(isocyanatomethyl) cyclohexane
I	isoprene
IMR	internal mold release
Ind	indenyl
IPDI	isophorone diisocyanate
IPN	interpenetrating polymer network
IPP	diisopropyl peroxidicarbonate
LCP	liquid crystal polymers
LDPE	low density polyethylene
LLDPE	linear low density polyethylene
LP	lauroyl peroxide
MDA	4,4′-diaminodiphenyl methane
MDI	4,4′-diphenylmethane diisocyanate
MFI	melt flow index
MMA	methyl methacrylate
M_n	number-average molecular mass
MOCA	4,4′-methylene-bis(2-chloroaniline)
MW	molecular mass

MWD	molecular mass distribution
NBR	butadiene–acrylonitrile copolymer, nitrile rubber
NDI	1,5-naphthalene diisocyanate
NMR	nuclear magnetic resonance
PAA	poly(acrylic acid)
PAI	poly(amide-imide)
PAN	poly(acrylonitrile)
PB	polybutadiene
PBT	poly(butylene terephthalate)
pbw	parts by weight
PC	polycarbonate
PCTFE	polychlorotrifluoroethylene
PE	polyethylene
PEI	poly(ether-imide)
PEO	poly(ethylene oxide) glycol
PEP	polyether polyol
PESP	polyester polyol
PET	poly(ethylene terephthalate)
PF	poly(phenol–formaldehyde) polymers
PFA	poly(tetrafluoroethylene-co-perfluoropropyl vinyl ether)
PGE	phenyl glycidyl ether
phr	per hundred
PI	polyisoprene
PMMA	poly(methylmethacrylate)
PMDI	polymeric diphenylmethane diisocyanate
PMP	poly(4-methyl-1-pentene)
PP	polypropylene
PPO	poly(propylene oxide) glycol
PTFE	polytetrafluoroethylene
PTHF	poly(tetrahydrofuran)
PTMG	poly(tetramethylene oxide) glycol
PU	polyurethane
PVA	poly(vinyl alcohol)
PVAc	poly(vinyl acetate)
PVC	poly(vinyl chloride)
PVDF	poly(vinylidene fluoride)
PVF	poly(vinyl fluoride)
PyHGC	pyrolysis–hydrogenation gas chromatography
RIM	reaction injection molding
RRIM	reinforced reaction injection molding
RTV	room temperature vulcanizing
S	styrene
SAM	styrene–methylacrylate copolymer
SAN	styrene–acrylonitrile copolymer

SB	styrene–butadiene copolymer
SBR	styrene–butadiene rubber
SBS	styrene–butadiene–styrene block copolymer
S–DVB	styrene–divinyl benzene copolymer
SEC	size exclusion chromatography
SIS	styrene–isoprene–styrene block copolymer
SM	methyl methacrylate–styrene copolymer
SMC	sheet molding compounds
TDI	toluene diisocyanate
TDMAMP	tri(dimethylaminoethyl) phenol
TEA	triethylamine
T_g	glass transition temperature
T_m	melting temperature
TMC-BP	trimethylcyclohexanone-bis-phenol
TMP	trimethylolpropane
TPE	thermoplastic polyester elastomer
TPU	thermoplastic polyurethane
TREF	temperature-rising elution chromatography
UF	urea–formaldehyde resins (aminoplasts)
UHMWPE	ultrahigh molecular weight polyethylene
UV	ultraviolet
VAc	vinyl acetate
VC	vinyl chloride
VCD	vinylcyclohexane dioxide
VLDPE	very low density polyethylene

PART ONE

Chain Growth Polymers

Polyolefins

1

1.1 POLYETHYLENE (PE)

1.1.1 Introduction

Polyethylenes are olefin polymers manufactured in the largest tonnage of all the thermoplastic materials. They are produced by the addition polymerization of ethylene to give the basic structure $-(CH_2-CH_2)_n-$. Based on the constitutional repeating unit in the structure, the name of polymethylene is recommended by the IUPAC Commission on Macromolecular Nomenclature for this family of polymers. This name has already been used before for the strictly linear polymer obtained from diazomethane $-(CH_2)_n-$. In common use, the term 'polyethylene' is still attributed to the polymers obtained by the polymerization of ethylene and will be used in this text.

Several well established families of polyethylenes can now be found on the market, each having a different structure and very different behaviors, performances and applications:

- low density polyethylene (LDPE) 0.915–0.935 g/cm^3
- high density polyethylene (HDPE) 0.941–0.967 g/cm^3
- linear low density polyethylene (LLDPE) 0.910–0.925 g/cm^3
- very low density polyethylene (VLDPE) 0.880–0.912 g/cm^3
- high and ultrahigh molecular weight polyethylene, 200 000–500 000 for HMWPE and 3 000 000 for UHMWPE, respectively.

Other categories of PE polymers are crosslinked PE, chlorosulfonated PE and the copolymers of ethylene.

In 1994 the global demand for polyethylenes of all types was of the order of 37 million metric tons per year [1] of which 8.2 million metric tons was produced in the US alone [2]. LDPE and HDPE each account for 40.5% in the global production and LLDPE for 19% [1].

The sources of ethylene and of other olefins are petroleum and natural gas [3]. Petroleum is first separated by distillation into broad fractions which are subsequently 'cracked' by heating (thermal cracking or pyrolysis). High yields of olefins are obtained and they are further separated into individual products by fractional distillation. Natural gas, the second major source of ethylene, contains methane (60–99%) and C_{2-5} paraffins which are converted to olefins by thermal cracking. Carbon monoxide, acetylene, oxygen, moisture and other impurities contained in raw ethylene must be carefully removed as they affect the polymerization process and the properties of the PE.

The first commercial production of PE started in 1939 at Imperial Chemical Industries Ltd in the UK and was based on the high pressure polymerization of ethylene experimented by Fawcet and Gibson. It was found that the new material had excellent electrical insulation properties; the first application was as underwater cable insulator and it also greatly aided in the development of radar. The needs of World War II significantly accelerated research and further developments in the field of PE [4].

In 1955, a plant using the Ziegler catalyst polymerization process began production at Farbwerke Hoechst AG Germany, almost simultaneously with Phillips Petroleum Co. (1957) and Standard Oil Co. (1961) of Indiana, both in the USA. These three processes were all based on the catalytic polymerization of ethylene in solution or in slurry at low pressures. The prepared PEs had higher densities and were more linear in structure than those produced under high pressure. They became known as high density polyethylene (HDPE) and the older material as low density polyethylene (LDPE).

In 1968, Union Carbide Corp. of USA developed a new low pressure polymerization process producing HDPE in the gas phase that required no solvents. This process led to the next major stage of development, the advent in 1977 of low density polyethylene produced by a low pressure, gas phase process and known as linear low density polyethylene (LLDPE). The production was less expensive and had lower energy demands than the conventional high pressure process [5].

The last twenty years have been marked by a huge research effort and accumulation of scientific knowledge on catalyst systems and by new developments of super high activity Ziegler catalysts and especially of the new class of metallocene catalysts. They have brought about not only improved efficiencies of the polymerization processes, but the manufacture of a new generation of tailor-made PE with an unprecedented control of MW, MWD and of comonomer incorporation [6,7]. In 1989, Exxon Chemical unveiled the so called Exxpol technology. They produced a wide range of PE grades with densities from 0.865 to 0.935 g/cm^3 by solution polymerization of ethylene with 'single site catalysts', a new class of catalysts of the metallocene family [7]. At present, resins prepared with single site catalysts are produced on a precommercial scale and around 20 companies are examining the possibility of using metallocene catalysts for the production of polyolefins.

Many application areas were found for PEs such as electrical insulators, packaging films, different domestic goods, toys, and in chemical plants. Each particular application requires the best resin choice which implies the knowledge of the fundamental properties of the polymers such as the melt index, the density, the MW, the polydispersity (MWD) and the degree of branching [8].

1.1.2 Low density polyethylene (LDPE)

LDPE is prepared by a typical free radical polymerization process at high temperatures and extremely high pressures. Azoisobutyronitrile (AIBN), benzoyl peroxide (BP) or oxygen (10–100 ppm) are free radical initiators commonly used. The free radical polymerization of ethylene has a critical dependence on the monomer concentration: growing polymer radical have a very limited life to react with the monomer and high monomer concentrations are necessary to obtain high MW polymer. To meet this requirement, the process is operated at very high pressures in the range of 150–350 MPa (1500–3000 atmospheres) in order to maintain the necessary concentration of monomer and to control the reaction rate [9]. The temperature is in the range 80–300°C, the lower the temperature (with active initiators only) the higher the MW.

Under these conditions, ethylene is above its critical pressure and temperature (of 9.7°C) and is in the gas phase. The polymerization is highly exothermic, the heat of polymerization of ethylene in the gas phase is 92–96 kJ/mol (3.35 kJ/g) and the temperature would rise by about 12–13°C for each 1% of polymerized ethylene (the specific heat of ethylene at these conditions is 2.5–2.8 J/g·°C). If the reaction temperature becomes too high, exothermic decomposition reactions of ethylene (142 kJ/mol) may occur leading to a mixture of methane, hydrogen and carbon and to increasing explosive pressure. Very efficient temperature control must be maintained to prevent runaway reaction. High cooling surface/volume ratio reactors are used as well as different inert liquid diluents such as benzene or water to dilute the reaction heat. For the same reason the initiators are sometimes dissolved in a solvent and are injected in the reactor at different points that minimize high local concentrations of the initiator in the reactor [10].

The process is operated continuously and two main types of reactor are in use, namely, a narrow bore tubular reactor and a stirred autoclave. The tubular reactor has obvious advantages at high pressures. Because there is no mechanical agitation, a plug flow through the tube is produced by continuously pumping the ethylene into the tube and by releasing the product periodically from the other end [11]. This helps to eliminate the buildup of polymer on the walls.

In a typical process, 10–30% monomer is converted per cycle to polymer. The obtained polymer is dissolved in ethylene and the whole resembles

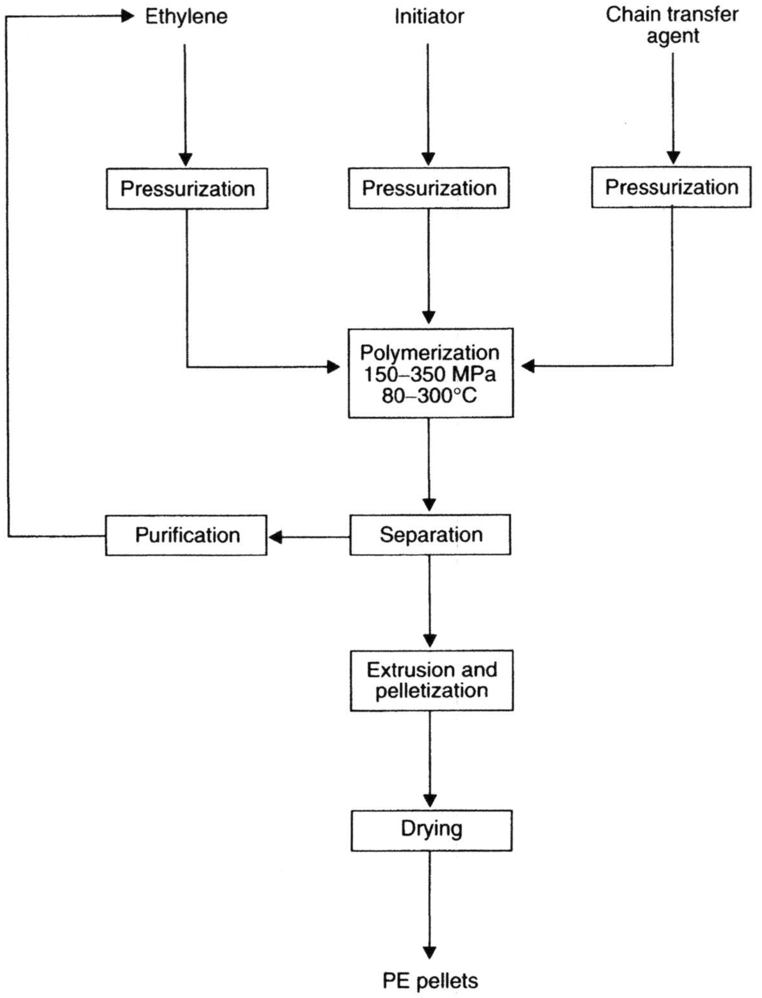

Figure 1.1 Continuous high pressure polymerization of ethylene.

conventional bulk polymerization. The molten polymer is separated from unreacted ethylene and from diluent by releasing the pressure, and then is extruded into a ribbon and granulated. Unreacted ethylene is purified and recycled.

A flow chart of the high pressure continuous polymerization of ethylene is presented in Figure 1.1.

Under the vigorous conditions of high temperature and pressure, side reactions occur leading to chain branching that determines the properties of LDPE. Both short and long branches are formed [12]. The presence of 20–30 short chain branches per 1000 carbon atoms in the main chain was evidenced by ^{13}C NMR, FTIR and pyrolysis–hydrogenation gas chromatography (PyHGC) [13–16]. They are mainly butyl branches, but also ethyl, methyl and amyl. Using size exclusion chromatography (SEC) and temperature-rising elution chromatography (TREF) combined with the methods mentioned above, it was found that the lower MW species have more short chain branches [12]. The level of short chain branching can be modified by the adjustment of pressure and temperature.

As for the long branches, there are on the average about three branches per 1000 carbon atoms and they have a range of lengths with an upper limit approaching the length of the main polymer chain. Long branches have been evidenced by combining SEC fractionation of the polymer with subsequent solution viscosity, light scattering and sedimentation measurements [17] and more recently with low angle laser light scattering [18] and ^{13}C NMR measurements [16]. The distribution of long chain branches with the MW depends strongly on the polymerization conditions [12].

The microstructure of LDPE may be represented as:

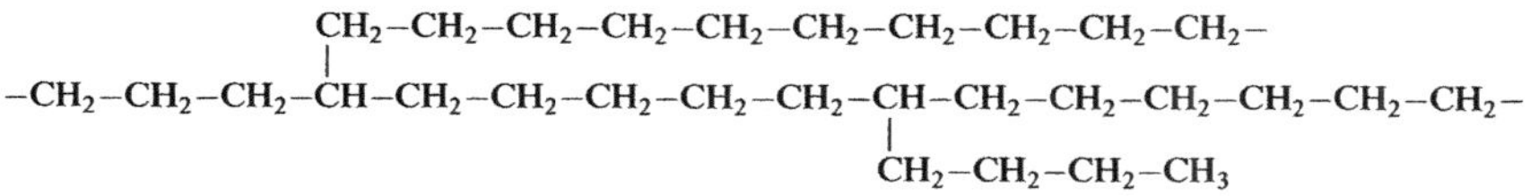

See also Figure 1.5.

The formation of branches is due to chain transfer reactions that occur both by intramolecular and intermolecular mechanisms [19]. Short chain branches are formed during the propagation stage by a 'backbiting' intramolecular mechanism. The end of the chain coiled backward extracts a hydrogen radical from a carbon atom in the chain, a six-membered ring transition state being favored. This results in butyl short branches. The propagation of the chain is thus transferred to the carbon atom from which the hydrogen was removed and the polymerization can continue in the normal way:

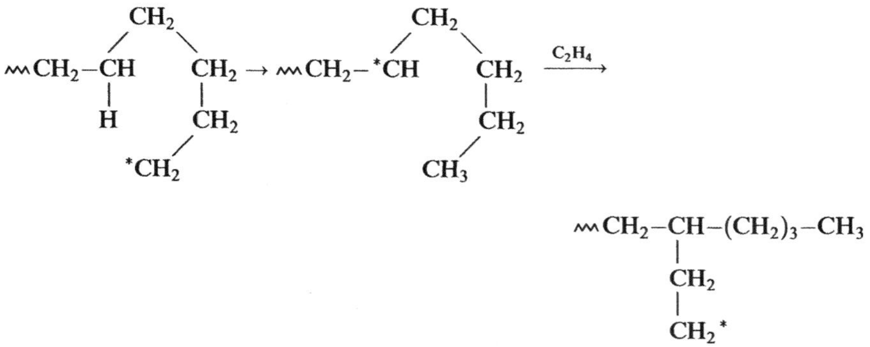

A second backbiting may occur either to (a) the butyl (Bu) group leading to ethyl (Et) side branches or (b) again to the main chain resulting in more branched side chains:

(a) 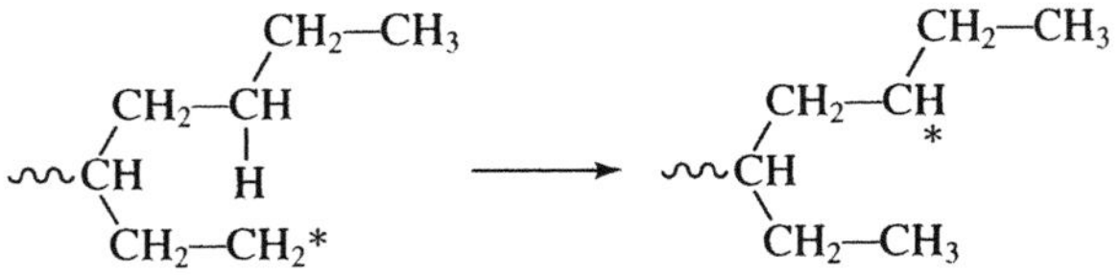

(b) 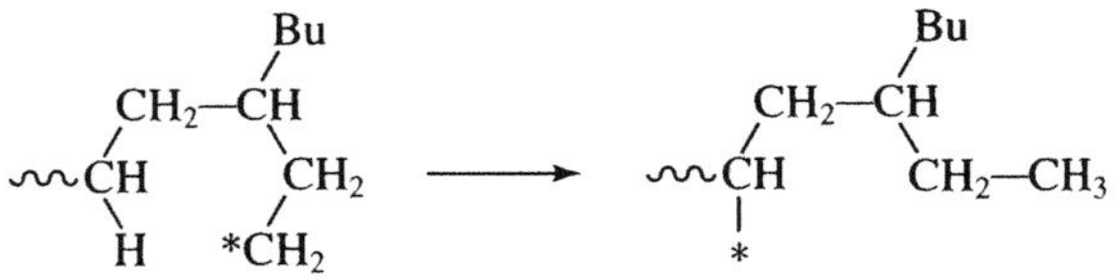

A third backbiting is also possible with the formation of a tertiary radical which could depolymerize by β-scission. This generates vinylidene groups which have been observed and found to provide about 50% unsaturation in LDPE, the rest being about evenly divided by vinyl and in-chains double bonds (there may be up to about three double bonds per 1000 carbon atoms).

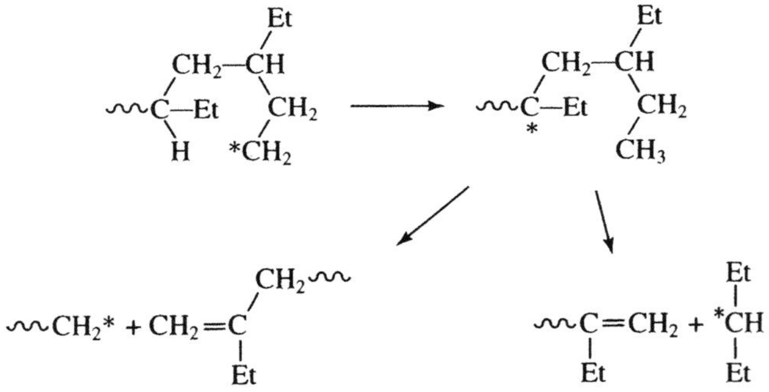

Long chain branches are formed through intermolecular transfer which arises when a growing chain extracts a hydrogen radical from a dead polymer molecule on which the propagation will continue leading to a long branch:

$-CH_2-CH_2^*$	+	$H-P$	$\longrightarrow$	$-CH_2-CH_3$	+	P^*_-
Propagating chain		Dead polymer		Dead polymer		Macro radical

$$P^*_- + CH_2{=}CH_2 \rightarrow P{-}CH_2{-}CH_2^* \xrightarrow{+CH_2=CH_2} \text{etc.}$$

On a number-average basis, a typical low density polyethylene may contain one long branch for 50 short branches. The distribution of long branches is very uneven, most of the branches are concentrated on a few very large molecules, while many molecules contain no long branches.

Chain transfer agents such as light paraffins, propylene or hydrogen are frequently employed to adjust the MW to the desired values as they terminate propagating polymeric chains and start other chains by an intermolecular chain transfer mechanism.

The characteristics of LDPE, such as branching and its distribution, MW and MWD, may be varied over a wide range by changing the conditions of the process (temperature, pressure) and the nature, the composition and the amount of initiator and of chain transfer agent.

Proprietary mathematical models have been developed for the operation of reactors used for the production of LDPE. They are very complex and serve to optimize the processes for desired types of PE.

A major trend in the high pressure technologies is their continued decline in favor of low pressure, gas phase processes. This is primarily due to the high operating costs and lack of product flexibility [6].

1.1.3 High density polyethylene (HDPE)

HDPE is produced by three basic types of process. They are all catalytic processes run at relatively low pressures. The details on methods of preparing the catalysts are closely guarded secrets.

In some of these processes a solvent is employed but, at the process temperature, the resultant polymer is not soluble so that a slurry of granular PE is formed. The slurry-type processes are the most important ones for the production of HDPEs. On the other hand, in the true solution processes, a higher temperature is employed permitting a complete dissolution of the PE in the solvent. The third type is a gas–solid process in which the powder of solid PE that forms is suspended in the gaseous ethylene.

HDPE can be broadly divided into two categories: high MW bimodal/broad MWD polymers used mainly for blow molding and pipe, and low MW grades with narrow MWD used for injection and rotomolding [6].

(a) Ziegler process

In the Ziegler process, the polymerization proceeds by a coordination mechanism which involves a catalyst–monomer coordination complex that controls the way in which the monomer approaches the growing chain.

The coordination catalyst is a complex formed by the interaction of (a) a compound of a transition metal in Groups IV–VIII of the Periodic Table (the catalyst) with (b) an organometallic compound of Groups I–III metals (the cocatalyst). The ligands in the catalyst compounds can be halide, oxyhalide, alkoxy, acetylacetonyl and phenyl. The metals in the catalyst are commonly titanium, vanadium, chromium, molybdenum, hafnium or zirconium. The cocatalysts are hydrides, alkyls or aryls of metals such as aluminum, lithium, zinc, tin, cadmium, beryllium or magnesium. One of the most common Ziegler–Natta catalysts is the complex $TiCl_4$–AlR_3 where R = alkyl.

Designing the chemical composition and the morphology of the heterogeneous catalyst particles is of paramount importance for both the polymerization efficiency and the polymer properties. New generations of super high activity catalysts have been prepared as result of the scientific understanding of their complex structure and intimate role in the polymerization. These catalysts are based on $MgCl_2$-supported titanium compounds in combination with trialkyl aluminum cocatalyst and, in some cases, with an electron donor [20,21]. The $MgCl_2$ support maximizes the number of active sites on the catalyst surface. Catalysts based on a single transition metal are used for polymers with narrow MWD, while those based on two or more transition metals lead to polymers with bimodal/broad MWD which are easier to process; each metal forms polymers of average MW that differ from each other by more than one order of magnitude [22].

Of the various mechanisms that have been proposed, the two that are generally accepted are the so called monometallic and bimetallic mechanisms, the former being favored in heterogeneous processes [23]. In both mechanisms, the monomer is pictured as being incorporated into the polymer by an insertion reaction between the transition metal atom and the terminal carbon of the coordinated polymer chain:

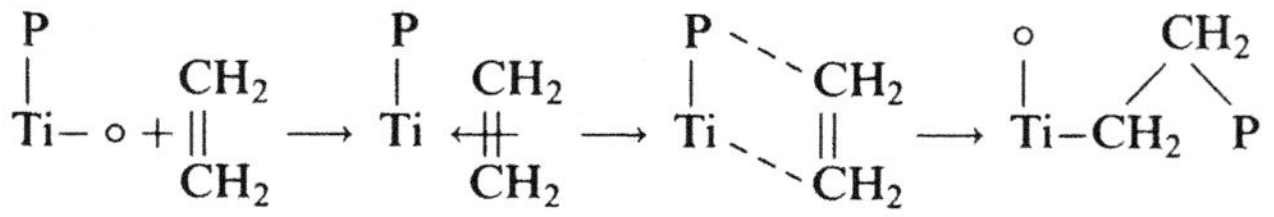

where P = polymer chain; o = vacant coordination site.

A very large number of patents claim different ways of catalyst preparation. Basically, the catalyst may be prepared in two general ways both conducted in an inert atmosphere (since oxygen and water reduce its effectiveness): (a) *in situ* by feeding the components separately and directly into the main reactor as solutions in diluents (heptane, toluene) or (b) by preparing the catalytic complex separately and then adding the resulting slurry to the reactor.

Loop reactors and autoclave reactors are both used for slurry processes.

Polymerization is conducted continuously by introducing into the reactor a liquid hydrocarbon diluent, the catalyst and the ethylene. The pressures are slightly above atmospheric, 0.2–0.4 MPa (2–4 atmospheres) and the temperatures are 50–75°C. In these conditions, the polymer is insoluble and forms a slurry which progressively thickens as the reaction proceeds and is continuously removed from the reactor.

Each catalyst particle forms one polymer particle. During the growth of polymer particles, the catalyst particle disintegrates and the fragments become dispersed throughout the whole polymer particle. With less efficient catalyst systems, the decomposition of the catalyst is very important particularly where the polymer is intended for use in electrical insulation. The catalyst is destroyed by the action of an alcohol (methanol, ethanol or isopropanol) and then extracted by alcoholic hydrochloric acid. By using extremely active catalysts, so little catalyst residue is in the polymer that the onerous step of removing the catalyst is no longer necessary. As an example, a typical $TiCl_4$–AlR_3 catalyst yields about 20–200 g of PE per gram of catalyst per hour per atmosphere of ethylene, whereas as much as 22–25 kg PE per gram may be produced using $MgCl_2$-supported catalysts [7,23]. Values as low as 2–4 ppm of titanium have been reported in some PEs [10]. The polymer is finally separated by centrifugation and then dried, extruded and granulated.

Another improvement in the economics of the polymerization process brought about by the new $MgCl_2$-supported catalysts is the possibility to eliminate the pelletizing step in the manufacture of the polymer. The polymer particle morphology depends on catalyst particle morphology, namely its shape, size, porosity, texture of particles and homogeneous distribution of active centers both in the core and on the surface layer (replication phenomenon). Tightly tailoring catalyst morphology makes it possible to control the morphology of the polymer particle in size, shape and bulk density, and thus avoids the pelletizing step [20].

The MW of the polymer may be controlled by the polymerization temperature, by the ratio Al–Ti in the catalyst or by using chain transfer agents. Hydrogen is the most commonly used chain transfer agent. A new multistep technology based on successive different hydrogen concentrations is used in order to broaden the MWD and thus to improve the processability of certain PE grades.

Ziegler HDPEs have densities of about 0.945 g/cm^3 and branching of the order of five to seven ethyl groups per 1000 carbon atoms. Butyl and other branches seem to be absent.

A typical flow chart of the process is presented in Figure 1.2.

(b) Phillips process

This process is based on the use of a heterogeneous catalyst consisting of transition metal compounds (chromium oxides). A finely divided support of silica or silica–alumina (75–90% silica) is impregnated with an aqueous solution of a chromium compound and heated at 400–800°C. The resulting catalyst contains 5% chromium oxides, mainly CrO_3. Very pure reactants must be employed as a number of impurities such as oxygen, nitrogen, acetylene and chlorine are catalyst poisons.

Two main variations of this process are known: the process may be conducted in solution or in a slurry depending on the temperature. The reaction temperature is highly important for controlling the MW of the PE. In the solution process, reactors are operated at 130–160°C, which is above the melting point of the PE and which makes the polymer more soluble. The solvent is a liquid hydrocarbon such as cyclohexane which dissolves the PE as it is formed and also serves as a heat transfer agent. In the slurry variant

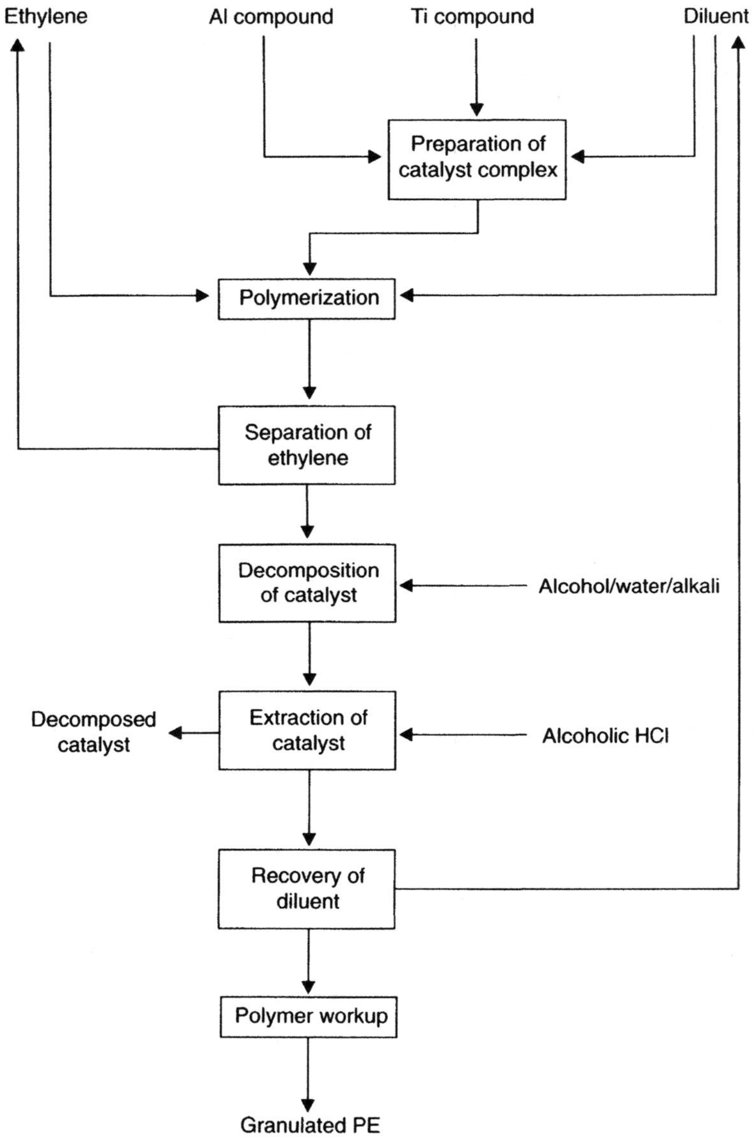

Figure 1.2 Ziegler process for ethylene polymerization.

of the process, temperatures are 90–100°C, which is below the crystalline T_m of the PE and so its solubility in the diluent is low.

Pressures are generally 1.4–3.5 MPa (14–35 atmospheres) which are between those used in high pressure and in Ziegler processes.

After reaction, the mixture is continuously withdrawn from the reactor into a gas–liquid separator where the unreacted ethylene is flashed off.

From the hot solution the catalyst is removed by filtration and the polymer is separated from the solvent either by flashing off the solvent or by precipitating the polymer by cooling. In the slurry process, after flashing off the ethylene, the polymer is separated by centrifugation.

In the slurry process, the granules of PE are each formed around individual catalyst particles which remain in the polymer. To reduce the level of contamination of the polymer, the polymerization is conducted to high conversion rates [10]. Higher MW polymers are prepared by the slurry method.

HDPEs prepared by this process have the highest density (about 0.96 g/cm^3) of any commercial PE and are almost completely linear with up to three methyl groups per 1000 carbon atoms and no ethyl or butyl groups detected.

The flow chart of the Phillips process is presented in Figure 1.3.

(c) Standard Oil Company (Indiana) process

Many similarities are found between the Phillips and Standard Oil processes. The Standard Oil process is conducted preferably in solution in the presence of a metal oxide catalyst at 200–300°C and 4–10 MPa (40–100 atmospheres). The catalyst may be molybdenum trioxide, titanium dioxide or zirconium dioxide in combination with sodium or calcium hydrides as promoters and supported on alumina [24]. The catalyst must be activated before use by reducing it with hydrogen or with a metal hydride (in the Phillips process, the catalyst is activated by the ethylene itself). A hydrocarbon solvent is used as the reaction medium, serving both to dissolve the polymer and to remove the heat of reaction. The same marked effect of the temperature on the MW of the PE as in the Phillips process is evident.

Polymerization mechanisms by metal oxides are not well understood. The polyethylenes obtained have densities of about 0.96 g/cm^3.

(d) Union Carbide gas phase process

HDPE is produced by a gas phase process developed at Union Carbide and known as Unipol process [25,26]. Ethylene serves at the same time as fluidizing gas and as reactant. No solvent is employed and hence there are no solvent recovery costs, the consumption of energy is lower and the process is simpler to operate.

The reaction is carried out at 75–100°C and 0.7–2 MPa (7–20 atmospheres). The Unipol technology employs highly effective transition metal catalysts of the Phillips-type (chromium oxide) or Ziegler-type (mostly titanium). Phillips catalysts consist of bis-tri-phenylsilyl chromate or of chromacene, $(C_5H_5)_2Cr$, on dehydrated silica support. The granules of PE

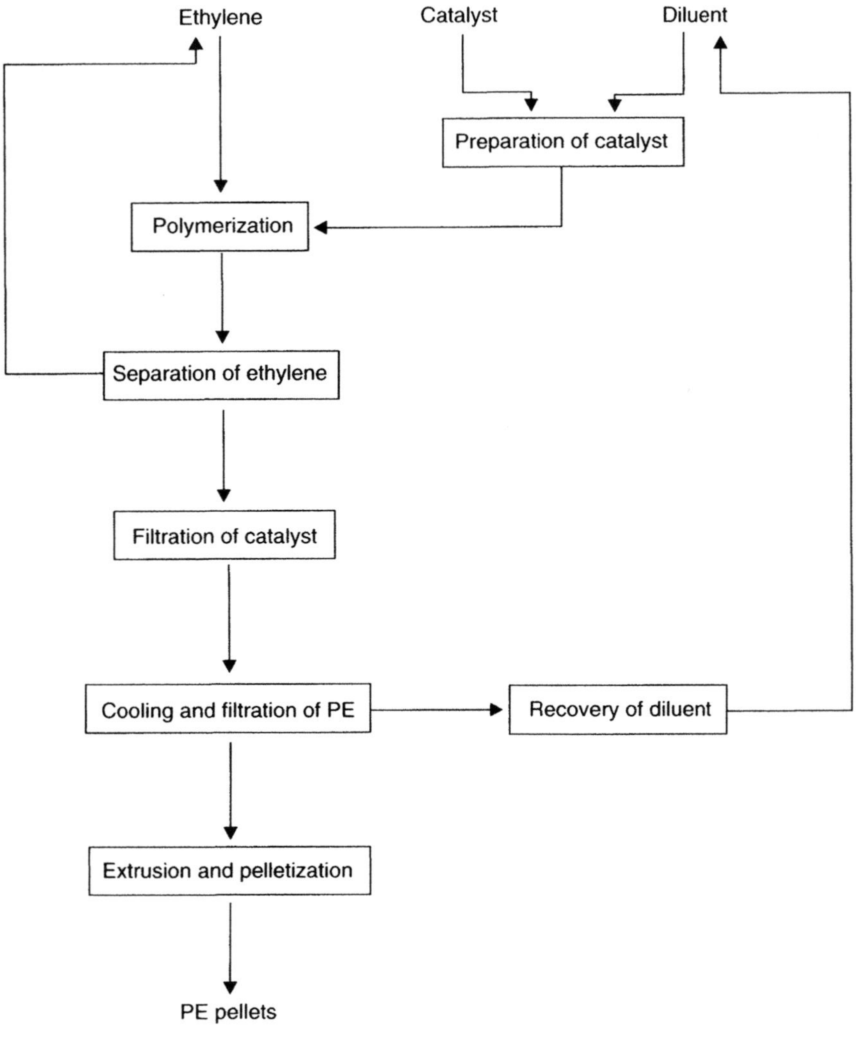

Figure 1.3 Phillips process for ethylene polymerization.

formed on the catalyst particles are in fluidized bed. Only about 2% of ethylene is polymerized at each pass and large flows are to be recycled but the overall conversion of ethylene is 97% [10].

PEs produced by this process have densities of 0.94–0.96 g/cm^3. They can be obtained with any desired MW and MWD using hydrogen as the MW regulator and multistep process technology where bimodal or broad MWD are required.

Since only small amounts of very effective catalysts are used in the process, the PE is not separated from the catalyst.

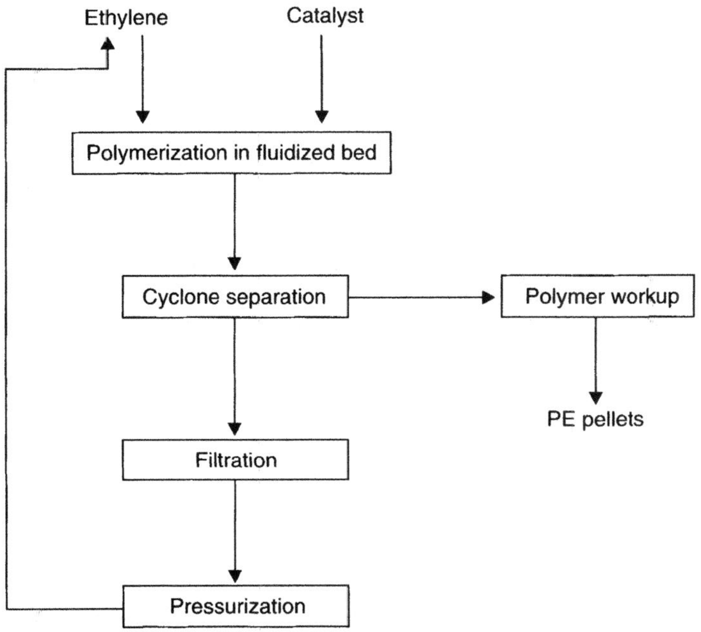

Figure 1.4 Gas phase process for ethylene polymerization.

Additional treatment for PE is used in conjunction with the main reactor. It consists of the incorporation of slip, antiblock and antioxidant additives into the granular particles of PE and thus the need for pelletizing is eliminated [10].

The flow chart of the gas phase process for the production of HDPE is presented in Figure 1.4.

Of all the methods for manufacturing PEs, the gas phase process is overwhelmingly preferred and still in expansion, due to its low cost and product flexibility (or 'swing' capacity) [6].

(e) Metallocene-based process

The newest type of catalysts for the polymerization of olefins has variously been termed Kaminsky-type, single-site, constrained-geometry or metallocene catalysts. The novelty of these soluble catalyst systems is that they allow a very precise control of the uniformity of chain length, degree of branching and stereoregularity leading to tailor-made polymers with predictable properties. In copolymerization processes, these catalysts show the capability of a rigorous control of the incorporation of comonomers into the polymer.

Metallocenes are extremely efficient catalyst systems, allowing yields in excess of one ton of polymer per gram of catalyst per hour [7]. Efficiencies as high as 25 tons of PE per gram have been reported [27].

Hundreds of published papers and patents describe an extremely large number of metallocene complexes in the polymerization of olefins but their detailed presentation is beyond the frame of this book. Various methodologies have been utilized to synthesize metallocenes with well defined electronic and steric structures [28].

In principle, a metallocene derivative has the formula L_2MX_2, where M is a Group IV transition metal such as titanium, zirconium or hafnium; X may be a halogen or an alkyl, phenyl, benzyl or a trimethylsilyl group; and L is a ligand associated to the metal atom by π-bonds. The ligands may be the cyclopentadienyl (Cp), the indenyl (Ind) or the fluorenyl (Flu) anions. The ligands may be unsubstituted (I) or may bring one or several substituents (II). The ligands may be independent of each other in non-stereorigid metallocenes (I and II) or may be linked by an ethylene or isopropylene bridge in stereorigid metallocenes (III). Some examples are given below:

Cyclopentadienyl (Cp)

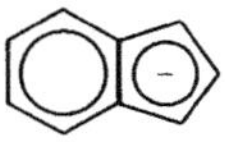

Indenyl (Ind)

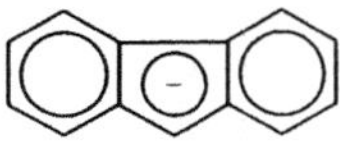

Fluorenyl (Flu)

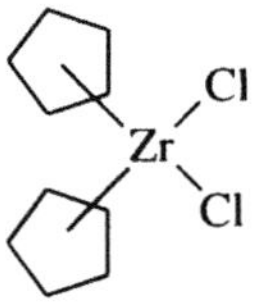

I

Bis(cyclopentadienyl) zirconocene dichloride Cp_2ZrCl_2

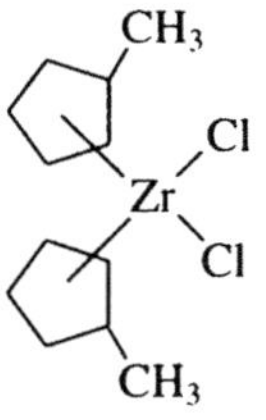

II

Bis(methylcyclopentadienyl) zirconocene dichloride $(MeCp)_2ZrCl_2$

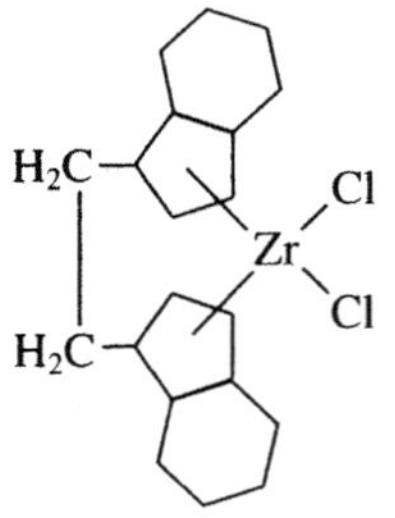

III

Ethylene bis(indenyl) zirconocene dichloride $Et(Ind)_2ZrCl_2$

Two categories of metallocene complexes have been used, bicomponent systems and single component systems. Besides the metallocene, the bicomponent complexes contain an alumoxane compound as cocatalyst. There are some uncertainties concerning the structure and the role of the alumoxane despite the numerous physicochemical measurements. Methyl alumoxane $(AlMeO)_n$ contains alternating aluminum and oxygen atoms and it is considered to be an oligomeric chain or ring compound whose molecules are associated in supramolecular aggregates [29]:

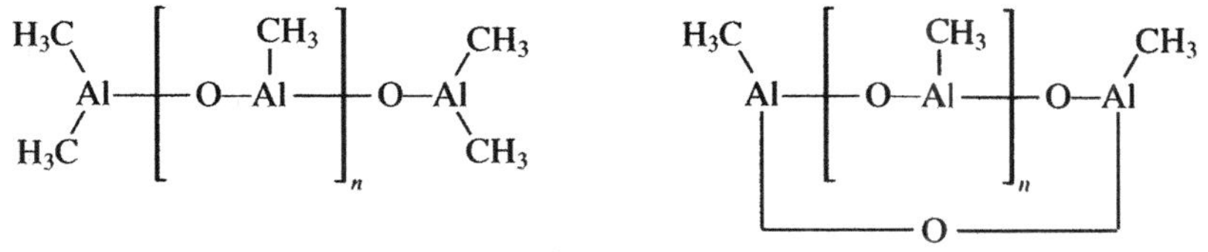

The alumoxane acts as a soluble carrier–activator producing and stabilizing metallocene cations $[Cp_2ZrCH_3]^+$ that are actual active species [30,31]

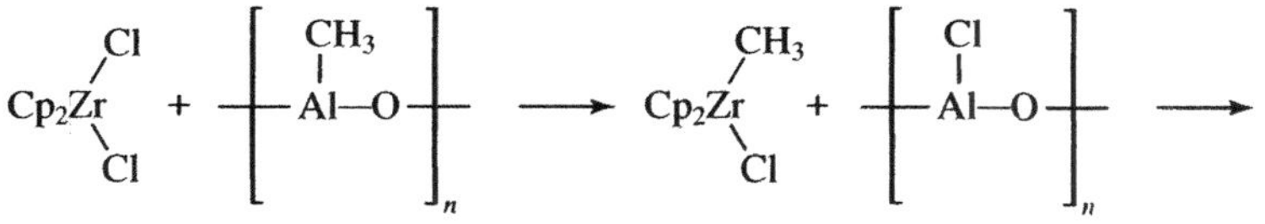

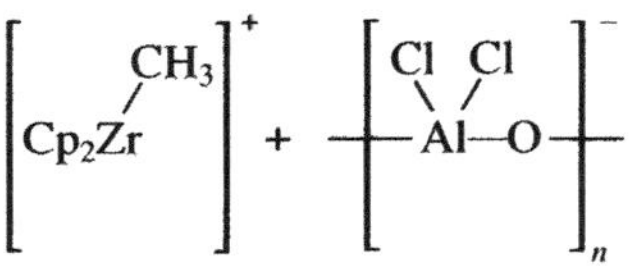

During the polymerization, the monomer is coordinated with the highly electrophilic and coordinatively unsaturated cationic complex followed by the insertion of the monomer into the metal–carbon bond that occurs through a four-center transition state. The polymer chain migrates and a vacant coordination site is reformed at the site originally occupied by the polymer chain [32]:

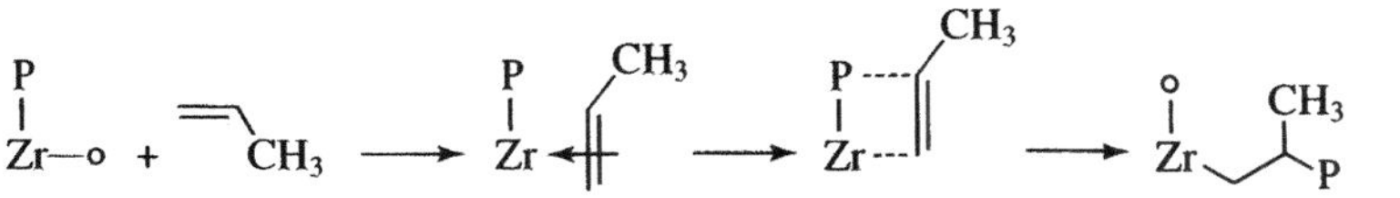

where P = polymer; o = vacant coordination site.

In metallocene catalyzed olefin polymerization, the termination of the polymer chain may occur through chain transfer to aluminum, to monomer or to hydrogen or to β-H or β-methyl eliminations depending on the conditions [28].

A judicious balance of electronic and steric effects is necessary for designing metallocenes of high productivity and strict control on the MW and MWD. The productivity in the homopolymerization of ethylene decreases by changing the metal in the order Zr > Ti > Hf. In the metallocene Cp_2ZrR_2 the nature of the sigma ligand, R, increases the productivity and the MW in the order $R = Me < Ph < CH_2Ph < CH_2SiMe_3$ [33]. Monosubstituted cyclopentadienyl ligands with electron donating groups enhance the productivity whereas polysubstituted and steric crowded cyclopentadienyls lower the productivity. Monomodal or bimodal MWD = 5.5–10 have been obtained by changing the composition of the metallocene complex [28].

Factors such as the composition of the cocatalyst, i.e. the ratio methylaluminum/alumoxane as well as the ratio catalyst/cocatalyst, influence the productivity, the MW and the evolution of the rate of polymerization in time (kinetic profile). A minimum ratio Al/Zr of 3000 represents a threshold value for the activity of the catalyst [34].

An optimum temperature for maximum productivity has been found for each metallocene system. The yield of PE increases between 0 and 40°C but higher MWs are obtained at lower temperatures. Temperatures between 40 and 60°C lead to bimodal distributions. Hydrogen is used as a MW regulator. Toluene has been found to be the best solvent for maximum catalytic activity of the metallocene.

Supported metallocenes are under study in order to control polymer morphology better [28]. They require several-fold less alumoxane cocatalyst compared to unsupported metallocenes. The silica and $MgCl_2$ support help to adsorb Cp_2MR^+ active species and thus to prevent their precocious deactivation [35].

The second category of metallocenes are the single-component systems which are ionic complexes with the general formula $[Cp_2MR]^+[B(C_6F_5)_4]^-$. They polymerize olefins without the utilization of alumoxane cocatalyst. The active species is the metallocene cation $[Cp_2MR]^+$ which is stabilized by the use of counteranions such as tetraphenylborate, $B(C_6H_5)_4^-$ or fluorinated tetraphenylborate, $B(C_6F_5)_4^-$. The MW, MWD and yield of PE depend on the nature of the ligands, metals and counterions in the ionic metallocene complex as well as on the solvent polarity and temperature. The polymerization productivity increases in the following order: $[(Ind)_2TiMe]^+ > [Cp_2TiMe]^+ > [Cp_2ZrMe]^+$ [36]. The optimum temperature is between −20°C and 15°C.

The exploitation of metallocenes is on a precommercial scale due to their prohibitive cost but it is fully expected that within a few years they will begin to be applied to the lower-price, larger-volume commodity markets [7]. Metallocene systems have been devised that are compatible with all existing polyolefin manufacturing technologies, and in most cases, they could be used in existing units to replace catalyst systems [7].

1.1.4 Linear low and very low density polyethylenes (LLDPE and VLDPE)

Different grades of PE are the linear LLDPEs which do not have long branches. Their density is kept in the low range (0.880–0.925 g/cm^3) by the deliberate introduction of a controlled amount of short branches of the desired length [37]. The word 'linear' refers to the absence of long branches. The molecules of LDPE, HDPE and LLDPE are represented schematically in Figure 1.5.

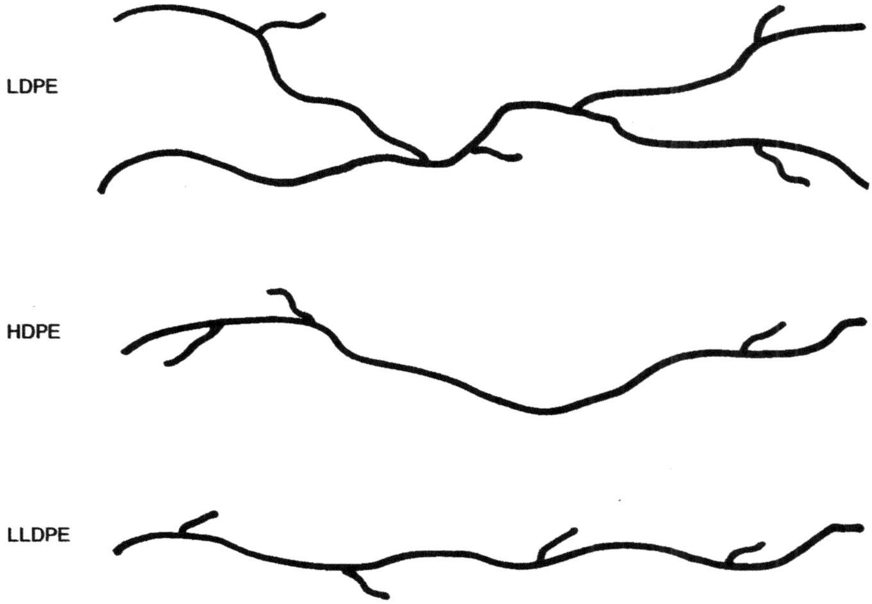

Figure 1.5 Schematic representation of the structures of various polyethylenes.

In fact, LLDPEs are copolymers of ethylene and another 1-olefin, such as 1-butylene, 1-hexene, 4-methyl-1-pentene or 1-octene. New generations of 'super strength' grades have been prepared most recently using higher olefin comonomers [1,6]. The length and position of the side chains are controlled by the comonomers used and by the catalyst system. Both the comonomer type and the polymerization conditions affect the properties of the copolymer.

The distribution of comonomers may be statistical or irregular (blocks). Statistical copolymers have lower density and crystallinity than the block copolymers.

LLDPE is produced basically by two types of low pressure processes, the gas phase and the solution processes [38]. Both the processes have been adapted from HDPE and many companies have swing capacities.

The gas phase fluidized bed process, initially developed at Union Carbide for the production of HDPE, has been modified for the production of LLDPE [39]. The reactor is operated at 80–100°C and 2.1 MPa (21 atmospheres) in the presence of a titanium-containing catalyst. It is now possible to obtain a complete array of LLDPE by the gas phase polymerization process.

Du Pont, Canada, was the first company to produce LLDPE. They have used a solution process at 150–300°C, 3–5 MPa (30–50 atmospheres) and a Ziegler-type catalyst [37].

DSC and TREF studies on LLDPE manufactured by different processes (gas, slurry, bulk and solution) have shown a bimodal distribution of the MW and a heterogeneous distribution of side branches [12]. The explanation is based on the existence of two groups of active sites in Ti-based heterogeneous Ziegler catalysts that produce different copolymerization rates of comonomers. In order to obtain a proper distribution of the side branches, the developed catalysts have to provide controlled rates of the copolymerization of comonomers.

Limited studies have been conducted on the copolymerization of ethylene with 1-butylene, 1-hexene and 4-methyl-1-pentene using metallocene–alumoxane catalyst systems [28]. The composition of the catalyst, the temperature and the concentration of comonomers determine the MW, MWD and the reactivity ratios. The reactivity ratios product (r_1r_2) of less than 1 suggests an alternating sequence for the insertion of the monomers.

VLDPE ranges in density from 0.880 and 0.912 g/cm^3. Their low degree of crystallinity imparts outstanding low temperature impact properties, flexibility and increased permeability to gases. These characteristics made them suitable materials for geomembranes, agricultural film, packaging for fresh produce and impact modifiers.

1.1.5 High and ultrahigh molecular weight polyethylenes

High molecular weight (HMW-HDPE) and ultrahigh molecular weight (UHMW-PE) polyethylenes are both HDPE with MW in the range 300 000–500 000 and 3–6 million respectively [40].

HMW-HDPE is produced by processes similar to those used for HDPE. The commercial process is based on a modified Ziegler catalyst system and the polymer is produced in the form of fine powder. The higher MW of these resins requires a broader polydispersity for acceptable processing characteristics. Typically these resins have bimodal MWD, the very high MW fractions impart strength and toughness while the low MW fractions facilitate extrusion [6].

UHMW-PE has essentially no flow above melting point due to its exceptionally high MW. Fabrication techniques are similar to that of metalworking, more intricate parts being machined from the solid resin.

UHMW-PE have a unique combination of physical and chemical properties such as exceptional impact strength at cryogenic temperatures, outstanding stress-crack resistance and the highest slurry-abrasion resistance of any thermoplastic. It therefore finds application in machine parts for service under severe conditions such as ore handling and food processing.

UHMW-PE is used to produce extended chain fibers (ECPE), the highest modulus and highest strength fibers ever made [41]. This enables PE to enter the field of specialized, high performance materials. The polymer is dissolved in a suitable solvent where the chains are disentangled and become extended.

Table 1.1 Comparison of some tensile properties of high performance fibers

Property	*ECPE*	*Aramid*	*Glass*	*Carbon*
Density (g/cm^3)	0.97	1.44	2.49	1.86
Tensile strength (MPa)	3000	2760	4590	2590
Tensile modulus (GPa)	172	131	90	393
Specific strength (cm × 10^6)	31.5	19.8	18.8	13.7
Specific modulus (cm × 10^6)	1813	927	355	2160

Then the solution is spun on conventional melt spinning equipment. The fibers are subsequently dried and postdrawn. The fibers acquire a specific structure in which all the macromolecules have fully extended chains oriented in the direction of the fiber so that the fiber is similar to a bundle of rods. The degree of crystalline orientation of the fibers is 95–99%.

Tensile properties of ECPE fibers are similar to other high performance fibers; however, because the density of PE is approximately two-thirds that of high modulus aramid fibers and half that of high modulus carbon fiber, ECPEs possess the highest ever obtained specific strength and specific moduli as can be seen from Table 1.1 [41].

1.1.6 Properties of polyethylenes

(a) Mechanical and thermal properties

There are available many hundreds of grades of PE which differ in their properties in one way or another. The main structural factors that determine PE properties are the degree of short and of long chain branching, the average MW and the polydispersity.

One of the most important characteristics that determines in the highest degree the properties and the behavior of different grades of PE is their branching. As has been mentioned in preceding sections, the number and the length of branches are controlled by the type and parameters of the polymerization process (temperature, pressure, type and amount of catalyst or initiator and of transfer agent, etc.). Branches prevent the polymer chains from packing together regularly and closely and have a predominant effect on the density of PE. The highest density for this family of polymers is about 0.98 g/cm^3 and is reached by the strictly linear polymer – the polymethylene – obtained from diazomethane and which is a useful reference. The density can be considered a first indication of the degree of branching – the lower the density the higher the degree of branching [37]. The effect of density on some general properties of PE is illustrated in Figure 1.6.

The presence of branches interferes with the ability of the polymer to crystallize. The degree of crystallinity of LDPE is usually of the order of 55–70% compared with that of HDPE which is 75–90%. The crystalline

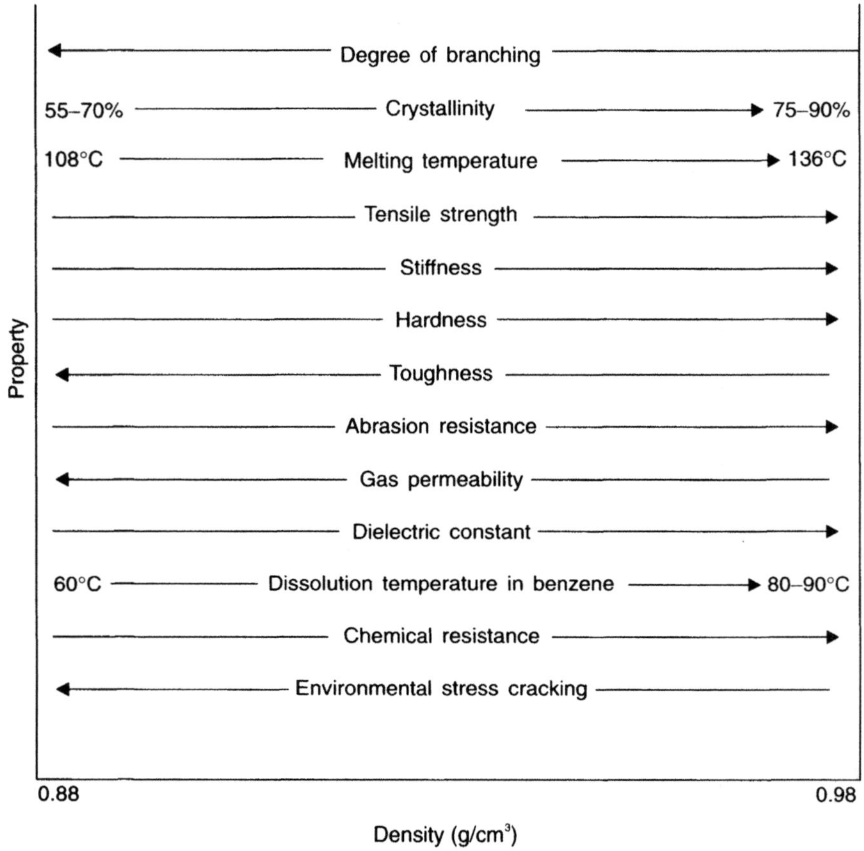

Figure 1.6 Effect of density on polyethylene properties.

T_m for PEs ranges from 108 to 132°C depending on the degree of crystallinity. Such low melting points are characteristic for structures with flexible C–C bonds and without strong intermolecular forces. T_g, which is associated with the motion of longer segments in amorphous matter, is reported in the literature as ranging between −60 and +130°C, the most likely value being considered −20°C [19].

The specific heat (C_p) of PE is higher than for most thermoplastics and strongly depends on the temperature. Values of 2.3 J/g are specific at room temperature and 2.9 J/g at 120–140°C for LDPE. This is an important feature for the processing of polyethylene.

Other properties depending on crystallinity, such as stiffness, hardness, tear strength, yield point, Young's modulus in tension and chemical resistance, increase with increasing degree of crystallinity (HDPE) whereas permeability to liquids and gases, flexibility and toughness decrease under the same conditions.

Since PE is crystalline non-polar hydrocarbon polymer it has no solvents at room temperature and dissolution takes place only on heating in solvents of similar solubility parameter (hydrocarbons and halogenated hydrocarbons). The higher the degree of crystallinity, the higher the dissolution temperature. LDPE dissolves at 60°C compared to 80–90°C for high density, more crystalline polymers.

The effect of branching also depends on the size of side chain branches. While short branches have a predominant influence on the degree of crystallinity and therefore on the density of the polymer, long branches affect more pronouncedly the polydispersity. The side chains may be as long as the main chain and like it may have a wide distribution of lengths. The higher the MW of the resulting polymer the wider the MWD, as chain transfer reactions may occur as well on side chains. Such a polymer may be made up of short chains grafted onto short chains, long chains onto long chains and a vast range of intermediate cases.

The ratio of weight- to number-average molecular weight (MWD) varies from 1.9 (for polymethylene) to 100; values of 20–50 are considered typical. High density polymers have MWD values of 4–15 [37]. The very high values of MWD for LDPEs are, in part, the result of long chain branching and sometimes it is not clear if an effect on the properties is due to branching or to MWD. It is generally considered, however, that when the other structural factors are constant, a narrower MWD leads to an increase in impact strength, tensile strength, toughness, softening point and resistance to environmental stress cracking.

Long chain branches also affect the flow properties. Long branched molecules are more compact and tend to entangle less with other molecules resulting in lower solution and melt viscosities as compared with unbranched polymers.

Another factor that influences the properties of the melt, as well as those properties that involve large deformations, is the weight-average MW. Ultimate tensile strength, tear strength, low temperature toughness, softening temperature, impact strength and environmental stress cracking increase as the MW increases; on the contrary, the fluidity of the melt and the coefficient of friction (film) decrease.

For technological purposes, the MW is characterized by the melt flow index (MFI) that is the weight in grams extruded under a standard load in a standard plastometer at 190°C in 10 minutes [42]. The higher is the MFI, the lower is the MW. However, MFI has to be considered carefully as the viscosity of the melt is strongly affected by the presence of long chain branches as has already been mentioned.

The influence of the density on some properties of different grades of PEs is illustrated in Table 1.2 [10,11,19,38,40].

The electrical insulating properties of PEs are excellent. The dielectric constant increases linearly with increasing density. As it is a non-polar material,

Table 1.2 Some properties of different grades of polyethylene

Property	*LLDPE*	*LDPE*	*HDPE*	*UHMW-PE*	*Poly-methylene*
Density (g/cm^3)	0.910–0.925	0.915–0.935	0.941–0.967	0.93	0.98
Melting temperature (°C)	125	106–112	130–133	132	136
Tensile strength (MPa)	14–21	6.9–17.2	18–30	20–41	34.5
Elongation at break (%)	200–1200	100–700	100–1000	300	500
Flexural modulus (MPa)	248–365	415–795	689–1654	–	–
Izod impact strength (J/m)	–	0.67–21	27–160	no break	–
Hardness (Shore D)	41–53	45–60	60–70	–	–

the dielectric constant and the power factor are almost independent of temperature and frequency. The presence of impurities such as metallic catalyst fragments or carbonyl groups formed by the oxidation of the PE has an adverse influence on the power factor.

(b) Degradation

PE has a chemical resistance similar to that of paraffins. Non-oxidizing acids, alkalis and aqueous solutions do not attack PE. On the contrary, PE is sensitive to oxidation that occurs on exposure to strong oxidizing agents (concentrated nitric acid, hydrogen peroxide or potassium permanganate), or even in air under ultraviolet light as well as at elevated temperatures during processing or service life. Resistance to oxidation increases with increasing density because a less branched polymer has a diminished permeability to gases and a smaller number of tertiary carbon atoms in the molecule (which constitute sensitive points of attack). Traces of metals (catalysts) enhance the oxidation process.

The oxidation proceeds by a free radical mechanism through the formation of hydroperoxides which decompose to new free radicals and the reaction is autocatalytic. Two kinds of reactions take place on oxidation. (a) The scission of polymer chains that leads to a deterioration of the mechanical properties as a result of the reduction of the MW:

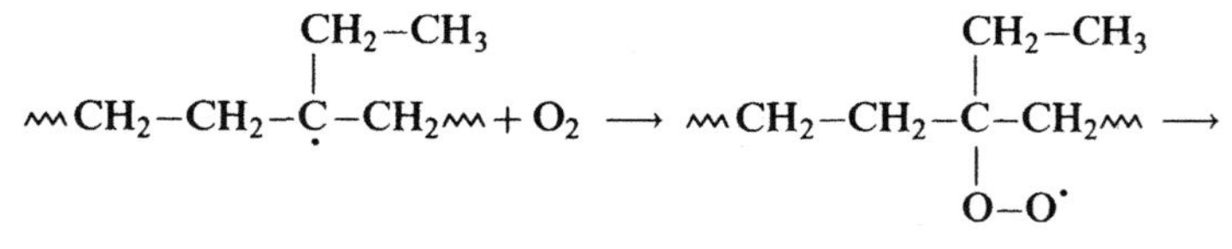

$$\sim CH_2-CH_2-\underset{\underset{O-OH}{|}}{\overset{\overset{CH_2-CH_3}{|}}{C}}-CH_2\sim \longrightarrow \sim CH_2-CH^{\bullet} + \underset{\underset{O}{\|}}{\overset{\overset{CH_2-CH_3}{|}}{C}}-CH\sim + {}^{\bullet}OH$$

(b) At higher levels of oxidation, combination reactions between existing radicals result in crosslinking of polymer chains and the formation of brittle products.

As a result of oxidation, there is a serious deterioration of the power factor and of electrical properties. The incorporation of antioxidants in PE is necessary for electrical insulation applications in order to reduce this effect.

Exposed under a certain stress or strain to some liquids such as alcohols, sulfonated alcohols, liquid hydrocarbons, esters, ketones, metallic soaps, silicone fluids, greases and oils, PE undergoes a fracture in a much shorter time than in the absence of this environment. This phenomenon is known as 'environmental stress cracking'. Some stress may remain in the material as a result of processing or may be due to distortion during use. The resistance to environmental stress cracking may be much improved by good processing conditions and good design of specific products.

(c) Additives and processing

PE is often used without the addition of any additive, but different types of additives may be incorporated into the polymer in order to impart various properties [43]. For long-term applications, especially as an electrical insulator, PE is blended with antioxidants which provide protection against oxidative degradation. Hindered phenols in small amounts (up to 0.1%) are used almost exclusively as chain breaking antioxidants. They block the hydroperoxide radicals formed by oxidation and thus interrupt the propagation of the oxidative process: the new aromatic radicals formed from phenols are too stable and are unable to propagate the oxidative chain.

Reinforcing fillers such as carbon black or still better silane and titanate coupling agents are used sometimes with PE. Pigments have to be chosen with special care to ensure that they do not catalyze the oxidation and do not bring about the rise of the power factor. Common pigments are compounds of cobalt, cadmium, manganese, iron or chromium.

As PE burns readily, halogenated compounds and antimony trioxide are used as flame retarders. Among other additives, antistatic agents of glycol alkyl ester type are widely used to reduce dust attraction. Antiblocking agents, such as fine silicas, are used to prevent the high cohesion between the layers of LDPE films and, in addition, slip agents are added in order to reduce the friction between film layers [19].

The addition of 10–40% starch to LDPE results in a biodegradable material used for carrier bags and for mulching of some crops. Starch decomposes in moist soils in one to five years and the remaining PE framework breaks up and is absorbed in the soil [44].

Products with a dielectric constant of about 1.45, useful as dielectric materials, can be obtained from cellular PE. Blowing agents such as 4,4-oxy-bis-benzenesulfonohydrazide and azocarbonamide are incorporated into the polymer, they decompose on extrusion and the evolved gas gives rise to a cellular extrudate.

PEs are processed almost exclusively by melt processing techniques. Extrusion, blow molding and injection molding are widely used with low, intermediate and high density polyethylenes. Mold shrinkage is 1.5–5.0% with LDPE and 2.5–6.0% with HDPE, the rather high values being due to the crystallization of the material.

There are basically two processes of making PE film, the tubular process and the flat film process. In the tubular process, used in larger proportion, a thin tube is extruded in a vertical upward direction and the tube is inflated into a bubble by blowing air through the die head. The bubble is cooled, flattened and wound up. In the flat film process, the film is extruded from a slit die either into a water bath or onto chilled rolls.

Coating with polyethylene is done by fluidized bed powder coating, by spraying, or by powder molding and powder rotational molding.

1.1.7 Applications

The widespread use of PE is due to its excellent electrical insulation properties and chemical resistance, easy processability and low cost.

A general overview of PE applications according to density and melt index is illustrated in Figure 1.7 [6]. The major applications of PEs (LDPE, LLDPE and HDPE) have been as film for general packaging (bags, sacks, food wrapping) and building and agricultural industries. HDPE film has also been used in a net-like form (or pseudofiber) [19].

PE (especially HDPE) is an important injection molding material for a wide range of products including toys, electrical fittings, seals, chemical plants, containers, cases, crates, housewares and many other applications. Specialized uses include the disposable syringes for medical purposes [19].

HDPE has been widely used for blow molding of bottles for milk and other foodstuffs, household chemicals and drug packaging. Squeeze bottles are made from LDPE.

Other areas for PE applications have been domestic water and gas piping, and agricultural piping [45]. PE has also been used as filament for ropes, fishing nets and fabrics.

Based on its excellent electrical insulation properties, PE has been extensively used in cable and wire covering (undersea cables) and air-borne

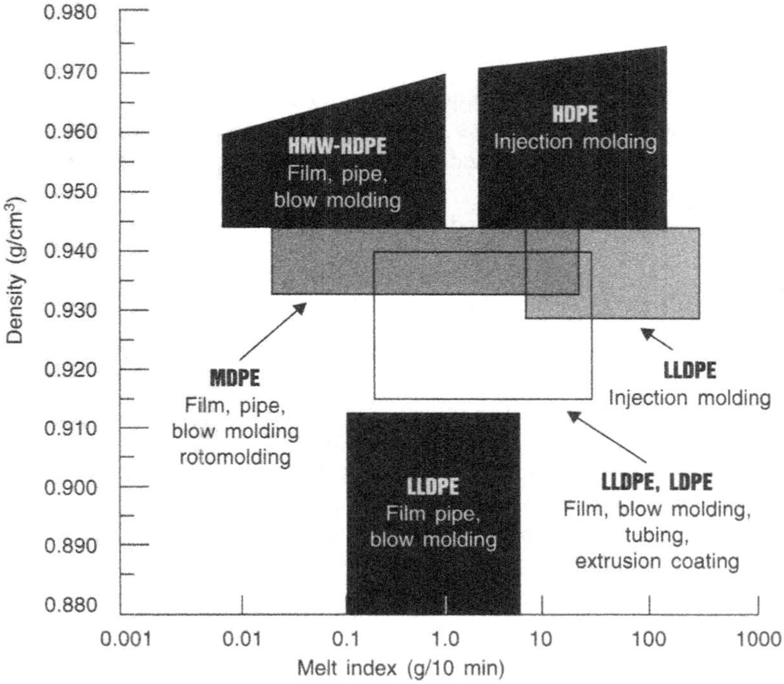

Figure 1.7 Polyethylene applications [6]. (Reprinted by special permission of McGraw-Hill, Inc., from *Modern Plastic Encyclopedia*, Vol. 71 (12), 1994. Copyright McGraw-Hill, Inc., New York, NY 10020.)

radar. Cellular PE with a very low dielectric constant (1.45) is used as insulator for television downlead aerials.

Synthetic wood pulp with cellulose-like morphology (fibrides) made from HDPE are used as battery separators and teabags [46].

Low MW oxidized PE (in the range of MW = 1000) is a wax-like product, water and chemical resistant. It constitutes an anticorrosion additive in lubricating wax formulations [47].

With the arrival on the market in 1985 of the ECPE fibers made from UHMW-PE, PE entered the specialized high performance composite industry. ECPE fibers are available in roving, fabric, continuous mat and chopped fiber. They have been used with a wide variety of resin systems imposed by the end-use application. Among the resin matrix are epoxies, polyesters, silicones, urethanes and polyethylene. For higher level of adhesion, specific surface treatments such as corona discharge or plasma treatments are applied to fibers [41]. They are used as reinforcing material for ballistic armor, radar protective domes, aerospace impact shields, sail cloth, marine ropes, sports equipment and concrete reinforcement [41,45].

1.1.8 Crosslinked polyethylene

A thermosetting PE has been obtained by a controlled process of crosslinking (vulcanization) [19]. The material has enhanced stability of the form at elevated temperatures, can withstand prolonged aging at 105°C or short-term exposures at 230°C and has a better resistance to environmental stress cracking. Crystallization is reduced by crosslinking and consequently modulus and hardness are lower.

Polyethylene may be crosslinked by three different processes [48] as follows.

1. Peroxide crosslinking is done by the incorporation of a peroxide with higher decomposition temperature such as dicumyl peroxide or di-*t*-butyl peroxide. The peroxides have to withstand temperatures during compounding and shaping operations and decompose only in a post-processing curing reaction at higher temperatures. The peroxide free radicals abstract hydrogen from the polymer leading to polymeric free radicals which combine to a network structure. The incorporation of carbon black into the PE before vulcanization improves the mechanical properties of the material.
2. Radiation crosslinking (X-rays, γ-rays or fast electrons) is done commercially on thin sections such as films. Small quantities of hydrogen, methane, ethane and propane are formed during irradiation and it seems that they result from the short side chains of the PE. Some insaturation is introduced in the PE by irradiation.
3. The third way for crosslinking is done by grafting onto the PE chain of trialkoxyvinylsilane compounds. The obtained trialkoxysilane PE is typically extruded onto a wire which is then treated with low pressure steam. The alkoxy groups hydrolyze with condensation leading to siloxane crosslinks:

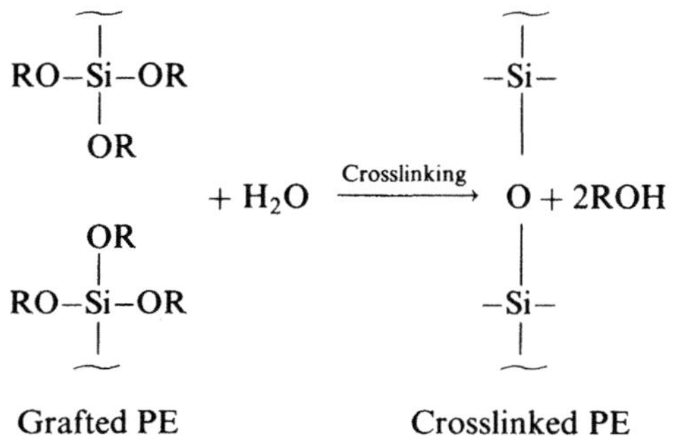

Crosslinked PE has found application for insulating electrical power cables (which are subjected to elevated temperatures), transformers, motors and generators. Crosslinked cellular PEs, produced particularly in Japan, have been used for carpeting in the automotive industry (sound deadening), for pipe insulation and as flotation media.

1.1.9 Chlorinated and chlorosulfonated polyethylenes

Chlorinated polyethylene (CPE) is produced commercially from both LDPE and HDPE. CPE has significant advantages compared to parent PE, such as an increased flexibility and toughness, ignition resistance, better filler acceptance and excellent compatibility with a large variety of polymers [49].

Chlorination may be effected in solution, in aqueous suspension or in fluid bed at temperatures in the range of 45–75°C in the presence of peroxides. Secondary hydrogen atoms are substituted by chlorine atoms that are distributed unevenly and are concentrated in certain regions of the macromolecules resulting in a block CPE.

The properties of CPE depend on the properties of the starting material and on the amount and distribution of introduced chlorine. The end products can be thermoplastics or elastomers.

CPE with 25–30% Cl by weight is a thermoplastic material used as an impact modifier additive for rigid PVC and styrenic polymers leading to composite materials with improved impact strength for piping and sheathing. CPE with 36–48% Cl are elastomers that do not contain chain unsaturation and have improved oil, heat and aging resistance. CPE elastomers can perform continuously at 150°C and intermittently at 162°C, they retain their flexibility at −40°C and are used for cable sheathing and different technical goods [49]. Above 50% chlorine, the material becomes stiff.

CPEs have good resistance to most acids, bases, ozone, oil and fuel, and to many solvents.

CPE elastomers are used for high performance applications such as wire jacketing, industrial and automotive hoses, tank linings and others.

When the chlorination of PE is conducted in the presence of sulfur dioxide, both chloride and sulfonyl chloride groups are introduced into the polymer, their ratio depending on the conditions of the process. Common commercial polymers contain about 1.5% sulfur and 30% chlorine, i.e. that some 25–42% of the monomer units are chlorinated. The polymer may be represented as follows:

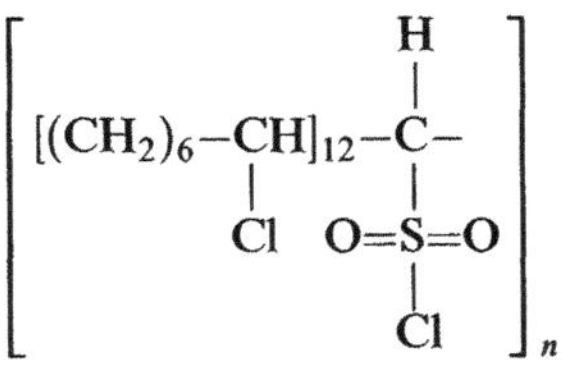

The sulfonyl chloride groups serve to subsequently vulcanize the material. The vulcanization is effected in the presence of metal oxides (litharge, MgO)

and of water and results in a crosslinked structure:

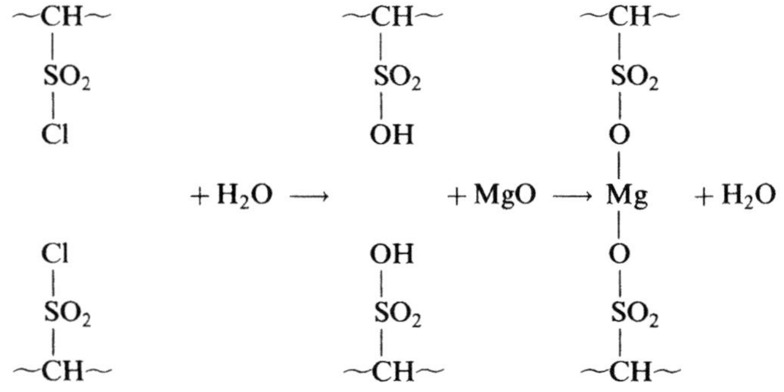

Chlorosulfonated PE   Crosslinked PE

The chlorosulfonated PE is a rubber with good resistance to oxidizing agents, such as oxygen and ozone, and excellent mechanical properties at elevated temperatures. It is used as a cable coating for special applications.

In the building industry, chlorosulfonated PE has been used as a rubber and sealant and also where resistance to weathering, corrosion or abrasion is required, e.g. flooring, lining for chemical plants, roofing, etc. [45].

1.1.10 Ethylene copolymers

Ethylene is copolymerized with different olefinic or non-olefinic comonomers, in order to obtain more flexible and tougher materials, by the reduction of the degree of crystallinity and by the introduction of some polarity. Ethylene copolymers with higher olefins such as 1-butylene, 1-hexene and 1-octene have already been discussed as they are known as LLDPE. The copolymers with propylene will be discussed further.

In this section are presented the copolymers of ethylene with non-olefinic comonomers such as vinyl acetate, methyl and ethyl acrylates, acrylic and methacrylic acids. All these copolymers are prepared by free radical, high pressure polymerization technologies.

(a) Ethylene–vinyl acetate copolymers (EVAc)

EVAc is a random copolymer with the general formula:

$$-(CH_2-CH_2)_x-(CH_2-\underset{\displaystyle OCOCH_3}{\underset{|}{CH}})_y-$$

The content of vinyl acetate varies between 5 and 50% and controls the crystallinity and the flexibility of the material. EVAc copolymers have improved clarity, low temperature flexibility, adhesion and stress cracking resistance but are more permeable to oxygen and vapors.

Depending on the composition and the MW of EVAc copolymers, they may have a low softening point and are used as wax additives or may be rubbery materials resembling LDPE in appearance. EVAc copolymers have good compatibility with bitumen and these mixtures are used for roof sheeting with improved properties [45].

EVAc copolymers with 5% VAc are used for film applications in packaging; with 10% VAc for extrusion and injection molding of flexible tubing and hose, footwear components, gaskets and squeeze bulbs; with 20% VAc they enter into hot-melt-adhesive formulations [48].

By the hydrolysis of the acetate moiety in EVAc, the copolymer ethylene–vinyl alcohol is prepared. This copolymer has excellent gas barrier properties associated with high moisture absorption. It has found use in multilayer film for food packaging.

(b) Ethylene–methyl acrylate (EMA) and ethylene–ethyl acrylate (EEA) copolymers

They are copolymers containing up to 20% acrylic comonomer [50] and may be represented as:

$$-(CH_2-CH_2)_x-(CH_2-\underset{\displaystyle COOR}{\underset{|}{CH}})_y-$$

where $R = CH_3$ or C_2H_5

Compared to EVAc copolymers, they have better thermal and abrasion resistance but lower clarity and toughness. They are used for extrusion and injection molding of different items.

(c) Ethylene–acrylic acid (EAA) and ethylene–methacrylic acid (EMAA)

They may be represented as:

$$-(CH_2-CH_2)_x-(CH_2-\underset{\displaystyle COOH}{\underset{|}{CH}})_y- \qquad -(CH_2-CH_2)_x-(CH_2-\overset{\displaystyle CH_3}{\overset{|}{\underset{\displaystyle COOH}{\underset{|}{C}}}})_y-$$

EAA EMAA

These copolymers exhibit very good adhesion to different substrates and particularly to metals due to their increased polarity compared to PE. Their current usage is as adhesive layers between aluminum foil and plastic films for applications such as toothpaste tubes.

(d) Ethylene–methacrylic acid ionomers

The term 'ionomer' refers to any copolymer of 1-olefins containing acid groups (carboxylic or sulfonic) but only ethylene–methacrylic acid ionomers have achieved significant commercial production.

They are produced by the high pressure copolymerization of ethylene with 1–10% methacrylic acid followed by partial neutralization of the carboxylic groups with a metal cation (sodium, magnesium, zinc). They may be represented as:

$$\sim(\mathrm{CH_2{-}CH_2}-)_x-(\mathrm{CH_2}-\underset{\mathrm{COO^-}}{\overset{\mathrm{CH_3}}{\mathrm{C}}}-)_y-(\mathrm{CH_2{-}CH_2}-)_v-(\mathrm{CH_2}-\underset{\mathrm{COOH}}{\overset{\mathrm{CH_3}}{\mathrm{C}}}-)_z\sim$$

$$\mathrm{Me^+} \qquad \mathrm{Me^+}$$

$$\sim(\mathrm{CH_2{-}CH_2}-)_x-(\mathrm{CH_2}-\underset{\mathrm{CH_3}}{\overset{\mathrm{COO^-}}{\mathrm{C}}}-)_y-(\mathrm{CH_2{-}CH_2}-)_v-(\mathrm{CH_2}-\underset{\mathrm{CH_3}}{\overset{\mathrm{COOH}}{\mathrm{C}}}-)_z\sim$$

An ionic crosslink is formed which is stable at ambient temperature, but which reversibly breaks down on heating. The ionomers behave like thermosets at room temperature, but like thermoplastics on heating and they can be melt-processed by usual techniques.

The structure of ionomers consists of three molecularly interconnected different phases: an amorphous hydrocarbon phase, a fringed micellar crystal

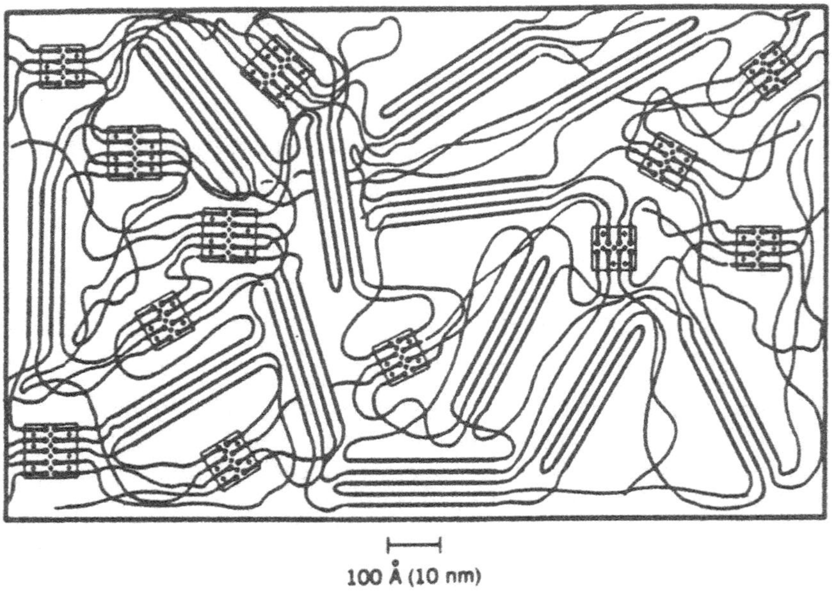

Figure 1.8 Schematic structure of ionomers [52]. (Reprinted by permission of J. Wiley & Sons, Inc., from *Handbook of Plastic Materials and Technology* by I.I. Rubin (ed.), 1990. Copyright J. Wiley & Sons, Inc.)

of PE and a layered ionic cluster [51]. Figure 1.8 represents a schematic structure of a ionomer [52].

Compared with LDPE, the ionomers have higher toughness, tensile strength and modulus, greater clarity and better abrasion and oil resistance. Recently, glass fiber ionomer composites have been introduced that have superior low temperature properties (at −29°C) and higher modulus (1034–2241 MPa) than unreinforced ionomers (172–345 MPa).

The ionomers are used as film and sheet with excellent optical properties and heat sealability for various food packaging. Injection moldings are used in footwear and ski boots. Ionomers' cut-through resistance and toughness make them superior materials for golf ball covers.

(e) Ethylene–carbon monoxide copolymers

Alternating copolymers are prepared by gas phase catalyzed copolymerization of ethylene and carbon monoxide [53]. The copolymers contain ketone groups in the main chain which absorb light that makes them photodegradable materials by chain scission [54]. They are used as packaging and sheeting film.

1.2 POLYPROPYLENE (PP)

1.2.1 Introduction

PP is a major volume thermoplastic material ranked third in the production of plastics, after PE and PVC. Due to its cost effectiveness, versatility, overall competitiveness and excellent environmental aspects, PP is the fastest growing commodity plastic. Advanced polymerization processes in conjunction with highly developed compounding techniques make it possible to target PP for engineering applications. The world production volume was about 19 million tons in 1995, including 4.7 million tons in the US alone [55].

PP is produced by the stereospecific addition polymerization of propylene. The common name of the polymer is polypropylene, which corresponds fairly well with the name recommended by IUPAC where the repeating propylene unit is in brackets.

Commercial production of PP started in 1957 at Montecatini, Italy, at Hercules, USA, and at Farbewerke Hoechst AG, Germany, and was the result of the discovery of heterogeneous stereospecific catalysts by Ziegler in Germany that effected the polymerization of ethylene in 1953. Using these catalysts, Natta in Italy obtained in 1954 crystalline polymers of propylene. Earlier efforts to polymerize propylene had only resulted in amorphous polymers without any use at that time. Natta classified the three geometric forms adopted by the polymer chain of PP as follows.

I. Isotactic; all methyl groups are situated on the same side of the chain.

I

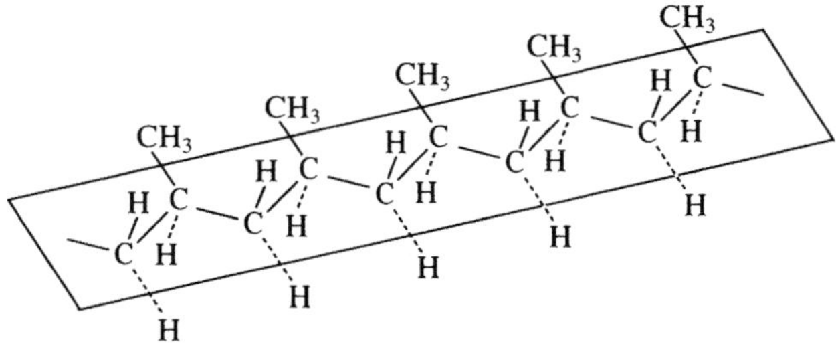

II. Syndiotactic; methyl groups are on alternating sides.

II

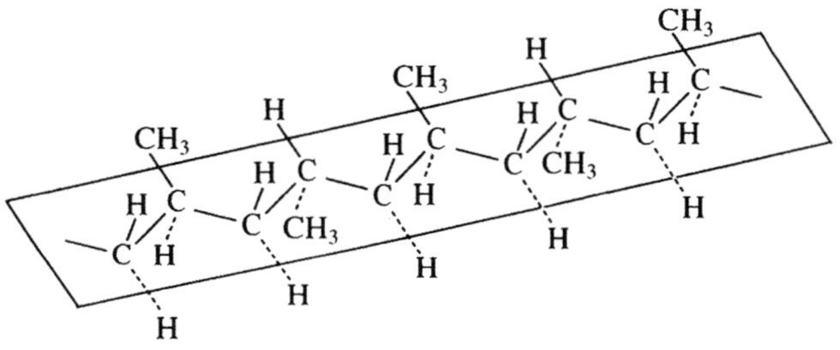

III. Atactic; methyl groups are situated at random.

III

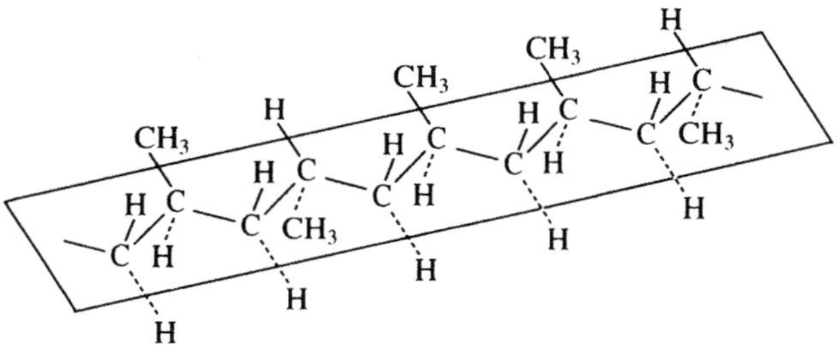

Isotactic and syndiotactic PPs are crystalline, whereas atactic polymer is amorphous.

As for ethylene, the raw materials for propylene are petroleum and natural gas from which the C_3 fraction is obtained after cracking and fractional distillation. Impurities such as water and methylacetylene must be carefully removed and the purity of propylene brought up to 99.7% [56].

Three different categories of PP materials are produced commercially: homopolymers, copolymers and blends.

1.2.2 Polymerization

The polymerization of propylene is effected in the presence of Ziegler-type catalysts which are composed of the reaction products of titanium trichloride with aluminum compounds such as aluminum triethyl or tributyl or aluminum diethyl monochloride. These compounds form a 10% slurry in a diluent such as naphtha. The preparation of the catalyst for propylene polymerization is the most important step in the overall process. The composition of the catalyst as well as the shape and size of its particles are determinant for the properties of the polymer [57,58]. A good catalyst should produce high yield of polymer per part of catalyst, with a high content of isotactic PP, and with controlled MW and morphology. MWD are generally broader when insoluble catalysts are used and much narrower with soluble catalysts [59]. Hydrogen is used for MW control.

The different technologies that are used for the polymerization of propylene using Ziegler-type catalysts are (a) the slurry processes including the bulk process, (b) the solution process and (c) the gas phase process.

(a) The slurry processes

The oldest and the major processes are the slurry processes [60]. The polymerization resembles the Ziegler process described for PE. It may be carried out either in stirred batch reactors or in continuous loop reactors. The suspending liquid (C_4–C_6 paraffins), the catalyst and the propylene are metered separately into the reactor and the polymerization is conducted at 50–80°C and 0.5–3.0 MPa (5–30 atmospheres). The required isotactic polymer develops around the catalyst particles and forms a slurry since it is insoluble in the diluent. Some atactic amorphous polymer that is also formed is largely soluble in the diluent. The addition of the monomer is continued until the slurry reaches 30–60% solids. At this point the reaction mixture is transferred to a flash tank for the removal and recycling of unreacted monomer. The catalyst is then solubilized and deactivated by the addition of an acidified alcohol. The slurry is centrifuged for the separation of the crystalline isotactic polymer from the diluent which contains the soluble atactic polymer and the solubilized catalyst. The polymer is washed and dried. The flow chart of the slurry process is presented in Figure 1.9.

The newer, high activity and high stereospecificity magnesium-supported catalyst systems provide high enough yields of polymer per unit of catalyst so that the removal of the catalyst becomes unnecessary. Thus, 40 kilograms of PP are produced with today's catalysts compared to only 4 kg/g produced in the 1970s [61]. The ratio of isotactic/atactic PP is also improved and the separation of atactic polymer is no more required [60]. With some catalysts, polypropylene may be produced as dense spherical granules so that a pelletizing operation is avoided [62,63].

A variant of the slurry polymerization technology is the bulk process in which liquid monomer serves both as reactant and as suspending liquid for the resulting PP [56]. Temperature and pressure are adjusted to 45–80°C

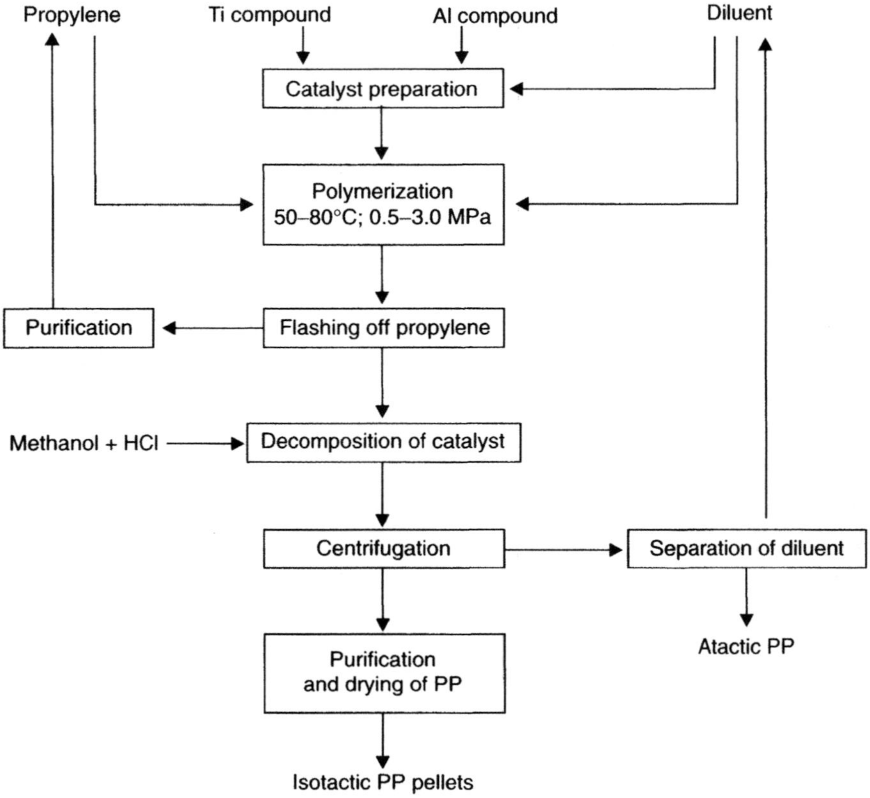

Figure 1.9 Polymerization of propylene by the slurry process.

and 1.7–3.5 MPa, respectively, in order to maintain the propylene in the liquid phase. PP results as a slurry in liquid monomer and is continuously withdrawn from the reactor when the slurry reaches 30–50% solids. In the flash chamber the unreacted monomer is eliminated and the polymer forms a powder that is subsequently dried and pelletized with stabilizers.

The slurry processes, which have been dominating the field for a long time, are increasingly giving way to superior gas phase technologies and to a combination of bulk and gas phase techniques [55].

(b) The solution process

In solution processes that are more expensive and hence used only for specialty products, higher temperatures (160–250°C) and pressures (2.7–8.0 MPa) are employed in order to dissolve the native polymer in the liquid monomer. PP produced in this process has higher impact strength and lower brittleness temperatures than conventional PP.

(c) The gas phase process

The gas phase fluidized bed process introduced by Union Carbide for the production of LLDPE has been adapted for the polymerization of propylene to homopolymers or block copolymers [64,65]. The plants are equipped with swing capacities for the transition from PE to PP and back to PE [66]. Fluidization is maintained by the high rate of gaseous monomer recycle. The reactor has two zones: the reaction zone at the bottom where the catalyst and the growing polymer particles are fluidized by the gas stream of monomer and the disengagement zone at the top where the unreacted monomer and the solid particles separate. The gas is recycled while the polymer particles fall down and are continuously withdrawn. The polymer may be obtained in granular form with spherical morphology which eliminates any pelletizing operation [55]. The flow chart of the fluidized bed process is presented in Figure 1.10.

Compared to conventional PP, the property spectrum of the new PP types makes it possible to rate them as advanced materials designed for engineering applications. The technologies of the preparation of highly active catalysts and the state-of-the-art polymerization processes are highly proprietary and little information on any details can be found in the literature.

Totally new metallocene-based catalyst systems are entering the PP manufacturing technology. The general description of metallocene catalysts is given in Section 1.1.3(e). Such factors as the temperature, the transition metal (Ti, Zr, Hf), the ratio alumoxane/metallocene, the substitution in

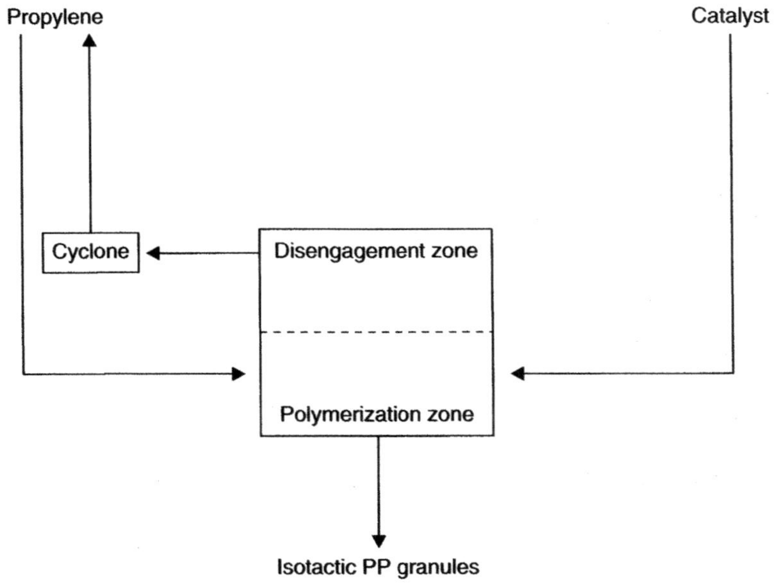

Figure 1.10 Gas phase fluidized bed polymerization of propylene.

the ligands (position, number and type of substituents) directly control the catalytic activity, the average MW, the MWD, the degree of tacticity, the sequence distribution and the microstructure of PP [28].

Stereoregulation in the polymerization of propylene is one of the most important capabilities of metallocenes. The possibility of steric design of the π-ligands (cyclopentadienyl, indenyl and fluorenyl) allows the synthesis of metallocenes with different symmetries, each giving rise to different stereoregulated PPs. Four types of PP can be produced [28]: isotactic, syndiotactic, atactic and hemiisotactic. In hemiisotactic structures every other methyl group is in isotactic placement and the remaining methyls are placed at random [28]:

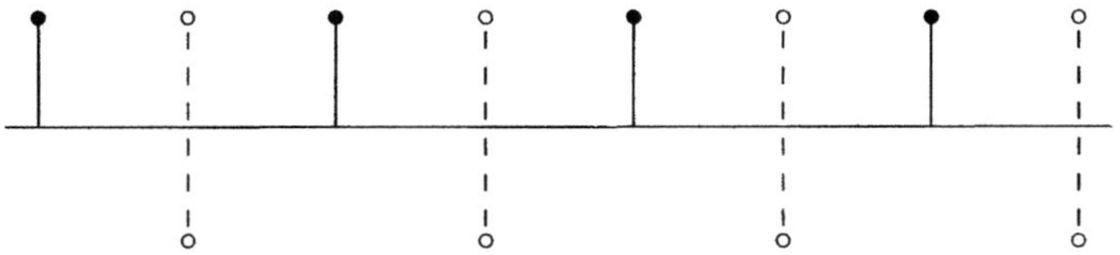

where • = isotactic placement; ○ = random placement.

The stereoregularity of the PP is controlled in two different ways: (a) by the chirality of the catalyst which governs the insertion of the monomer into the metal–carbon bond (enantiomorphic site control) and (b) in the case of achiral metallocenes, the tacticity is controlled by the configuration (chirality) of the last inserted monomer unit (chain end control) [28]. The tacticity of the PP is mostly affected by the symmetry of the metallocene [68].

Spherical megaparticles of PP are formed with metallocene systems that are quite different as compared with the particles obtained with conventional Ziegler catalysts [28].

Supported metallocenes on silica or $MgCl_2$ have also been studied.

Cationic metallocenes give productivities that are six times higher than metallocene/alumoxane systems and the stereospecificity is higher at low temperatures (96.3% isotactic PP at −55°C) [69].

1.2.3 Properties

The tacticity of PP plays a major role in determining its properties. The isotactic PP has the most regular structure with the carbon chains arranged into right-hand and left-hand helices. Each turn of the helix is formed by three monomer units. The methyl groups are on the outside of the helix. These regular helices can pack closely together to form crystalline regions. Commercial polymers are 75–98% isotactic and may contain some atactic and syndiotactic forms as separate macromolecules or as blocks. Stereoblock chains may also be present in which right-hand and left-hand blocks alternate. When the inversion of these blocks is not frequent the polymers behave like isotactic structures. The greater the amount of isotactic structure, the higher the degree of crystallinity.

PP is a crystalline polymer and the rate and manner in which the crystals have been formed from the melt influence the crystalline structure. This in turn largely determines both physical properties and processing characteristics of the polymer. Nucleation agents (crystalline organic acids or metal salts) are added in amounts below 0.1% to provide additional crystallization sites and the formation of smaller and more numerous polymer crystals. This controlled morphology of the polymer, referred to as 'high crystallinity PP', results in higher bulk properties of the material, namely the softening point, the stiffness, the tensile strength, the modulus and the hardness [56].

The melting temperature, T_m, of isotactic PP is 176°C, some 40°C higher than that of HDPE and this fact makes possible steam sterilization of many PP articles. Strain-free articles retain their shape even at 135–140°C. The glass transition temperature, T_g, of PP is situated at about 0°C and below this temperature the impact strength drops down and the polymer becomes brittle. A way to improve this behavior is by block copolymerization with ethylene that will be discussed later.

Some properties of PP are:

density (g/cm^3) = 0.903;
tensile strength (MPa) = 35.5;
tensile modulus (MPa) = 1380;
flexural modulus (MPa) = 1690.

The non-polar nature of PP provides the material with excellent electrical properties similar to those of HDPE. Chemical resistance to most chemicals and solvents is exceptionally high. The polymer is insoluble at room temperature and dissolves in hydrocarbons and chlorinated hydrocarbons only above 80°C. Unlike PE, PP does not suffer environmental stress cracking [70].

PP differs from PE by the presence of methyl side branches and of relatively labile tertiary hydrogen atoms at every second carbon atom on the backbone. This makes the polymer more susceptible to oxidation by oxidizing agents and by air at elevated temperatures. Antioxidants must be used in all commercial compounds. The general practice is to incorporate small quantities of stabilizers even before the drying operation or before a long-term storage. Additional stabilizers are blended with the polymer during pelletizing [56]. The general antioxidants are hindered phenols, hydroperoxide decomposers or various phosphites and they are synergistic. For special applications more specific stabilizers are added.

Unlike PE which may be crosslinked by high radiation, by heating with peroxides or by chlorosulfonation, PP is degraded to lower MW products by similar treatments.

PP is produced in a large variety of melt flow index (MFI) from 0.3 to 800 g/10 min [55]. Weight-average MW of commercial polymers is in the range 220 000–700 000 and MWD is 5.6–11.9. Both MW and MWD are dependent on many factors such as the polymerization process, the temperature, the pressure, the concentration of the monomer, the composition of the catalyst system, and the addition of hydrogen as chain transfer agent. A common practice to narrow the MWD is a post-reaction treatment usually with peroxides that also produces some chain scission. The products are known as 'controlled rheology grades' (CR) and they manifest a better balance of physical properties and processing characteristics [70].

The specific heat of PP is lower than that of PE allowing an increase in the plasticizing capacity of the injection molding machines compared to PE.

The higher the degree of induced crystallinity by the processing conditions the higher will be the molding shrinkage of the material. The reduction of the shrinkage can be realized by coupling a higher temperature of the melt with a lower temperature of the mold.

Besides antioxidants, other additives have been used with PP. Different mineral fillers such as talc, calcium carbonate or glass fibers confer an improved range of properties [56]. For example, 25% by weight of glass fibers leads to a fourfold increase of the tensile strength and to an eightfold increase of the rigidity of PP at 100°C. Rubbers, especially ethylene–propylene rubbers taken in amounts of up to 50% by weight, have been used as fillers in order to improve the impact strength of PP. Pigments, flame retarders, carbon black and ultraviolet absorbers have been used for specific applications. Foaming agents contribute to improve the ratio stiffness/weight and to eliminate the shrinkage [71].

Processing methods such as injection molding, blow molding, fiber extrusion and film manufacture (tubular bubble process or flat process) are the general processing techniques used with PP. The bulk of the film is used in biaxially oriented form which provides higher strength and better barrier properties.

1.2.4 Polypropylene copolymers, rubbers and blends

Besides homopolymers, PP materials are produced as copolymers and as blends.

PP copolymers account for 30% of PP products. The minor, second monomer is ethylene taken in 5–30% in order to improve impact properties especially at lower temperatures. Both random and block copolymers are produced.

(a) Random copolymers

Random copolymers contain 2–5% ethylene which is mixed with propylene in the reactor. The copolymerization processes are basically the same as those described for the homopolymers. Random copolymers have lower strength and crystallinity but higher clarity and improved impact strength than homopolymers.

(b) Ethylene–propylene rubbers (EPRs or EPMs)

Ethylene–propylene rubbers are a special type of random copolymers. They are completely amorphous and are obtained when the propylene content varies from 30–60 mol%. They are prepared by Ziegler solution polymerization techniques using soluble catalysts such as vanadium oxychloride and aluminum trihexyl mixtures. In order to develop good physical properties, these linear copolymers have to be vulcanized. The vulcanization may be done in two ways. For ethylene–propylene copolymers without unsaturated groups, only peroxide vulcanization is possible which occurs by the abstraction of hydrogen atoms from the chain followed by the combination of resulting macroradicals.

(c) Ethylene–propylene–diene terpolymers (EPDMs)

EPDMs are important commercial rubbers. They are prepared by the introduction of a small amount (no more than 1.2%) of a non-conjugated diene as the third comonomer, besides ethylene and propylene. The diene unit provides the necessary amount of unsaturation, usually 2–15 double bonds/1000 carbon atoms. These diene units allow the terpolymer to be crosslinked to a vulcanized rubber network. The diene comonomers commonly used for EPDM rubbers are:

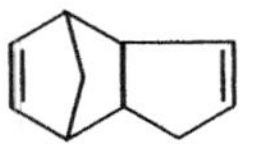

Dicyclopentadiene

$CH_2{=}CHCH_2{-}CH{=}CHCH_3$

1,4-Hexadiene

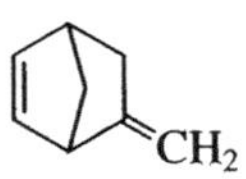

5-Methylene-2-norbornene

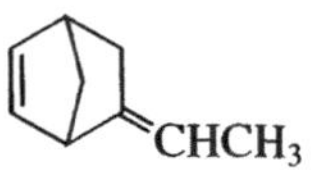

5-Ethylidene-2-norbornene

One of the double bonds copolymerizes with ethylene and propylene entering the main chain, while the other remains like a pendant unsaturation. The vulcanization is subsequently done using common sulfur vulcanizing systems from the rubber industry.

EPDM rubbers have low polarity, good mechanical strength, ozone resistance, abrasion resistance, weatherability and thermal stability. EPDMs are widely used as roofing materials in the building industry [45].

(d) Block copolymers

Block copolymers of different types are produced by means of two or more reaction steps using slurry, bulk or gas phase polymerization techniques. The selection of the preferred process is based on both the desired properties of the copolymer and the operating costs.

Numerous combinations and modifications have been used for the multistep operation. In the first reaction zone, only homopolymers of propylene are formed. They are then transferred to the second reactor where ethylene is fed in and it polymerizes to form the second block. The order of block formation may also be reversed.

(e) Polyallomers

A special type of block copolymers are the polyallomers which contain 5–15% ethylene. They are highly crystalline materials that combine the most valuable properties of both crystalline PP and HDPE. Their impact strength is three to four times that of PP and they withstand temperatures from −40 to 149°C. Strength properties of polyallomer film in the transverse direction are three to four times higher than those of PP or PE films [72].

(f) Reactor-made blends

Reactor-made blends may be prepared by the homopolymerization of propylene in the first reactor, followed by random copolymerization to a rubbery copolymer in the second reactor into which ethylene and propylene are fed. During this step, propylene homopolymer and the random ethylene–propylene copolymer form an intimately mixed blend. The common blends are prepared with less than 50% of the rubbery copolymer.

(g) Polyolefin thermoplastic elastomers

Polyolefin thermoplastic elastomers are blends of PP with up to 65% ethylene–propylene rubber. Thermoplastic elastomers are a class of specialty elastomers that behave as vulcanized elastomers at room temperature but can be processed as thermoplastics by extrusion and injection molding. These characteristics are obtained with linear macromolecules that are linked together by physical crosslinks resulting in physical network structures. Polyolefin thermoplastic elastomers represent about 23% of the global volume of thermoplastic elastomers, which also include polyester, polyurethane and styrenic thermoplastic elastomers [73]. In the case of polyester and of polyurethane the physical crosslinks have a polar nature. In polyolefin thermoplastic elastomers that are non-polar, the physical crosslinks have another nature. It is supposed that the PP homopolymer in the blend co-crystallizes with propylene short blocks of ethylene–propylene rubber and the resultant microcrystalline regions act as virtual crosslinks at room temperature. At higher temperatures, these microcrystalline regions reach their softening temperature and disappear so that the blend behaves like a thermoplastic.

These elastomers are low cost, general use rubbers with excellent resistance to ozone, oxygen and polar solvents and with good mechanical properties. Their main use is in the automotive field.

1.2.5 Applications

Around 30% of PP and its related copolymers is used as fibers and filaments produced by extrusion. Excellent wear and resistance to staining associated with low cost have led to uses in woven and non-woven form for carpeting in indoor and outdoor constructions, for clothing, inner liners, drapes, tea bags, wall coverings, furniture and automotive upholstery, and for ropes and netting [56].

Coatings for wire, piping or ducts, sheeting and film are also produced by extrusion.

Another large market for PP and for its copolymers is as film (both oriented and cast) for packaging of food and tobacco products and for textile goods.

Injection molded objects represent about 30% of PP applications. These are toys, laboratory and medical supplies, housewares, closures, appliances, luggage, automotive and truck battery cases, washing machines, agitators, pump housings, different parts in dishwashers, refrigerators and freezers.

In addition to its light weight, PP provides outstanding resistance to creep, high temperature rigidity, impact strength and resistance to corrosion. Disposable syringes and other medical applications are based on the possibility of applying sterilization either by autoclaving or by radiation [56].

Ethylene–propylene rubbers have found applications in the building industry for outside uses as sheeting and roofing [45]. Elastomer modified blends are widely used as bumpers, radiator grilles, fascia panels and protective strips in the car industry as well as weather strips for windscreens.

Atactic PP, which was an unwanted by-product for many years, is now specially produced for some applications. The material, which is intermediate between a wax and a rubber, is compatible with mineral fillers and bitumens. It has found use for roofing materials, as asphalt additive, for the formulation of sealing and adhesive compounds and for paper laminating [45].

Syndiotactic PP, which is not yet commercially available, has lower T_m, better impact resistance and more clarity than isotactic PP.

An obvious advantage of PP is its ease of recycling; this is especially important in many of the packaging and automotive applications [55].

1.3 POLYISOBUTYLENE (PIB)

In the 1980s, the USA's yearly consumption of this elastomer was 125 000 tons, compared with 150 000 tons in Western Europe and 30 000 tons in Japan.

The monomer isobutylene is obtained from petroleum sources through the following procedures: thermal and catalytic cracking of various petroleum and natural gas fractions, catalytic dehydrogenation of isobutane, depolymerization of di-isobutylene and alcohol dehydration.

Isobutylene is used in the petroleum industry for the synthesis of high octane number gasoline by alkylation, as an additive (polyisobutylene with low MW) for mineral oils, and in the synthetic elastomer industry.

Isobutylene polymerization occurs through a cationic mechanism at extremely low temperature (i.e. below −100°C) due to the low energy of activation [74]. Estimation of the activity of different acidic catalysts is evaluated based on the reaction rate, conversions and the MW of the obtained elastomer. The activity of Friedel–Crafts catalysts used for isobutylene polymerization decreases from BF_3 to $SnCl_4$ according to the following series: $BF_3 > AlBr_3 > AlCl_3 > TiCl_4 > SnCl_4$.

In the production of synthetic oils based on polyisobutylene production, a butane–butylene fraction, which contains 17–30% isobutylene, is used. For this application, polymerization is catalyzed by $AlCl_3$ and the polyisobutylene obtained has a low MW in the range 10 000–12 000. For the synthesis of a high MW polymer (above 100 000) a pure monomer is needed.

Isobutylene polymerization with BF_3 is one of the fastest known reactions in organic chemistry; even at very low temperatures, the reaction takes place violently, almost explosively. In some measure, the reaction rate can be regulated by adding diluents.

According to the modern conceptions, Friedel–Crafts catalysts form with alkenes complex molecular combinations able to activate polymerization reactions. Isobutylene is considered by J.P. Kennedy [75] as an ideal cationic monomer. It is an extremely reactive species under all kinds of acidic conditions and it is one of the few cationic monomers, indeed one of the few monomers, that can be readily polymerized from the lowest oligomer, through medium MWs up to the highest polymers with MWs in the millions.

Immediately after catalyst introduction (i.e. $AlCl_3$ in methyl chloride at −78°C) tens or hundreds of thousands of monomer molecules are mobilized, and polymer chains of MWs in the millions are formed within a fraction of a second.

One of the commercial procedures used to produce high MW polyisobutylene uses BF_3 as a catalyst diluted with ethylene. Liquid isobutylene and ethylene in a 1 : 1 ratio are sprayed on a cold, moving, stainless-steel trough. Polymerization is almost instantaneous and the considerable heat of polymerization is removed by the partial evaporation of the internal

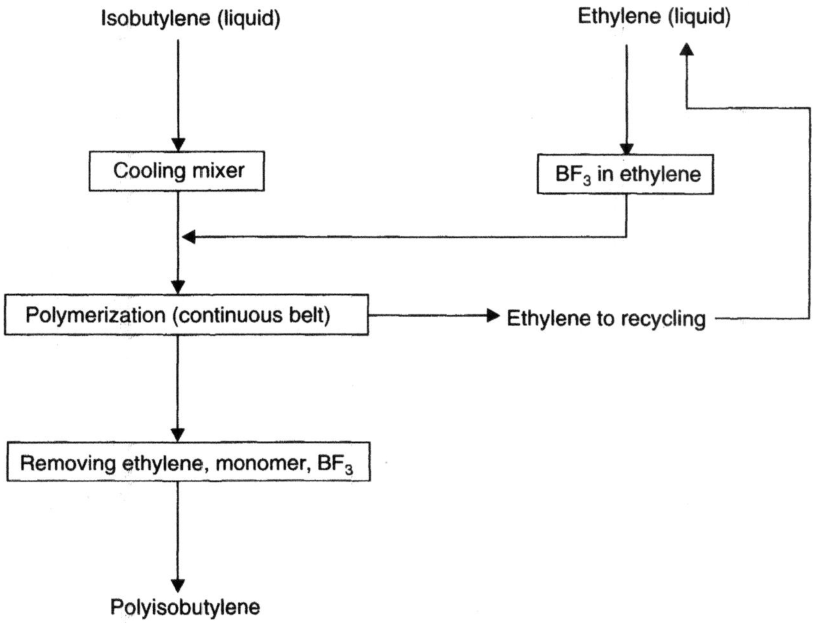

Figure 1.11 Solution polymerization of isobutylene.

coolant ethylene. The whole apparatus is enclosed to avoid loss of gases. The thick polyisobutylene layer transported by the moving belt is scraped off, and falls on steam heated compacting rolls to remove from the polymer BF_3, unreacted isobutylene, and entrapped ethylene. This process is shown in Figure 1.11.

Other variants use methyl chloride as the diluent, and $AlCl_3$ as the catalyst in a polymerization at −100°C. It is known that the energy of cationic polymerizations is lower than 15 kcal/mol, and even negative in some cases. The high polymerization rate, e.g. that observed for isobutylene in methyl chloride with $AlCl_3$ co-initiated with ethyl chloride, cannot be reduced by lowering the temperature to −180°C [74]. Isobutylene forms copolymers with conjugated dienes [76]. Also known is the graft copolymer (poly vinyl chloride-*g*-isobutylene) [77].

(a) Properties

Polyisobutylene has good low and high temperature properties; maximum continuous service temperature is +65°C and minimum continuous service temperature is −50°C. It is tacky and has a very low T_g (−70°C). It has good aging resistance, excellent sunlight resistance, resistance to chemical agents and very good electrical insulating properties [78]. Measurements of complex shear modulus and complex shear compliance of relatively high

MW over a high range of frequencies show clearly the glassy region, the viscoelastic region, the rubbery and flow regions, characteristic for amorphous polymers [79].

(b) Applications

Lower MW polymers are viscous liquids, but the more important polymers with medium MW (1×10^5–4×10^5) are highly extensible, tough, rubber-like products. These medium MW products are used in unvulcanized adhesive rubber compositions, in caulking and sealing compositions, to plasticize, tackify and impermeabilize polyethylene (for melt coating and heat sealing applications), as an additive to asphalts and waxes, and in vulcanizable rubber mixes and as a chewing gum base [45,80]. For these applications, the polymer is compounded with a wide variety of other materials, i.e. talc, calcium carbonate, asbestos, clays, etc. A large amount of PIB is used in blends with waxes because it improves wax low temperature brittleness and flaking, scuff resistance, tensile strength, flexibility, heat-sealing, strength and barrier characteristics. Wax–PIB blends are used in the paper industry to reduce water vapor transmission. Tank lining is an important application of PIB and its mixtures, especially with polyethylene. PIB–asphalt blends are useful in roofings, ditch and reservoir linings, and in general where moisture barrier characteristics are necessary [81]. The use as an additive for oils is related to PIB change in shape with increasing temperature. Since lubricating oil is not able to dissolve this polymer, its macromolecules will exist as coils when mixed at room temperature with the oil. As temperature increases the macromolecular chain tends to uncoil and the oil becomes a good solvent under these circumstances. This effect tends to counteract the decrease in the oil viscosity with the increasing temperature [80].

1.3.1 Butyl rubber

There is an abundance of information in the patent literature on the copolymerization of isobutylene with other monomers. The most important comonomers for isobutylene copolymerization are aromatic olefins like styrene and dienes, for example, isoprene.

Styrene and preferably α-methylstyrene were used as the second monomer in the two-stage synthesis of isobutylene-containing polymers. In this way, α-chloro-poly(isobutylene-block–styrene) was synthesized from the asymmetric prepolymer, but it was contaminated by homopolymers, the amount of which was dependent on the experimental conditions [82].

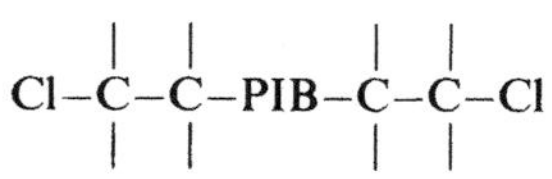

Among the different copolymers discussed in technical literature, only one product, an isobutylene–isoprene copolymer known as butyl rubber, has gained industrial importance. Its annual consumption in the USA is *c.* 150 000 tons.

In this isobutylene and isoprene copolymer, the latter is present in relatively small amounts, usually 0.6–3 mol% necessary to provide the unsaturation required for vulcanization. This unsaturation is only 0.6–3% of that of natural rubber, and many of the properties of butyl rubber are dependent on its very low unsaturation. This copolymer has the following microstructure:

$$\left[(-\underset{\displaystyle CH_3}{\overset{\displaystyle CH_3}{\overset{|}{\underset{|}{C}}}}-CH_2-)_{100}(-CH_2-\underset{\displaystyle CH_3}{\underset{|}{C}}=CH-CH_2)_x\right]_n$$

where x varies from 0.6 to 3.0 and n from 350 to 1000.

Isoprene units are bonded almost exclusively as 1,4 units; the amount of 1,2 and 3,4 units is less than 1% of the unsaturated groups.

$$[\text{PIB block}]-CH_2-\underset{\displaystyle CH_3}{\underset{|}{C}}=CH-CH_2-[\text{PIB block}]$$

Butyl rubber is produced by copolymerizing isobutylene with small amounts of isoprene, using $AlCl_3$ catalyst in methyl chloride solution at *c.* −100°C. Although the catalytic system is not essentially stereospecific, the isobutylene units bond head-to-tail, and the copolymer chains contain a very regular structure, owing to the symmetrical nature of isobutylene.

The process is continuous and the copolymerization reaction is cooled by circulating liquid ethylene in the mantle of the reactor. The cooled mixture of comonomers and solvent has a 27–30% amount of comonomers. The slurry resulting from copolymerization contains 15–20% copolymer and 10–12% unreacted comonomers. Copolymerization is controlled through the catalyst feeding rate; it is a very fast reaction, and at the mixing of reaction medium with catalyst solution, each drop is covered with a copolymer thin film and further comonomer molecules diffuse in the already formed polymer–monomer particle.

Due to the copolymer low thermal conductivity, temperature in the polymer–monomer particles is higher than the reaction medium temperature, a fact which favors the formation of a low MW copolymer. This is the reason why thorough attention has to be given to a fast, fine and homogeneous distribution of the catalyst solution in reaction mass.

For an acceptable conversion (70%) the reaction mass has to be circulated in the copolymerization reactor for 90–120 minutes [83].

Copolymerization development can be monitored by observing the change of the amount of isobutylene in the slurry, or by observing the ethylene vapor

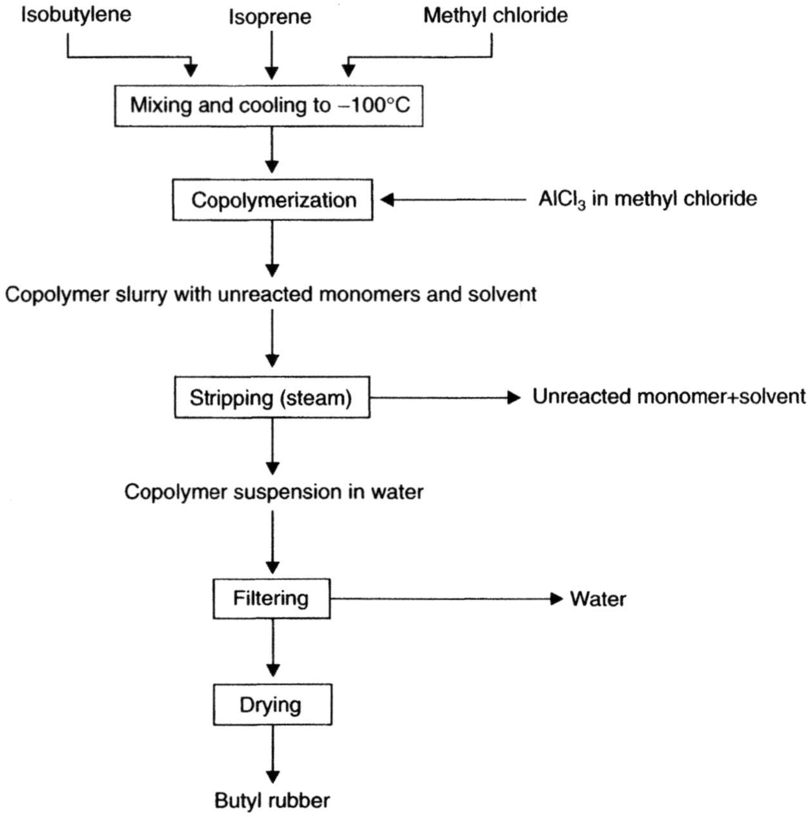

Figure 1.12 Butyl rubber synthesis.

pressure (ethylene which is circulating for the cooling of the system). The technical literature describes continuous plants for the synthesis of butyl elastomer, some of which have only one copolymerization reactor and some of which have several. The scheme for butyl rubber synthesis is shown in Figure 1.12.

During stripping, an antioxidant and zinc stearate are introduced. The antioxidant will prevent breakdown of the polymer in the subsequent operations. The stearate is added to prevent the agglomeration, or sticking together, of the wet crumb [84].

A solution process, developed in the former USSR, uses a C_5–C_7 hydrocarbon as the solvent and aluminum alkyl halide as the catalyst. The copolymerization is conducted in scraped surface reactors at −90° to −50°C. Producing the elastomer in solution avoids the use of methyl chloride and there is an advantage when butyl rubber has to be transformed in halogenated rubber [85].

Standard grades of butyl rubber have various mole per cent unsaturation. Butyl rubber is shipped in uncured condition and contains 1 part (by wt) of zinc stearate and 0.5 part (by wt) of an antioxidant. It is tacky and has a tendency to flow under its own weight and to adhere to storage racks. It does not require mastication before being compounded.

(a) Properties

Different types of butyl rubber have MWs in the range 40 000 to 85 000 and a density *c.* 0.91 g/cm^3.

The principal drawback of this copolymer is its low elasticity at the ambient temperature. Compared with natural rubber, its gas permeability is 10–30 times lower. Maximum continuous service temperature is 100°C [78].

Due to low unsaturation, vulcanization of butyl rubber needs a longer time and higher temperature than other synthetic rubbers. Butyl rubber is free from branching provided that the degree of unsaturation is low. As the unsaturation is increased to 5%, cure rates are faster and the vulcanizates have better thermostability and higher modulus but lower ozone resistance [86]. In the case of sulfur vulcanization, the correct choice of the accelerator is very important. Vulcanization introduces a chemical crosslink approximately every 250 carbon atoms, producing a molecular network. The vulcanization can be done with phenolic resins instead of sulfur [85].

Butyl rubber belongs to the crystallizable rubbers group. It possesses a good adhesion, is stable to oxygen action, to that of oxidants, to most inorganic acids, alkali, salts and oxygenated organic compounds like alcohols, esters, ketones, animal and vegetable oils, and has a good weatherability.

The combination of low absorption properties with a good ozone stability and with good electrical indices, opens great possibilities of using butyl rubber as an electrical insulator. Carbon black is a very useful ingredient and is recommended for this kind of rubber; for reinforcement in this case, the ratio 1 : 1 is used.

Butyl rubber can be processed on conventional equipment and compounded besides carbon black with conventional mineral fillers, plasticizers, mineral oils and processing aids. It is usually cured in the temperature range of 148–200°C, but fast curing formulations give satisfactory vulcanizates at temperatures as low as 100°C.

A commercially important butyl rubber derivative is chlorinated butyl rubber, known also as chlorobutyl. It is a high MW isobutylene–isoprene copolymer containing reactive (allylic) chlorine atoms. It is superior to butyl rubber since it has a faster rate of vulcanization and enhanced compatibility with many other elastomers. Both brominated and chlorinated butyl rubbers are available on the market [86]. Their structure, determined by

NMR, is

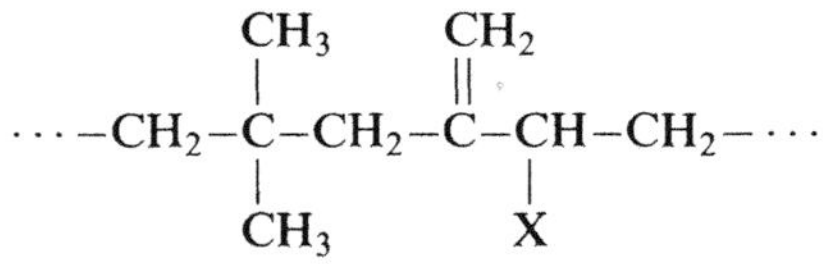

where X stands for bromine or chlorine.

Halogenated butyls are vulcanized by the same agents as butyl rubber, as well as by zinc oxide, diamines, peroxides and dithiols [85,86].

One of the drawbacks of butyl rubber is its poor adhesive properties to metals and to other rubbers. This is largely due to the lack of polar groups in the microstructure of this copolymer; halogenation remedies this deficiency [45].

G. Balasz *et al.* [87] reported on a DSC study which shows that butyl rubber is very sensitive to light but more stable to thermo-oxidation or to ozone attack.

(b) Applications

Due to its very low permeability to gases, butyl rubber is used extensively for inner tubes, and in the form of chlorobutyl rubber, in tubeless tires. The service life of inner tubes made of butyl rubber is 10–18% longer than that of those made of natural rubber.

Its stability to steam and relatively high temperature permits the use of butyl rubber in the production of diaphragms and conveyer belts for hot materials.

Excellent chemical stability determines its use for tanks and pumps for very aggressive liquids, lining, for the manufacturing of textiles impermeable to yperite gas, various gas masks, various impermeable textiles, rubber coated pelerine, pneumatic boats, windproof and waterproof clothing, belts, pipes, steam and firefighting hoses, gaskets, adhesives, sealants, and caulkings. Most butyl caulking compounds cure by solvent release. Some compositions are based on cured rubber solvated by mineral spirits. Some more solvent may be used to incorporate more filler into the compound [82]. Butyl rubber also proved suitable as a high quality roofing membrane [88].

Chlorobutyl rubber is widely used in applications where resistance to pollutants, barrier properties, speed and versatility of curing, ability to blend and covulcanize with other rubbers, and where high heat resistant vulcanizates are required.

Butyl rubber sealants can be obtained using butyl crude rubber, which can be dissolved in an appropriate organic solvent to give a solvent based butyl sealant. Non-shrinking caulking compounds can be obtained by using low MW polyisobutylene oils to soften the butyl rubber.

As shown in Table 1.3, the formulation of such a sealant is complex because of the numerous additives necessary for its compounding. In the

Table 1.3 Butyl rubber based caulking compound [89]. (Reprinted by permission of J. Wiley & Sons Inc. from *Construction Sealants and Adhesives* by J.R. Panek and J.P. Cook, 1984. Copyright © J. Wiley & Sons)

Ingredient	*% by weight*
Polyisobutylene	2.5
Butyl rubber, 50% solution in mineral spirits	20.5
Talc powder	30
Calcium carbonate filler	20
TiO_2 pigment	3
Adhesion resin	4
Polybutylene	10
Plasticizer	2
Thixotropic agents	2
Drying catalysts	0.05
Mineral spirits	6
Total solids content	84

formulation, butyl rubber is used for structure but polyisobutylene and an adhesive resin are used for tack. The trace driers are used to give a tack-free surface and slowly cure. The total butyl composition is approximately 22%, but the total solvent is 16%, which could result in approximately 30% volume shrinkage. Most butyl caulking compounds cure by solvent release. Some compositions are based on crude rubber solvated by mineral spirits. Some more solvent may be used to incorporate more filler into the compound [89].

Besides adhesives and sealants, butyl rubber is used for weather seals and gaskets in buildings. Butyls require some joint cleaning, but extensive surface preparation is not required. Laitance and loose concrete should be removed from the faces of the concrete joints, but priming is unnecessary. Metal surfaces may be wiped with an oil-free cloth wet with solvent to remove contaminants. Wood surfaces and glass should be wiped clean. Butyl caulks work best in relatively non-moving joints. They are used on the mating surface of heating and air-conditioning ducts and for sealing openings where pipes and ducts pass through roofs and partitions. Butyl caulks can be used below grade on masonry joints. They may be used in lap joints of sheet siding, metal roofs, glazing, wooden doors and windows, and protective sidings. Butyls adhere to aluminum, steel and vinyl siding. Some advantages and disadvatages of butyl caulks are presented in Table 1.4.

1.4 POLY(1-BUTYLENE)

Another polyolefinic thermoplastic material is poly(1-butylene) prepared by the stereospecific polymerization of 1-butylene in the presence of

Table 1.4 Advantages and disadvantages of butyl caulks [92]. (Reprinted by permission of J. Wiley & Sons Inc. from *Construction Sealants and Adhesives* by J.R. Panek and J.P. Cook, 1984. Copyright © J. Wiley & Sons)

Advantages	*Disadvantages*
1. Reasonable cost	1. Very slow cure
2. Availability	2. High shrinkage
3. Good flexibility	3. High compression set
4. Good adhesion to most substances	4. Limited to joints with ±7.5% joint movement
5. One-component	5. Not recommended for expansion joints
6. Little surface preparation	
7. Good water resistance	
8. Good color stability	
9. Four colors available	
10. Only material for capping neoprene gaskets	

Ziegler–Natta catalytic system [90,91]:

$$-(CH_2-\underset{\displaystyle |}{\overset{\displaystyle C_2H_5}{C}}H-)_n-$$

The polymerization of 1-butylene was also effected using metallocene–alumoxane catalysts [28].

Commercial products have high MW, in the range 700 000–3 000 000 and are 98–99.5% isotactic. The polymer can exhibit three crystalline forms [92]. The first form is obtained by crystallization from the melt, has $T_m = 124°C$ and the density $0.89\,g/cm^3$. This form is unstable and after several days changes to a stable form with $T_m = 135°C$ and the density $0.95\,g/cm^3$. The hardness, the stiffness and the yield strength all increase. This characteristic has to be considered and the polymer must be aged for a week after processing. The third crystalline form is obtained by crystallization from solution.

Polybutylene exhibits the general properties of a polyolefin. Its outstanding property is a very high creep resistance that was advantageously used for the manufacture of pipes having a much reduced wall thickness as compared to PE and PP pipes. The main use of polybutylene is for pipes for cold and hot water plumbing as well as for the transportation of abrasive and corrosive materials in mining, chemical and power generation industries [93].

Film applications include food packaging and industrial sheeting as abrasion resistant liner.

Atactic polybutylene is used for roof coverings and sealing compounds [45].

1.5 POLY(4-METHYL-1-PENTENE) (PMP)

PMP is a high performance specialty plastic without being a high volume material. The production volume is 6000 tons per year but the demand for PMP has grown steadily at a rate of 20% per year [45]. First produced by ICI Ltd of Britain in 1965, it is now manufactured and marketed solely by Mitsui Petrochemical Industries.

The repeating unit in PMP is:

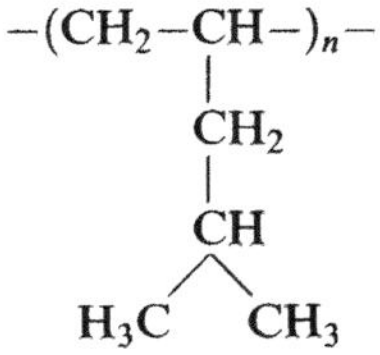

The monomer, 4-methyl-1-pentene, is obtained by the dimerization of propylene:

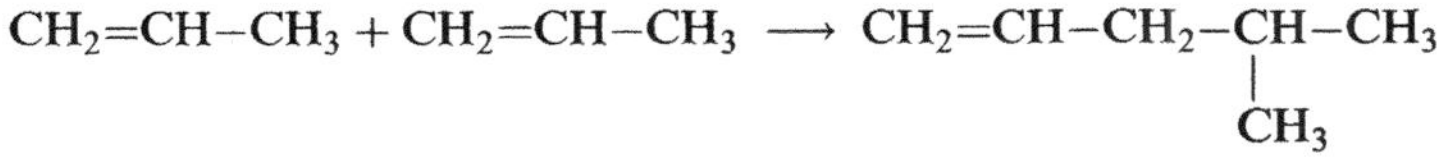

The polymerization of the monomer is effected in a hydrocarbon diluent at 30–60°C and atmospheric pressure in the presence of a Ziegler–Natta catalyst composed of titanium trichloride and diethyl aluminum chloride. At the end of the polymerization the catalyst residues are washed out with an alcohol and the polymer is dried and compounded with antioxidants and other additives.

Experimental studies have been conducted on the polymerization of 4-methyl-1-pentene with metallocene–alumoxane catalyst systems [28].

The commercial polymer is isotactic and about 40% crystalline. In the crystalline part the molecules are arranged in a helix with seven monomer units per two turns of the helix. Crystalline and amorphous zones have similar refractive indices and similar densities and this accounts for the outstanding high degree of transparency of the material. The white light transmission is 90%. Some properties of PMP are in Table 1.5.

Table 1.5 Some characteristics of PMP

Property	*Value*
Density (g/cm^3)	0.834
Melting temperature (°C)	240
Glass transition temperature (°C)	50–60
Tensile strength (MPa)	195
Tensile modulus (MPa)	800
Flexural modulus (MPa)	750

PMP has the lowest density of any available plastics material. The bulky side chains contribute to the stiffening of the chains resulting in high T_m and T_g. The high permeability to gases is due to the same structural factor. PMP is capable of separating nitrogen and oxygen due to the great difference in the permeability of the two gases [94].

The presence of relatively labile tertiary hydrogen atoms on the chain makes the material more sensitive to oxidation especially at high processing temperatures and in working conditions at elevated temperatures. Hence compounding requires the use of efficient antioxidant systems.

The applications of PMP take advantage of the high transparency and heat resistance of the material, which can be used at up to 150°C. It can be bonded to metals and other materials through its own shrinkage without the use of adhesives. Medical and laboratory sterilizable equipment such as syringes, blood test cells, animal cages as well as cosmetic containers account for one third of PMP uses. Other applications are for heat resistant houseware such as microwave trays, coffee maker parts, nursery bottles and many others.

In the industrial area, PMP has been used as material for resin molds designed for manufacturing thermosetting plastics such as polyurethanes, epoxies, phenolics, melamine and urea resins as well as for molding contact lenses. It is also used for covering heat resistant wire, power cables, fuse covers, lighting fixtures, optical memory and optical fibers.

REFERENCES

1. Anon. (1994) *C & EN*, 31 October, p. 10.
2. Bennett, R.A. (1991) *Polymer Preprints*, **32**(2), 121. Paper presented at the ACS Philadelphia, Pennsylvania Meeting, June.
3. Haley, J. and Turner, L. (1975) in *Modern Petroleum Technology* (ed. G.D. Hobson), Applied Science Publishers, London.
4. Restaino, M.C. (1990) in *Handbook of Plastic Materials and Technology* (ed. I.I. Rubin), J. Wiley & Sons, New York, p. 311.
5. Anon. (1979) *Chem. Eng.*, **86**, 26.
6. Manders, P.W. (1994) *Modern Plastics, Encyclopedia '95*, **71**(12), p. B-3.
7. Anon. (1994) *Plastics Engineering*, July, p. 14.
8. Romanini, D. (1982) *Polym. Plast. Technol. Eng.*, **19**(2), 201.
9. Symcox, R.O. and Ehrlich, P. (1962) *J. Amer. Chem. Soc.*, **84**, 531.
10. Albright, L.F. (1985) *Processes for Major Addition-Type Plastics and their Monomers*, Robert Krieger Publ. Comp., Florida, p. 52.
11. Bibee, D.V. (1990) in *Handbook of Plastic Materials and Technology* (ed. I.I. Rubin), J. Wiley & Sons, New York, p. 317.
12. Usami, T. (1989) *Handbook of Polymer Science and Technology. Vol. 2: Performance Properties of Plastics and Elastomers* (ed. N.P. Cheremisinoff), Marcel Dekker Inc., New York, p. 437.
13. Usami, T. and Takayama, S. (1984) *Macromolecules*, **17**, 1756.

14. Usami, T. and Takayama, S. (1984) *Polym. J.*, **16**, 731.
15. Ohtami, H., Tsuge, S. and Usami, T. (1984) *Macromolecules*, **17**, 2557.
16. Usami, T., Gotoh, Y., Takayama, S. *et al.* (1987) *Macromolecules*, **20**, 1557.
17. Wild, L., Ranganath, R. and Barlow, A. (1977) *J. Appl. Polym. Sci.*, **21**, 3331.
18. Axelson, D.E. and Knapp, W.C. (1980) *J. Appl. Polym. Sci.*, **25**, 119.
19. Al-Sammerai, D. and Al-Nidawy, N.K. (1989) in *Handbook of Polymer Science and Technology. Vol. 2: Performance Properties of Plastics and Elastomers* (ed. N.P. Cheremisinoff), Marcel Dekker Inc., New York, p. 341.
20. Zucchini, U. and Dall'Occo, T. (1991) *Polym. Sci.* (Symp. Proc. Polym. 91), **1**, 221.
21. Barazzoni, L. *et al.* (1993) *J. Mol. Catal.*, **82**(1), 17.
22. Floyd, S., Heiskanen, T., Taylor, T.W. and Mann, G.E. (1987) *J. Appl. Polym. Sci.*, **33**, 1021.
23. Stevens, M.P. (1990) *Polymer Chemistry*, 2nd edn, Oxford University Press, New York.
24. d'Ouville, E.L. (1960) in *Polyethylene*, 2nd edn (eds A. Renfrew and P. Morgan), Interscience, New York.
25. Batleman, H.L. (1975) *Plastics Eng.*, **31**(4), 73.
26. Rasmussen, D.M. (1972) *Chem. Eng.*, **79**, 104.
27. Kaminsky, W. (1986) in *Studies in Surface Science and Catalysis. Catalytic Polymerization of Olefins* (eds T. Keii and K. Soga), Vol. 25, Elsevier, Tokyo, p. 293.
28. Gupta, V.K., Satish, S. and Bhardwaj, I.S. (1994) *J. Macromol. Sci., Rev. Macromol. Chem. Phys.*, **C34**(3), 439.
29. Kaminsky, W. and Lindenhof, N.M. (1990) *Bull. Soc. Chim. Belg.*, **99**, 103.
30. Zucchini, U., Dall'Occo, T. and Resconi, L. (1993) *Indian J. Tech.*, **31**, 247.
31. Shishta, C., Hathorn, R.M. and Marks, T.J. (1992) *J. Amer. Chem. Soc.* **113**, 3623.
32. Castonguay, L.A. and Rappe, A.K. (1992) *J. Amer. Chem. Soc.*, **114**, 5832.
33. Giannetti, E., Nicoletti, G.M. and Mazzocchi, R. (1985) *J. Polym. Sci., Polym. Chem. Ed.*, **23**, 2117.
34. Reddy, S.S., Shashidhar, G. and Sivaram, S. (1993) *Macromolecules*, **26**, 1180.
35. Satyanarayana, G. and Sivaram, S. (1993) *Macromolecules*, **26**, 4712.
36. Bochman, M., Jaggar, A.J. and Nicholls, J.C. (1990) *Angew. Chem., Int. Ed. Engl.*, **29**, 780.
37. Short, J.N. (1981) in *Kirk–Othmer Encyclopedia of Chemical Technology*, 3rd edn, Vol. 16, J. Wiley & Sons, New York, p. 386.
38. Pate, T.J. (1990) in *Handbook of Plastic Materials and Technology* (ed. I.I. Rubin), J. Wiley & Sons, New York, p. 327.
39. Anon. (1979) *Chem. Eng.*, **86**, 80.
40. Hanna, R.D. (1990) in *Handbook of Plastic Materials and Technology* (ed. I.I. Rubin), J. Wiley & Sons, New York, p. 349.
41. Cordova, D.S. and Donnelly D.S. (1990) in *Handbook of Plastic Materials and Technology* (ed. I.I. Rubin), J. Wiley & Sons, New York, p. 919.
42. Standard Specification (1974), ANSI/ASTM D1238, Philadelphia.
43. Hawkins, E.L. (ed.) (1972) *Polymer Stabilization*, J. Wiley & Sons, New York.
44. Briston, J.H. (1989) *Plastics Films*, 3rd edn, Longman Scientific & Technical, Chapter 2.
45. Feldman, D. (1989) *Polymeric Building Materials*, Elsevier Applied Science, London.

46. Mark, H.F. (ed.) (1967) *Encyclopedia of Polymer Science and Technology*, Vol. 6, Interscience, New York, p. 324.
47. Al-Sammerrai, D. (1986) *J. Appl. Polym. Sci.*, **31**, 1.
48. Brydson, J.A. (1995) *Plastics Materials*, 6th edn, Butterworths, London.
49. Blanchard, R.R. (1990) in *Handbook of Plastic Materials and Technology* (ed. I.I. Rubin), J. Wiley & Sons, New York, p. 67.
50. Baker, G. (1980) *Plastics Engineering*, **36**(2), 45.
51. Earnest, T.R. and MacKnight, W.J. (1981) *J. Polym. Sci., Macromol. Reviews*, **16**, 41.
52. Longworth, R. (1990) in *Handbook of Plastic Materials and Technology* (ed. I.I. Rubin), J. Wiley & Sons, New York, p. 137.
53. Gautier, P.A. *et al.* (1993) Eur. Patent Appl. EP534,570; *CA*, **119**(7), 140026.
54. Xu, F.Y. and Chien, J.C.W. (1993) *Macromolecules*, **26**(14), 3485.
55. Anon. (1993) *Kunststoffe* (Engl.), **83**(10), 7.
56. Hanna, R.D. (1990) in *Handbook of Plastic Materials and Technology* (ed. I.I. Rubin), J. Wiley & Sons, New York, p. 433.
57. Henrici-Olive, G. and Olive, S. (1981) *Chemtech.*, **11**, 746.
58. Boor, J. (1970) *IEC Prod. Res. Dev.*, **9**, 439.
59. Zucchini, U. and Cecchin, G. (1983) *Adv. Polym. Sci.*, **51**, 101.
60. Vandenberg, E. and Repka, B. (1977) *Encyclopedia of Polymer Science and Technology*, Vol. 29, J. Wiley & Sons, New York, p. 337.
61. Anon. (1994) *World Plastics & Rubber Technology*, **7**, 53.
62. DiDrusco, G. and Luciani, L. (1981) *J. Appl. Polym. Sci., Applied Polym. Symp*, **36**, 95.
63. Cipriani, G. and Trischman, C.A. (1981) *J. Appl. Polym. Sci., Applied Polym. Symp.*, **36**, 101.
64. Crespi, G. and Luciani, L. (1981) in *Kirk–Othmer Encylopedia of Chemical Technology* (ed. M. Grayson), 3rd edn, Vol. 16, J. Wiley & Sons, New York, p. 453.
65. Anon. (1983) *Plast. Technol.*, December, 89.
66. Anon. (1983) *Plast. Eng.*, December, 33.
67. Castonguay, L.A. and Rappe, A.K. (1992) *J. Amer. Chem. Soc.*, **114**, 5832.
68. Burger, P., Hortmann, K. and Brintzinger, H.H. (1993) *Macromol. Chem. Macromol. Symp.*, **66**, 127.
69. Bochmann, M., Jaggar, A.J. and Nicholls, J.C. (1990) *Angew. Chem., Int. Ed. Engl.*, **29**, 780.
70. Brockschmidt, A. (1982). *Plastic Technology*, **29**(3), 67.
71. Heck, R. (1984) in *Modern Plastics Encyclopedia*, Vol. 60, McGraw-Hill, New York, p. 142.
72. Sharp, G. (1990) in *Handbook of Plastic Materials and Technology* (ed. I.I. Rubin), J. Wiley & Sons, New York, p. 215.
73. Rader, C.P. (1993) *Kunststoffe*, **83**(10), 36.
74. Marek, M., Pecka, J. and Halaska, V. (1984) in *Cationic Polymerization and Related Processes* (ed. E.J. Goethals), Academic Press, London, p. 17.
75. Kennedy, J.P. (1975) *Cationic Polymerization of Olefins, a Critical Inventory*, J. Wiley & Sons, New York.
76. Cesca, S. (1984) in *Cationic Polymerization and Related Processes* (ed. E.J. Goethals), Academic Press, London, p. 105.

77. Lucas, R. and Toman, L. (1990) *Polymer Bulletin (Berlin)*, **29**(3), 285.
78. Daniels, C.A. (1989) *Polymers: Structure and Properties*. Technomic Publ. Co., Lancaster, PA, p. 75.
79. Ward, J.M. and Hadley, D.M. (1993) *Mechanical Properties of Solid Polymers*, J. Wiley & Sons, New York.
80. Seymour, R.B. and Carraher Jr, C.E. (1992) *Polymer Chemistry*, 3rd edn, Marcel Dekker, New York.
81. Kennedy, J.P. and Kirshenbaum, T. (1971) in *Vinyl and Diene Monomers*, Part 2 (ed. E.D. Leonard), J. Wiley & Sons, New York, p. 734.
82. Jerome, R., Fayt, R. and Ohadi, T. (1984) *Prog. Polym. Sci.*, **10**, 87.
83. Feldman, D. (1974) *Macromolecular Compounds Technology*, Ed. Technica, Bucharest.
84. Fusco, J.V. and Hous, P. (1987) in *Rubber Technology* (ed. M. Morton), Van Nostrand Reinhold Co., New York, p. 284.
85. Mark, H.F., Bikales, N.M., Overberger, C.G. and Menges, G. (1987) *Encyclopedia of Polymer Science and Engineering*, Vol. 8, J. Wiley & Sons, New York.
86. Brydson, J.A. (1978) *Rubber Chemistry*, Appl. Science Publ. Ltd, London.
87. Balasz, G., Doszlop, S., Biro, D. and Palotas, L. (1990) *J. Therm. Anal.*, **36**(5), 1861.
88. Anon. (1982) *Chem. Week*, **133**(17), 16.
89. Panek, J.R. and Cook, J.P. (1984) *Construction Sealants and Adhesives*, 2nd edn., J. Wiley & Sons, New York.
90. Chatterjee, A.M. (1990) in *Handbook of Plastic Materials and Technology* (ed. I.I. Rubin), J. Wiley & Sons, New York, p. 247.
91. Rubin, I.D. (1968) *Poly(1-Butene). Its Preparation and Properties*, Gordon and Breach Science Publishers, New York.
92. Nakafuku, C. and Miyaki, T. (1983) *Polymer*, **24**, 141.
93. Lindegren, C.R. (1975) *Plast. Eng.*, July, 38.
94. Ohi, H. and Furuya, K. (1990) in *Handbook of Plastic Materials and Technology* (ed. I.I. Rubin), J. Wiley & Sons, New York, p. 375.

2 Vinyl polymers

2.1 POLYSTYRENE (PS)

2.1.1 Introduction

Although known since 1845, polystyrene became a commercial product only after the second world war, as a result of a high production of styrene (S) for the synthetic rubber. S, along with most of its substituted derivatives, can act as either an electron donating or an electron attracting center. Therefore, in addition to free radical polymerization, these monomers can be polymerized by all other primary propagation mechanisms: anionic, cationic and coordination (Ziegler–Natta) [1].

The world production of 3.3 million tons in 1970 is surpassed by 5 million tons in 1980. In USA, the 2.5 million tons production of PS and its copolymers in 1983 has reached 3.8 million tons in 1993 [2]. However, the production rate of PS is lower than that of total thermoplastics, due not only to the low number of applications, but especially to the fact that styrene production is oriented to obtaining copolymers and to the crosslinking of unsaturated polyesters.

S, like many other aromatic compounds, is toxic, and concentrations in the atmosphere should not exceed 10 ppm [3]. S polymerization takes place in most cases through the radicalic thermal mechanism, with or without initiators (peroxides). The system needs heat only for starting as the reaction is exothermic (71 kJ/mol). As the temperature or initiator concentration increases, reaction rate increases and in the meantime, polystyrene MW decreases.

At large scale, polymerization takes place in bulk, solution or the suspension technique [4].

2.1.2 Bulk polymerization

Discontinuous bulk polymerization with or without initiators is realized in glass forms where a solid PS is obtained. Because of the difficulties in removing the reaction heat and because of the low thermal conductivity of PS, a polydisperse product is obtained. In the case of thick blocks, where the polymerization is uncompleted, styrene traces lead to the polymer aging, a fact which may be visualized through crackings and opacification.

Continuous mass polymerization permits the formation of a PS without styrene traces, and with a high MW. The process has two stages. In the first one, a partial polymerization leads to the formation of a syrup (prepolymer). In the second stage, in a tower reactor, the polymerization is taken substantially to completion.

Prepolymerization in large aluminum kettles is completed at 80°C until the conversion reaches 30–35%. Continuously, the syrup is fed into the top of a tower reactor (8 m high). The temperatures in the tower are controlled in six zones, independent of each other, and increase from top to bottom from 100 to 200°C. This procedure takes into account a certain ratio between the main parameters, mainly time, tower height and reaction rate. The flow chart is presented in Figure 2.1.

In many plants, the polymerization phase is carried out in cylindrical reactors, which may be horizontally or vertically arranged. Besides this

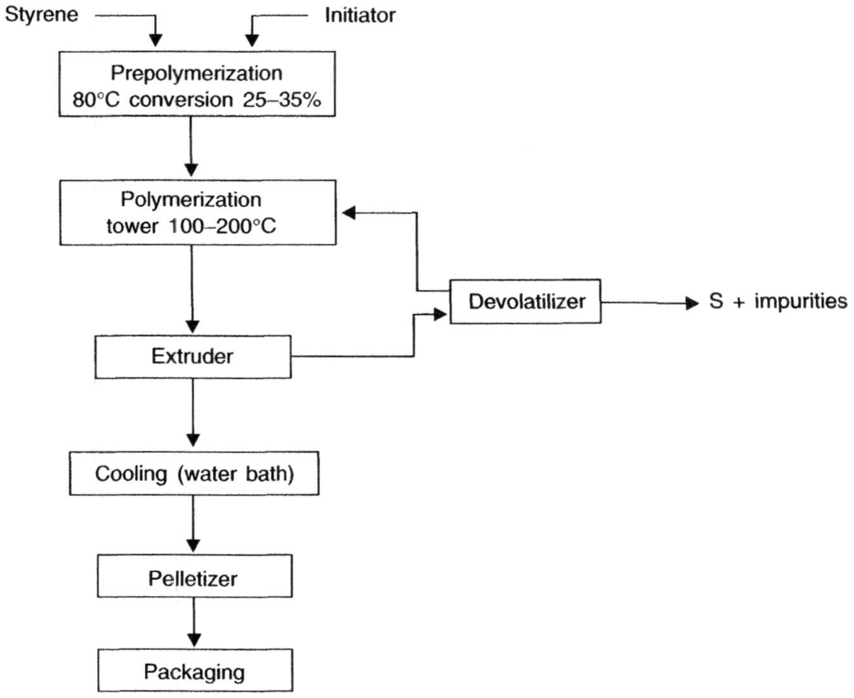

Figure 2.1 Styrene bulk polymerization.

modification before the melt extrusion, PS has to pass through a devolatilizer, where it is foamed at 10–20 mm Hg low pressure, for the release of unreacted styrene and impurities like ethyl-benzene and isopropyl benzene. The use of a recycle tubular reactor for the prepolymerization seems to be economically attractive due to its low energy consumption compared to conventional systems based on batch or continuous tank reactors [5]. A mathematical model for the continuous polymerization process has been developed to include three different types of polymerization reactors: stirred tank, plug flow reactor and tubular reactor [6].

In order to decrease the viscosity of the polymer, it is sometimes preferred to dilute the monomer with 3–12% ethyl-benzene as in the so called 'continuous mass polymerization with solvent' process. Polymerization is carried out in a series of reactors in similar thermal conditions as in the continuous mass process, i.e. under reduced pressure in the devolatilizer (225–250°C), the unreacted styrene and the ethyl-benzene are flushed off, condensed, and recycled.

This procedure may be easier to control because the medium viscosity is lower and the temperature is uniformly distributed. The solvent's presence affects the polymerization rate and the MW is lower because of the chain transfer activity of the solvent. For example, the chain transfer constant of ethyl-benzene is 6.7×10^{-5} at 60°C and 16.2×10^{-5} at 100°C.

2.1.3 Emulsion polymerization

Emulsion polymerization is less frequently used in the production of PS because the presence of the emulsifier traces in the final product affects some of its main characteristics. Some PS latices are obtained for special applications and the technology is identical with that of vinyl chloride emulsion polymerization. Different aspects of this polymerization are discussed in the literature [7–10].

2.1.4 Solution polymerization

This process provides a high purity PS. The principal technical challenges are the control of the exothermic heat of reaction and the handling of the highly viscous polymer.

A schematic diagram of the continuous solution S polymerization is presented in Figure 2.2.

This polymerization takes place at sequentially higher temperature from 90 to 180°C, with a final conversion at 60–90%. After the volatilization of the solvent and remaining S, the PS melt is extruded and pelletized. S and the solvent, usually ethyl-benzene, are condensed and recycled to the reaction feed [1].

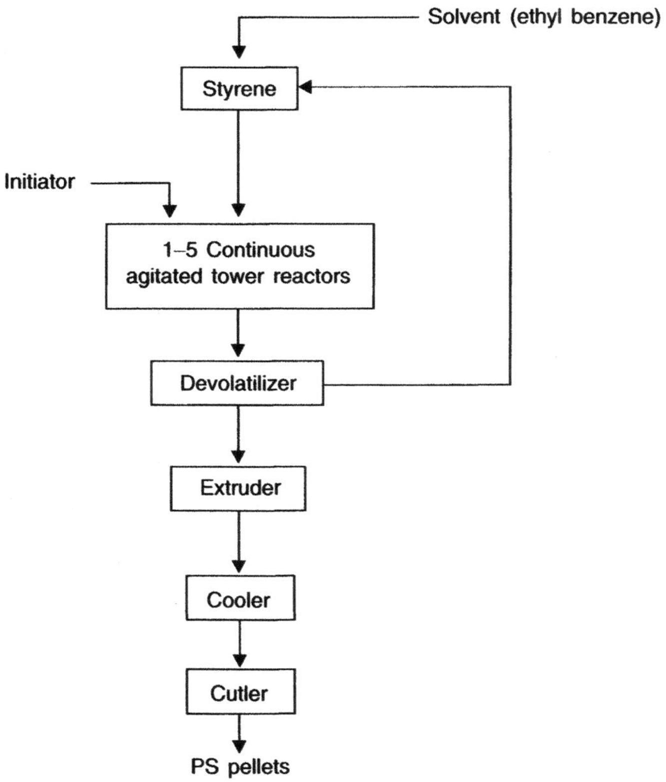

Figure 2.2 Styrene continuous solution polymerization.

2.1.5 Suspension polymerization

In this technology, styrene is dispersed in water with a dispersing agent (suspending agent) like poly(vinyl alcohol), calcium phosphate, magnesium silicates, etc. Generally, a combination of initiators like benzoyl peroxide and *t*-butyl hydroperoxide are used. The main phases of this procedure are:

- the removal of the inhibitor from styrene;
- preparation of the dispersing agent, initiator, etc.;
- polymerization;
- suspension processing;
- drying;
- pelletizing;
- stirring.

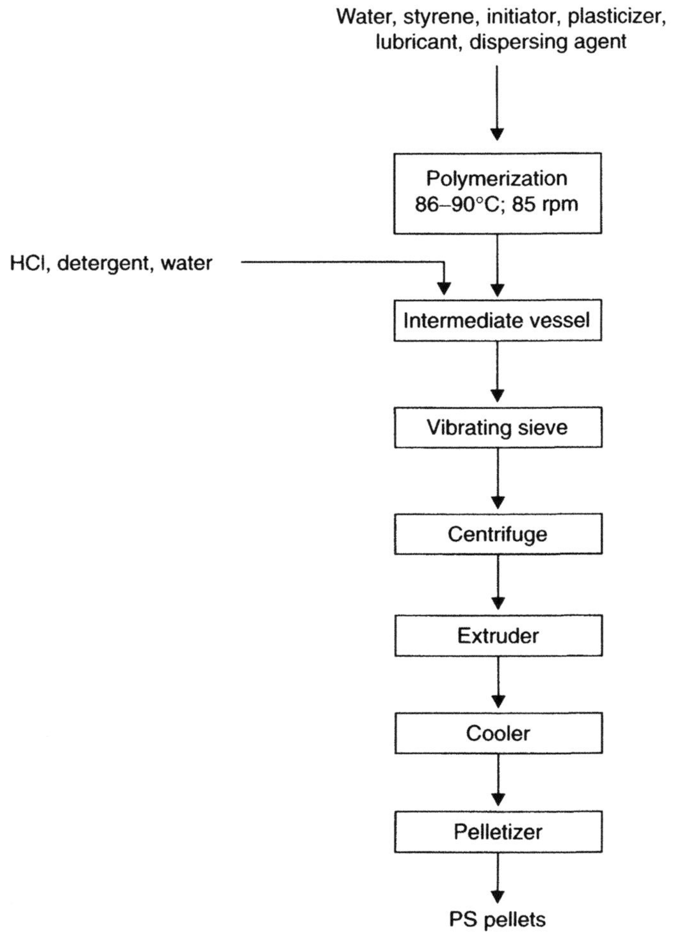

Figure 2.3 Styrene suspension polymerization.

The process uses demineralized water, depending on the PS applications. In some cases, plasticizer and lubricants are added. The styrene suspension polymerization process is presented in Figure 2.3.

For a better control of the bead's characteristics, the dispersing agent is introduced in the polymerization reactor in portions. After polymerization, an amount of 36% HCl is added to the already cooled suspension, to dissolve the phosphate on the bead's surface. This leads to the improvement of the bead's clarity. The pH is adjusted at *c.* 3.0. A detergent may improve the polymer's color; it is introduced with the water for washing the beads. After separation from the centrifuge, PS with 1.5% water is sent to the rotating dryer, where the drying takes place at 140°C, with air (counter-current).

For a colored PS, a colorant is added to the beads before extrusion, and to improve the processing of PS pellets, an external lubricant is added. Simulation of the conversion and average MW in bulk and suspension S polymerization process is reported [11].

Styrene suspension polymerization leads to a wide range of different grades, such as:

- general purpose PS;
- high MW PS;
- high softening point PS;
- low MW PS;
- high shock resistant PS;
- medium shock resistant PS.

2.1.6 Properties

Polymerization conditions determines PS purity, styrene content, MW, branching degree and polydispersity. PS obtained through continuous mass polymerization does not contain emulsifier, salts, initiator or foreign impurities, providing this polymer with superior dielectric properties.

The polymer with a MW of the order of 50 000 is used for coatings and a 200 000–300 000 MW polymer is suitable for processing through pressing or injection molding. PS is an amorphous polymer which has a high degree of branching. Roentgenographically, no orientation is detected after stretching.

The good dielectric properties are determined by its very low polarity ($n = 0.73 \times 10^{-18}$); in this regard, PS is inferior only to PE. It has a good chemical stability (towards acids, alkalis, etc.) and stability to water. It is soluble in chlorinated and aromatic hydrocarbons, dioxan, esters like ethyl and butyl acetate and insoluble in kerosene and alcohols. At room temperature it is hard, transparent, and has a density of 1.05 g/cm^3. Transmission of visible light is 80–90%. Pure PS without extraneous groups and analyzed in vacuum has no absorption in near UV light, i.e. at 290 to 400 nm. It shows strong absorption at 250–280 nm due to phenyl groups [12]. Above 150°C in air, it darkens and thermal degradation starts. PS with average MW depolymerizes at 300°C yielding styrene (45–65%), diphenylpropane, diphenylbutane, triphenylpentane, etc.

The most important weaknesses are its brittleness, low thermostability, low resilience, and flammability. This last characteristic might be improved by the introduction of halogen atoms in the benzene ring. Consequently, a PS of a 35.4 oxygen index was obtained [13,14].

2.1.7 Applications

PS blocks and sheets are processed especially for electrochemical plants and lenses. PS is very suitable to be processed through extrusion, injection and blowing. Due to PS flowing properties, the most widely used processing procedure is injection forming. For this proposal, the polymer is supplied as a powder or as pellets. Injection occurs at 180–200°C and 50–200 MPa pressure.

Throughout extrusion, PS is processed in bars, tubes, sheets and films. Films might be obtained also from PS solutions in binary mixtures like xylene and dioxan, or xylene and methylene chloride. Adding plasticizers leads to the improvement of PS flexibility; low polarity plasticizers are preferred because they do not affect the dielectric properties.

PS has a lot of applications in electro and radio technique laboratory equipment, packaging for food and pharmaceutical supplies, pipes for acids, etc. In polygraphy, PS replaces the lead letters. PS varnishers are applied in radiotechnology and the chemical industry. In construction, PS is used for wall tiles, light diffusing panels and electrical appliances. Expanded PS is extensively used as thermal insulation in buildings and acoustic tiles for ceilings. It is also used for cryogenic equipment, fridges, autoclaves and boilers.

More recently, in addition to and replacing PS, more and more polymers of styrene derivatives are used. Styrene substituted derivatives may be classified in the following three groups:

- ring substituted derivatives;
- vinylic group substituted derivatives;
- ring and vinylic group substituted derivatives.

The first group contains the most important derivatives which are able to polymerize as easily as styrene itself.

Among ring substituted PS, practical applications include the polymers of chlorostyrene. Poly(2,5-dichloro-styrene) is a transparent plastic with a density of (1.38–1.40 g/cm^3), superior thermostability and flame resistance. This polymer may be processed through injection, extrusion and also by machining. Due to its better thermostability compared to PS, its field of application is wider.

2.1.8 Styrene based copolymers

Styrene forms through copolymerization with vinyl and diene monomers, copolymers and terpolymers. Among these, the most important from a practical point of view are:

- styrene–acrylonitrile copolymer (SAN);
- methyl methacrylate–styrene copolymer (SM);
- styrene–α-methyl-styrene (SAM);
- styrene–butadiene copolymer (SB) (see Chapter 3);
- styrene–divinyl benzene copolymer (S–DVB);
- acrylonitrile–butadiene–styrene grafted copolymer (ABS).

(a) Styrene–acrylonitrile (SAN) copolymer

These kinds of copolymers might be obtained through all the well known polymerization procedures; copolymerization is radicalic and the copolymerization constants at 60°C are $r_{AN} = 0.04$ and $r_S = 0.4$. Due to the fact that styrene reactivity is higher than that of AN, styrene units will be consumed faster, a fact which will lead to a change in the composition. A wide range of copolymers containing different amounts of these two monomers are produced by continuous bulk, solution emulsion and suspension. The most used commercial SAN copolymers have a ratio S:AN of 76:24. It has been found that the optimum processing conditions are obtained when copolymerization occurs at 130–180°C, in a bulk system.

Copolymers produced through bulk, solution or those obtained through emulsion copolymerization are opaque. The flow chart of suspension copolymerization is shown in Figure 2.4.

Properties depend on %AN. With increasing AN content, SAN copolymers show continuous improvements in barrier properties, chemical and light, including UV resistance, but thermal stability decreases. The main properties of SAN copolymers are:

- specific gravity is 10% or more higher than that of some other polymers like polyolefins, PMMA, PVC, polyamides, etc.;
- tensile strength and hardness are higher than those of other thermoplastics;

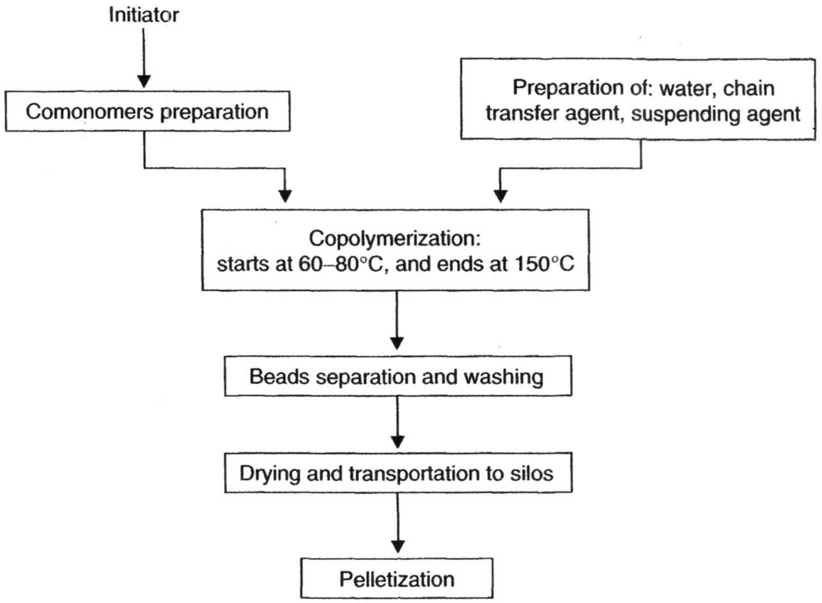

Figure 2.4 SAN production by suspension polymerization.

- possession of high transparency, but are not perfectly colorless and have a light yellow color;
- flammable;
- good thermal and acoustic insulation;
- stable to acids, bases and aromatic hydrocarbons;
- soluble in ketones, esters and chlorinated hydrocarbons;
- water absorption is higher than that of other styrene polymers.

SAN copolymer has the same applications as PS.

(b) Styrene–methyl methacrylate (SM) copolymer

SM copolymers are obtained through mass copolymerization with PB as initiator. The prepolymer is produced at *c.* 90°C and the end of copolymerization and the formation of SM copolymer blocks take place at 55°C.

SM copolymers possess good chemical and UV stability; thermal stability is 105–110°C, light transmission, 96.5–97%, Brinell hardness, 1.85–1.95 MPa. Their microstructure is discussed by A.M. Aerdts *et al.* [15].

(c) Styrene–α-methyl-styrene (SAM)

Styrene–α-methyl-styrene (SAM) copolymer is obtained throughout emulsion copolymerization. This copolymer possesses a better thermostability than PS; due to its electrical insulation properties, it is used as a dielectric for high frequencies.

(d) Styrene–divinyl benzene copolymer (S–DVB)

Suspension polymerization of styrene in the presence of several per cent of divinyl benzene (DVB) leads to a three-dimensional network. The copolymerization is produced in the presence of a suspending agent which together with the stirring conditions permits control of the bead size and bead size distribution. After the formation of the copolymer under controlled temperature, catalyst concentration, suspending agent concentration and stirring conditions, this product is washed to remove the suspending agent, dried and screened to obtain the desired particle size distribution [16–21].

As little as 0.01% of DVB can render the copolymer insoluble but swellable in benzene, while 1% renders it glass-like and virtually unaffected by organic liquids [18]. S–DVB beads are sulfonated to the extent of about one sulfonic group per benzene ring to provide cationic type ion exchangers. The commercial copolymer consists of small beads (0.1–0.5 mm in diameter) able to swell in a liquid medium, usually water, and form a gel-like product. The amount of swelling is controlled by the degree of crosslinking which depends on the amount of DVB used in copolymerization.

Conventional anion exchange polymers are produced by chloromethylation of the copolymer, followed by amination. Weak-base products have secondary amine (–NHR) or tertiary amine (–NH_2) functionality, and strong-base products have quaternary ammonium ion as the functional group [20].

If the resin is packed into a column, and if an aqueous solution containing ions whose removal is desired is allowed to permeate through the column, the eluate will contain the exchanged counter-ions from the polymer in place of the undesired ions [16].

H.Y. Erbil and B.M. Baysal [22] have used S–DVB copolymer for obtaining cation exchange membranes for the electrolysis of borax solutions. The copolymer may be introduced in PE in the form of film, powder or granules. In the last two cases, the resulting interpenetrating polymer network system is then processed into the desired shape. Sulfonation is subsequently applied. The system obtained is composed of three distinct phases, crystalline PE, amorphous PE and S–DVB copolymer, where the latter is formed in the amorphous PE phase. The amount of crosslinking agent (DVB), in the copolymer phase and the presence of the inert diluents affect various properties of the ion exchange membranes based on the PE/S–DVB system.

M. Steiberg [23] has produced mortars and concretes impregnated with S–DVB copolymer for the storage of nuclear waste materials. Such polymer impregnated products have shown radiation stability to 10^{10} rads which is the total integrated dose expected during a 1000 year exposure. Crosslinked PS is especially radiation resistant. Oxides incorporated in S–DVB copolymer for polymer–concrete composites also show promise.

(e) Acrylonitrile–butadiene–styrene (ABS) terpolymers

Over time, polymerization, grafting and blending have all evolved and ABS bulk polymerization in conjunction with bulk, emulsion and suspension polymerization are now available. Total worldwide sales in 1992 are estimated at 2.3 million tons and total capacity at around 3.6 million tons [24].

The ABS terpolymers have properties superior to high impact polystyrene and a wide range of applications.

The first technique used to produce ABS involves mechanical blending by mixing a butadiene–acrylonitrile elastomer with a styrene–acrylonitrile copolymer under such conditions that grafting is produced (Figure 2.5).

A variation of this process, illustrated in Figure 2.6, is based on mixing a SAN latex with a rubber latex (BAN or even polybutadiene).

These techniques are rather limited and have been largely superseded by a procedure based on a copolymerization system of styrene and acrylonitrile in the presence of a polybutadiene rubber latex.

As a result, a mixture of SAN copolymer, a graft copolymer of styrene–acrylonitrile and polybutadiene and unmodified polybutadiene homopolymer is obtained.

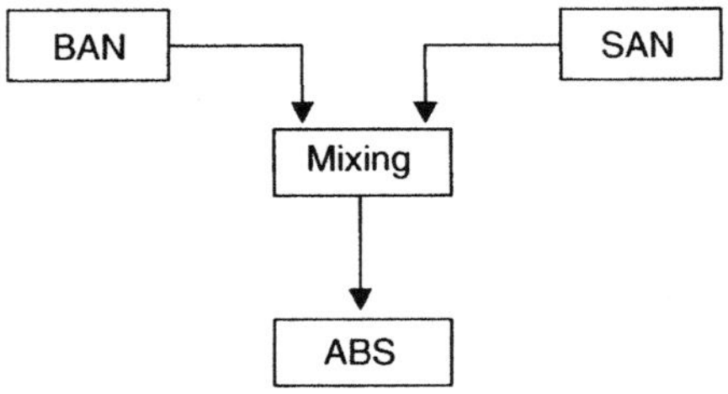

Figure 2.5 Mechanical mixing of solid copolymers.

Suspension polymerization
In the case of a suspension process, a solution of polybutadiene in styrene and acrylonitrile is first prepolymerized in a bulk system with agitation to obtain phase inversion. Then polymerization is continued in a suspension with water as the continuous phase (Figure 2.7) [25]. The bulk operation uses small amounts of solvents such as ethyl-benzene.

Another variation, based on two steps, uses the process presented in Figure 2.8. Initiation is done with PB, and *t*-dodecylmercaptan is used as chain transfer agent. The ratio between monomers, S : AN, is 3 : 1 [1].

Emulsion polymerization is widely used for improved varieties of ABS, in which copolymers SAN are grafted to polybutadiene. The styrene–acrylonitrile copolymerization can be carried substantially to completion, as can the grafting of the two monomers on polybutadiene in the form of a latex. Typical recipes include surfactants, a water soluble peroxide, a chain transfer agent, usually a mercaptan, e.g. the first step (PB latex formation) is shown in Table 2.1.

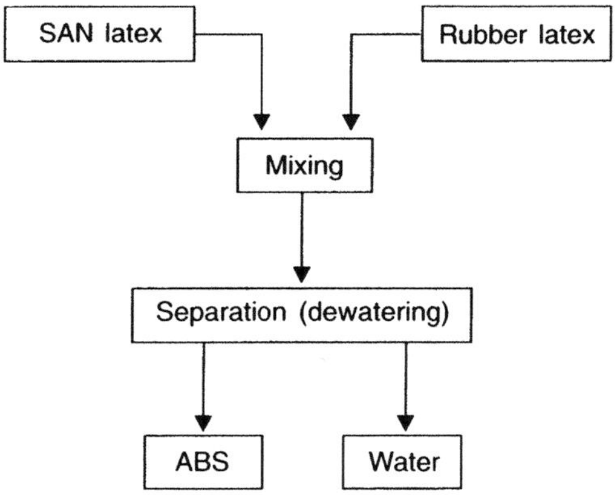

Figure 2.6 ABS production based on mixing latices.

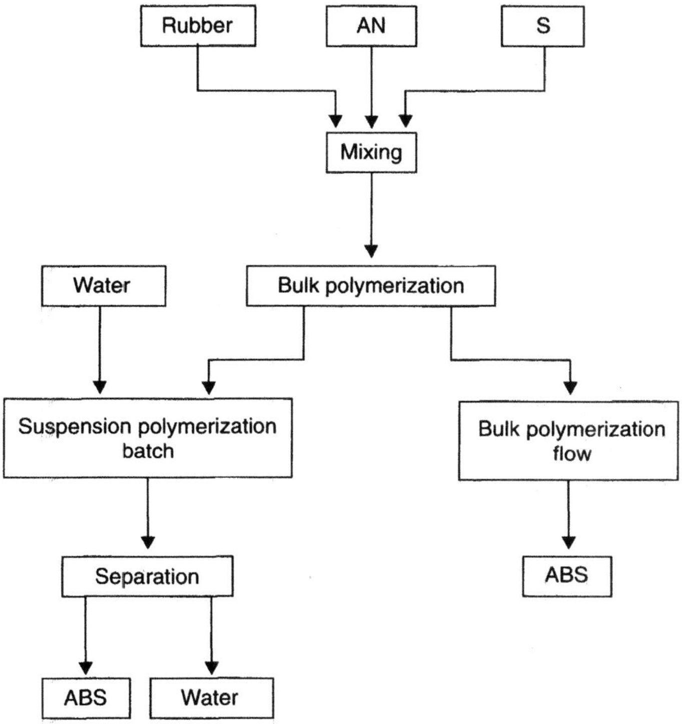

Figure 2.7 ABS production by bulk or suspension bulk polymerization.

The agitated batch reactor is stainless steel or glass lined, has a volume of 10–30 m^3 and must resist at 1000 KPa. The heat of reaction is 1350 J/g and the temperature is controlled at 10–85°C (Figure 2.9).

The second step is operated at 40–85°C and atmospheric pressure, and its typical recipe [1] is shown in Table 2.1.

ABS properties

These groups of copolymers possess a very good combination of mechanical, thermal and electrical properties, as well as chemical stability. Ease of processing, relatively low cost and the possibility of designing grades to meet the needs of some particular applications are other important advantages.

The three comonomers representing the three main characteristics like glassiness (styrene), elasticity (butadiene) and polarity (acrylonitrile) can be placed in a triangular phase diagram (Figure 2.10) [26].

ABS is superior to common polystyrene products. It can be compounded with a high degree of hardness or with great flexibility and toughness, and is commonly described as tough, hard or rigid.

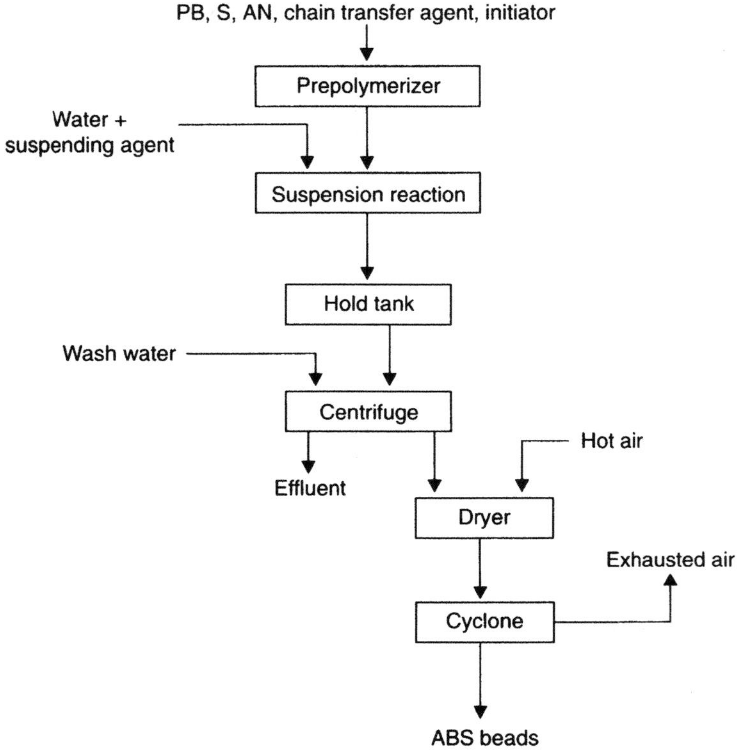

Figure 2.8 ABS production by suspension polymerization.

Table 2.1 Typical recipe for ABS

	Parts by weight
First step	
Water	200.0
1,3-Butadiene	175.0
Sodium oleate	4.0
Cumene hydroperoxide	0.3
Sodium pyrophosphate	2.5
Dextrose	1.0
Ferrous sulfate	0.05
Second step	
Water (including that of PB latex)	300.0
PB contained basis	30
S	50
AN	20
t-Hexadecylmercaptan	0.1
Sodium oleate	0.5
Potassium persulfate	0.5

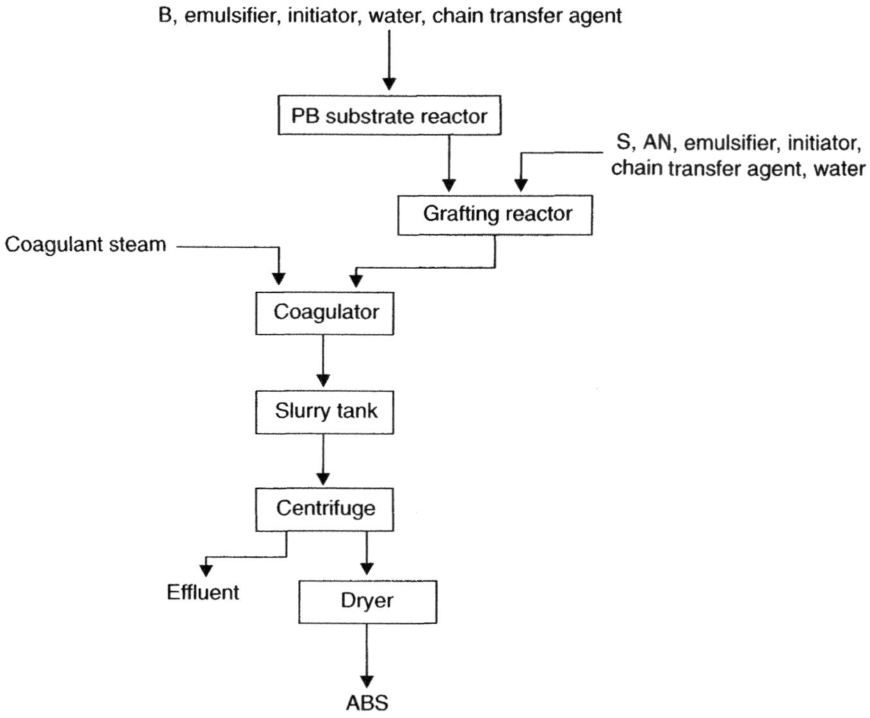

Figure 2.9 ABS production by emulsion polymerization.

The impact strength, which ranges from 107 to 534 J/m, may be increased by raising the proportion of polybutadiene, but this is accompanied by slight reductions in heat resistance, rigidity, the electrical properties and some of the other strength properties [27]. It is possible to avoid a decrease in the heat distortion temperature if styrene is substituted by methyl-styrene, but this will increase the cost.

Stiffness and dimensional stability are greatly improved by the addition of glass fibers. Its photo-oxidation resistance was recently studied [28].

ABS copolymers are used in applications that require impact resistance, colorability, hardness, electrical properties, moisture stability, limited thermal stability (105°C). Certain formulations are compounded with self-extinguishing additives to be used for electrical hand tools. ABS products are little affected by water, alkalis, weak acids and inorganic salts. Alcohol and hydrocarbon solvents may affect the surfaces and cause swelling if the exposure is extended.

ABS is processed by thermoforming, injection, blow, rotational, extrusion molding and calendering.

Examples of applications are business machines, camera housings, telephone handsets, electrical hand tools, automotive trim and hardware, pipes

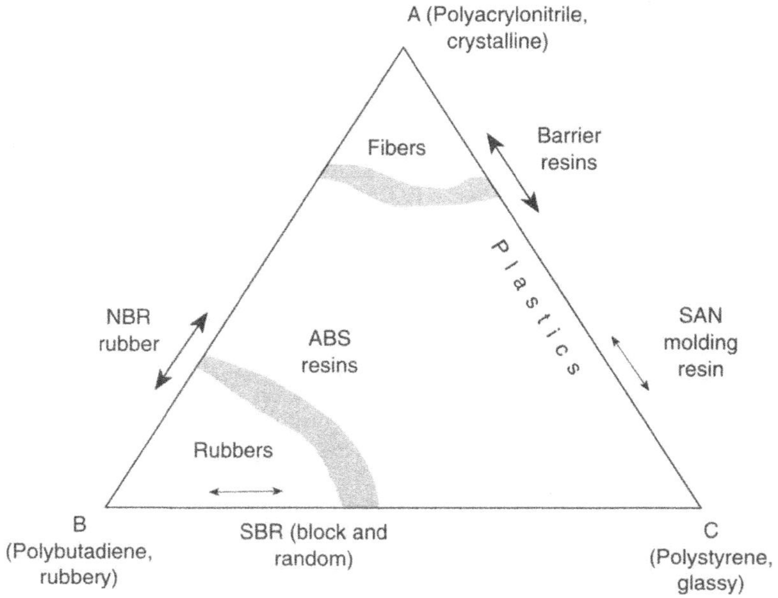

Figure 2.10 ABS compositions. Besides the homopolymers, interpolymers of three building blocks give rise to numerous modifications. Locations indicated are only approximate [26]. (From *Principles of Polymer Systems*, 2nd edn, 1982, by F. Rodriguez, Hemisphere Publishing Corp., New York. Reproduced with permission. All rights reserved.)

and pipe fittings. ABS may be metal plated, and such pieces are often used to replace die casings.

Polyurethanes proved useful for increasing the ultimate elongation, ductility and impact strength of ABS. This could be used to increase the modulus of thermoplastic polyurethane elastomers more easily and economically through reformulation of the polyurethane itself [29].

2.1.9 High impact PS (HIPS)

The polymerization of styrene in the presence of polybutadiene using PB or AIBN as radical initiators leads to HIPS, a rubber modified PS.

G. Riess and P. Gaillard [30] have studied the fabrication process of HIPS and the effects of the reaction parameters, e.g. molecular characteristics of polybutadiene, concentration of initiator and transfer agent, agitation. Morphological features of HIPS are discussed. The impact strength of HIPS increases with rubber particle size and concentration while gloss and rigidity decrease. The microstructure of polybutadiene has an important influence on HIPS properties and a 36% *cis*-1,4-polybutadiene provides optimal properties.

2.2 POLY(VINYL CHLORIDE) (PVC)

2.2.1 Introduction

PVC is one of the most important large volume thermoplastic polymers. In the last years, PVC production has reached 23% of the world total production of thermoplastic materials. Its global production in the 1980s was around 15 million tons per year, and has increased to 22.8 million tons in 1992 [31,32]. USA production of PVC and vinyl chloride (VC) copolymers has increased from *c.* 3 million tons in 1983 to 4.6 million tons in 1993 [2].

PVC is produced mainly by free radical initiated suspension polymerization, but emulsion, bulk and solution polymerization technologies are also employed to some extent.

In spite of its high volume production, PVC is one of the less stable thermally, chemically and with respect to light. For processing and uses it requires special stabilizers.

The first observation with respect to the possibility of polymerization of VC has been attributed to M.V. Regnault and dates from 1838. Important contributions regarding VC polymerization were brought by J.J. Ostromislenski through many studies done between 1912 and 1929. However, until 1927, PVC's industrial function was unknown. The beginning of some increase of its production is recorded only in 1936. PVC is processed in rigid or plasticized products. While rigid PVC is used for some items in which flexibility is not required, most PVC is plasticized.

2.2.2 Bulk polymerization

Polymerization of liquid VC yields the purest PVC, the polymer being insoluble in its monomer precipitates. About 10% of the world's PVC is produced by bulk polymerization. The process is carried out in an inert medium, because the presence of oxygen in the polymerization system leads to the formation of peroxides of VC which influence the process and the PVC properties [33].

With an initiator such as benzoyl peroxide (BP) the process can be completed in 17 hours at 55–75°C to 80–90% conversion, or with azo-bis-isobutyronitrile (AIBN) in 12 hours at 62°C to 62% conversion. Other initiators such as acetyl cyclohexane sulfonyl peroxide or peroxydicarbonates can be used. The pressure during the process varies from 0.5 to 1.2 kPa. Since the liquid monomer is always in equilibrium with its vapors, it is easy to eliminate the reaction heat due to the distillation and condensation of VC on the reactor walls.

The polymerization has two distinct stages (Figure 2.11). During the first one, the prepolymerization, polymer grains form; a strong agitation will

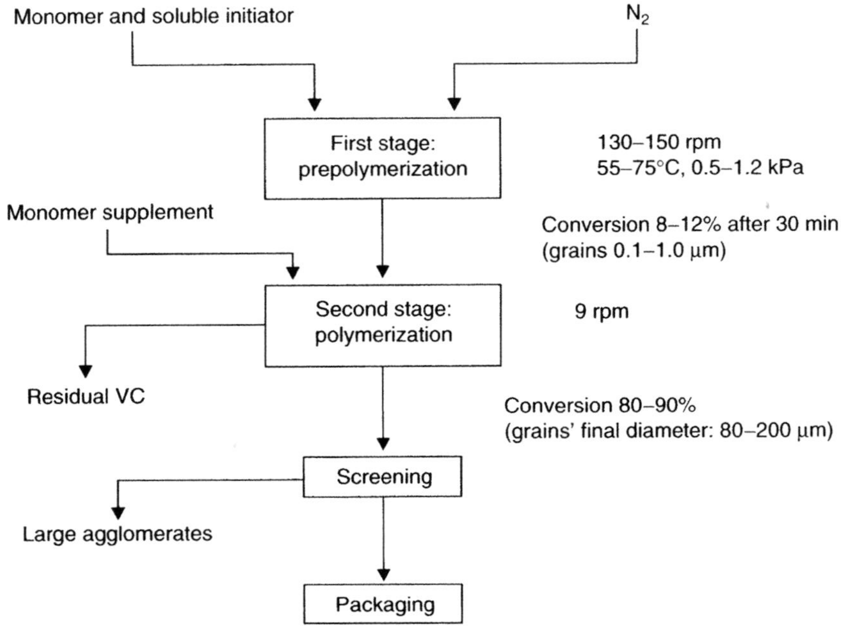

Figure 2.11 Bulk polymerization of vinyl chloride.

ensure grain uniformity and a correct morphology. Due to the grains' porosity, at the conversion obtained during prepolymerization (8–12%), much of the liquid monomer is absorbed and the low viscosity of the system becomes too high for an efficient stirring. Prepolymerization takes place in a vessel equipped with a reflux condenser. The second stage is conducted in a horizontal vessel having a ribbon blender type agitator. Additional monomer and initiator required for this stage are introduced during the transfer. At 20% conversion the system looks like a damp powder and at about 40% like a dry one. In this stage, the grains grow to sizes in the range of 0.1–1.0 μm and the conversion is allowed to reach about 80%, whereupon the unreacted monomer is vented off. Agitation history is one of the main factors which affect the grains' size. They appear as expanded beads, low in bulk density but very avid absorbers of liquid additives, especially plasticizers.

This polymerization takes place without protective colloids (surfactants, emulgators, suspending agents, etc.) and as a result PVC is purer and its grains do not have a skin like PVC obtained through suspension polymerization; therefore they have better heat stability and better dry-blending characteristics.

Bulk polymerization of a pure monomer offers the simplest process with a minimum of contamination of the polymer (only initiator fragments). No further drying is required, and after screening to remove agglomerates, the product is packaged.

The most important peculiarity of this polymerization is the separation of the polymer as a solid phase from the beginning of the process. One meets such separation of PVC in the case of vinyl chloride polymerization in organic solvents. These strongly affect the process's kinetics because the solvents behave as chain transfer agents. Figure 2.11 presents the flow chart of vinyl chloride bulk polymerization.

Because of this feature, the rate of vinyl chloride bulk polymerization is usually higher than that of heterogeneous polymerization in other media than the monomer. Another feature of vinyl chloride bulk polymerization consists in the *S* shape of the conversion–time diagram. Its activation energy is 21.8 kcal [34]. A fairly comprehensive but realistic model is described that has the ability to predict the rate of polymerization as well as the molecular properties as a function of the bulk or suspension polymerization conditions [35].

2.2.3 Solution polymerization

Solution polymerization is over 50 years old and only a small percentage of the produced PVC is obtained this way. Depending on the solvent capability to dissolve both the monomer and the polymer or only one of them, the system may be homogeneous or heterogeneous. This procedure is used almost exclusively for obtaining copolymers and terpolymers.

2.2.4 Emulsion polymerization

This polymerization type, which is carried out in a heterogeneous medium, is of particular interest because of the following characteristics.

- It permits a regulation of the process so that the heat transfer presents no difficulty for the normal development of the polymerization. This permits the production of polymers with good properties.
- It creates better technical possibilities for process regulation than the polymerization in the homogeneous mixture.
- It creates the possibility of the quantitative variation of the components while diminishing the role of secondary reactions such as chain transfer, branching, etc.
- Emulsion polymerization takes place at a high rate without affecting the molecular weight. The fact that the process has high rates at low temperature influences not only the reaction development but also the necessary equipment which becomes in this case more accessible and more productive.

The possibility of recuperating the unreacted monomer via simple vapor is another economic advantage due to the fact that water is in most cases the dispersion medium. The latex produced can be used for many applications as it is.

In the emulsion polymerization process, the monomer is dispersed in the form of drops in the aqueous phase which constitutes the dispersion medium. For better stability of the concentrated emulsions it is necessary that an absorption colloidal layer with a gel structure forms at the surface of the drops. This can be obtained with the aid of emulsifiers. An emulsion oil–water type is obtained, due to such additives, and the agitation effect. The emulsion is characterized by a high active surface.

Conventional emulsion polymerization involves the dispersion of a monomer in a continuous water phase with an oil-in-water emulsifier, followed by polymerization with a water soluble initiator. The product is a polymer colloid in which the particles are considerably smaller than the monomer droplets of the initial monomer dispersion. Hence the reaction mechanisms include particle nucleation and growth [36].

This process might be realized continuously or discontinuously, each of these two variations having advantages and drawbacks. The continuous one is more economical and permits the production of higher polymer amounts with less raw materials, less additives and less energy. However, through batch polymerization it is easier to control and regulate the process and to obtain a PVC with determined properties.

The latices obtained through the discontinuous polymerization have a better stability with less emulsifier than the continuous one; it is also easier to switch from the fabrication of one PVC type to another in the discontinuous process.

Emulsion PVC production usually passes through the following stages:

- preparation of the water emulsion and initiator solution;
- polymerization;
- demonomerization;
- polymer separation from the latex;
- drying.

The most important parameters of the process are the ratio of monomer to aqueous phase, the type and quantity of the emulsifier, the type and quantity of the initiator, temperature and pH.

Depending on factors such as the type of equipment, type of emulsifier, additives for latex stabilization and expected PVC properties, the ratio between phases (monomer–aqueous phase) varies in the range 1:1 to 1:2.3. If this ratio is brought to 1:1 or lower, the latex stability decreases and the polymerization slows down. The temperature affects the K value (Fikentscher constant) which characterizes the degree of polymerization of PVC.

In emulsion polymerization, the following components are used: water as continuous phase, VC, water soluble initiator and emulsifier. The emulsifying agents, either ionic or non-ionic, dissolve at low amounts in water. When the emulsifier concentration is higher, the surface tension between the phases decreases and the electrical conductivity of the system increases.

It is considered that emulsion polymerization takes place in the emulsifier-made micelles formed by the emulsifying agent and water, and not in the aqueous phase or in the monomer drops. Polymerization is produced in stages, some inside the micelles and some in the polymer–monomer particles. So, the first stages involve initiation and starting of propagation until a certain size occurs; termination of the growth of the macromolecular chains is produced in the polymer–monomer particles. The formation of such complex particles is considered to result from the existence of emulsified monomer drops able to swell or dissolve the polymer present in the reaction medium [37].

The active participation of the emulsifier in the form of micelles is supported by the fact that at high emulsifier concentrations one may produce, for the same polymer yield, higher amounts of polymer particles with lower sizes than those of monomer particles. In such latex systems the role of monomer drops is reduced to simple reservoirs from where the monomer diffuses first towards the micelles and after in the polymer particles. As the emulsifier disappears from the system by being absorbed by the polymer particles, it is possible that initiation be produced in the aqueous phase. Figure 2.12 represents the stages of an ideal emulsion polymerization.

Various emulsion polymerization theories are discussed by D.C. Blackley [38]. Difficulties are frequently encountered in reproducing polymerization rates in emulsion systems. These arise from a variation of the number of latex particles formed under supposedly identical conditions. Practically, the difficulty can be overcome by 'seeding' the process [39].

A typical recipe and the reaction conditions for the emulsion polymerization of VC are given in Table 2.2.

A simplified flow chart of VC emulsion polymerization is presented in Figure 2.13. The polymerization temperature is selected on the basis of decomposition kinetics of the particular initiator used and the desired molecular weight of PVC [40,41].

(a) Continuous technology

This technology permits the production of a coarse dispersion (with particles of 0.1–1.0 μm) with good mechanical and thermal stability superior to the fine dispersions 0.01–0.1 μm). For this reason, in addition to emulsifiers, surface active agents with very weak tensio-active hydrophilic groups (e.g.

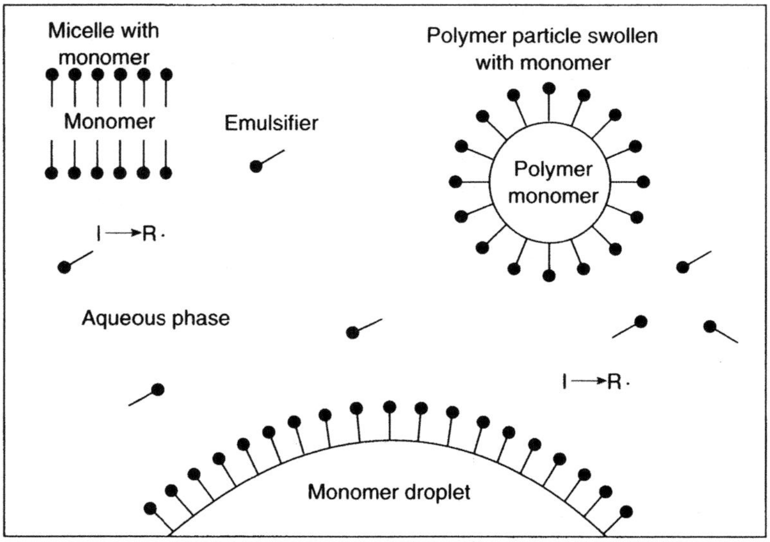

Figure 2.12 Simplified representation of an emulsion polymerization system. I = initiator; R· = free radical; •– = emulsifier molecule.

fatty alcohols C_{16}–C_{18}) are introduced into the system. The flow chart of the VC continuous polymerization is presented in Figure 2.14.

In a special vessel, a solution of 25% emulsifier in demineralized water is prepared. The emulsifier system is a mixture of sodium alkyl-monosulfates with C_{12}–C_{15}. In the same vessel, a 10% NaOH solution and a 70% phosphoric acid solution are introduced. The alkaline solution is necessary for the transformation of the free sulfur components and chlorides of the emulsifier into sodium salts. Phosphoric acid as a buffer, regulates the pH at 8.0 in the case of the use of persulfates and at 8.5 in the case of the initiation with hydrogen peroxide. With direct vapor under continous

Table 2.2 Typical recipe and reaction conditions for the emulsion polymerization of VC

	Parts by weight	*Conditions*
VC	100	
Water	100–200	
Sodium persulfate	0.4	
Sodium dodecylbenzenesulfonate	0.5–2.0	
Sodium bicarbonate		to pH of 4–9
Temperature		50–55°C
Reaction time		10–18 hours
Latex particle diameter		800–1600 Å

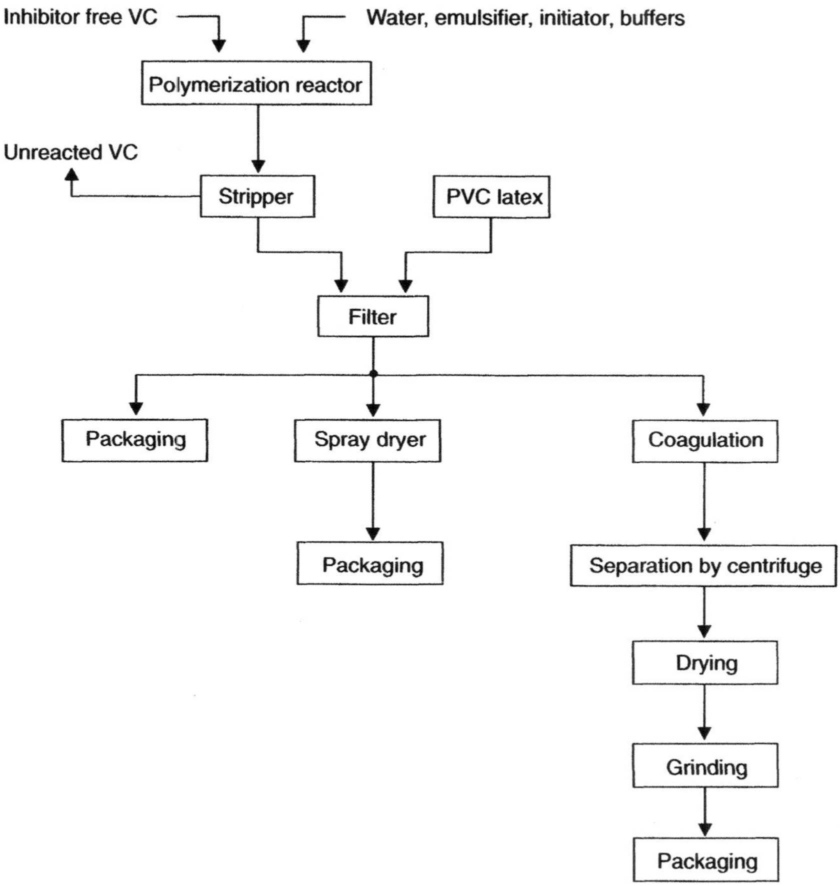

Figure 2.13 Discontinuous vinyl chloride emulsion polymerization.

stirring, this mixture is heated at 50–60°C and it is kept at these conditions until the complete dissolution of the emulsifiers (*c*. 5–6 hours). The aqueous phase is obtained by mixing the emulsifier solution, with the 3% persulfate solution and with solutions of the caustic soda and phosphoric acid. Polymerization is realized in vertical autoclaves (12 m^3, 6 m height). The volume of the polymerization medium is 10–10.5 m^3; the polymerization section usually has eight autoclaves. For each autoclave, the monomer is introduced with a flow rate of 300–360 l/h and the aqueous phase with 270–330 l/h. They vary depending on the type of polymer required. The temperature (43–58°C) and pressure (5–9 atm) are also chosen depending on the same factor.

At 43–45°C, the polymerization rate is slow and the PVC *K* value is 67–70; at 50°C and above this temperature, the rate is higher and *K* is 64. Agitation

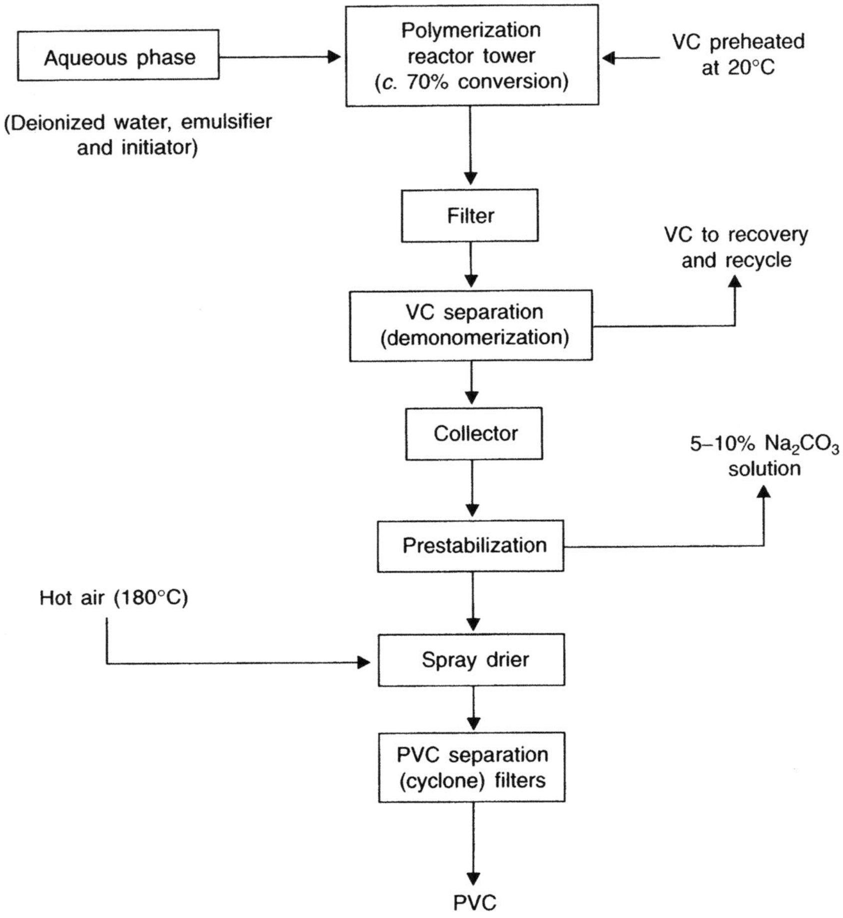

Figure 2.14 Continuous vinyl chloride emulsion polymerization.

is performed by stirring at 80 rpm. After prestabilization with Na_2CO_3, the latex has the following characteristics:

- specific gravity at 20°C, 1.138–1.165;
- solids content, 40–50%.

The drying thermal regime in the atomizer is:

- at the top, 175–180°C,
- in the middle, 80–90°C,
- at the bottom, 70–75°C.

The *K* value of PVC (powder) is 60–75 depending on the polymerization conditions mentioned above. Average specific consumptions per metric ton are given in Table 2.3.

Table 2.3 Average specific consumptions per metric ton for the production of PVC by continuous polymerization

Monomer	1050 kg
Emulsifier	35 kg
Initiator	0.1 kg
NaOH	2 kg
Na_2CO_3	4 kg
Demineralized water	2 m^3
Nitrogen	5 Nm^3
Vapor	5 tons
Recirculated water	150 m^3
Technological water	7 m^3
Electrical power	550 kWh
Compressed air	10 Nm^3
Brine	200 kcal

The principles of modeling the emulsion polymerization of vinyl chloride were reviewed by K. Tauer *et al.* [42] to show that various strategies for modeling the course, special aspects and special properties are available.

2.2.5 Suspension polymerization

Most plants for the production of PVC suspensions run discontinuous for polymerization and continuous for the drying operation, and are able to produce two different, especially from a morphological point of view, PVCs, one called 'compact normal', and the other one 'dry blend'. PVC compact normal is crystalline and composed of spherical particles with superficial adsorption of the plasticizer. The dry blend type is based on amorphous agglomerated particles with a capillary structure able to absorb the plasticizer very strongly so that the final compound looks dry. A mathematical model for the production of PVC suspension was developed by D.G. Kelsall and G.C. Maitland [43].

Suspension polymerization is generally easier to control and there is little, if any, loss in clarity or electrical insulation properties. Particle shape, size, size distribution and morphology may be controlled by varying the dispersing systems and the rate of stirring. Continuous suspension polymerization processes can use reactors with broad or narrow residence time distributions; in the second group, the residence time distribution is determined by forced or free convection [44].

The production of the two different types of suspension PVC implies some technological differences, especially in the polymerization stage.

A typical flow chart for the production of suspension PVC is presented in Figure 2.15. It is essentially similar to emulsion polymerization plants, apart from the substitution of the atomizer (spray dryer) for the centrifuge/rotary dryer system. Indeed, it is often possible for both methods of polymerization to be carried out in the same reactors. A numerical study on batch reactors

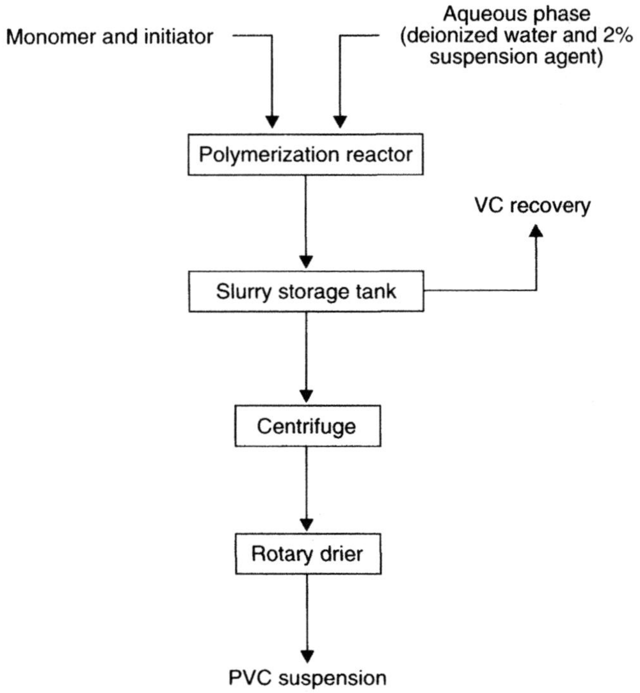

Figure 2.15 Vinyl chloride suspension polymerization.

was done recently by J.C. Pinto [46]. A typical basic formulation might be as shown in Table 2.4 [46].

Sometimes methylcellulose or other substances like gelatine, PVA, maleic-anhydride–vinyl-acetate copolymers, ethylcellulose, carboxymethyl cellulose and alginates, are used as a suspension agent [47–49]. Polymerization takes place under nitrogen; the stirring rate for the normal compact type is 80 rpm. Polymerization reactors are capable of withstanding pressures of 15 atm (1500 kPa) or above if low molecular weights are expected. For high MW, pressures are 5–10 atm (500–1000 kPa). The thermal regime and polymerization time depend on the kind of PVC wanted (Table 2.5).

Table 2.4 A typical basic formulation for the production of suspension PVC

	Parts by weight
Monomer	100
Water (demineralized)	100
Poly(vinyl alcohol)	0.04 (suspension agent)
Trichloroethylene	0.2 (transfer agent)
Lauroyl peroxide	0.2 (monomer soluble initiator)

Table 2.5 Conditions in polymerization reactors

K value	58	63	70	70 (dry blend)
Temperature (°C)	70	59–60	52–53	54
Time (h)	7	9	17	15

In some plants, reactors of 80–100 m^3 are used [50]. Figure 2.16 presents the variation of conversion and pressure with time in a typical suspension polymerization.

A.E. Hamielec *et al.* [51,52] have developed a comprehensive model for MW development during suspension radical polymerization of VC. With reasonable recipes and good quality of monomer and initiator, combined with correct operations, the polymerization is unlikely to get out of control. For a particular recipe, the shape of the reactor and baffles, and the shape and speed of the stirrer, are critical in determining the form of the final polymer particles, and indeed the success of the whole operation [53]. Maximum conversion reaches *c.* 80%.

For the production of dry blend PVC, polymerization implies the following differences compared to the process for compact normal.

- Dry blend uses only one half of the amount of the suspension agent used for compact normal.

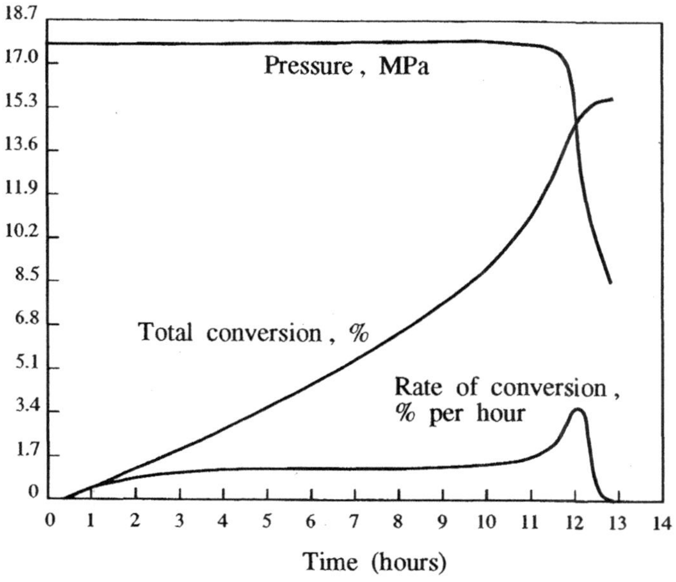

Figure 2.16 Variation of conversion and pressure with time in a typical suspension polymerization of vinyl chloride. (Reprinted with permission from *Applied Polymer Science*, 2nd edn, 1985, by R.W. Tess and G.E. Poehlein (eds), Copyright American Chemical Society.)

- Added to the suspension agent is a small amount of concentrated ammonia solution, which produces a partial precipitation of the suspension agent.
- Stirring conditions vary during the polymerization.
- During the process, the pH decreases from 9.5 to 8.0 and after polymerization is finished, the suspension is neutralized with concentrated HCl to adjust the pH at 7.

The increase of the absorption capacity of plasticizers can be obtained by the addition of soaps of superior fatty acids, especially those of stearic acid, during polymerization. In the meantime PVC density and thermal stability increases.

The specific consumption for 1 metric ton of suspension PVC are shown in Table 2.6.

Taking into account the different morphologies of the particles, the normal compact type is recommended to be processed into rigid items, and the dry blend type into plasticized articles.

The VC polymerization heat is 22.9 kcal/mol, so that an important parameter of the process is the removal of exothermic heat. This is balanced by adequate kinetic control which depends on the initiator to assist in obtaining uniform conversion rates.

Among the monomer soluble initiators often used are lauroyl peroxide (LP), di-2-ethyl hexyl peroxydicarbonate, AIBN, diisopropyl peroxydicarbonate (IPP). Oil soluble initiator and a partially saponified PVAc (DP *c.* 800) as dispersing agent led to a PVC with no scale formation, which had an average particle size of 140 μm, no coarse particles, void volume of 0.30 ml/g, and low fish eye formation in films [54]. The gradual dosage of the initiator is discussed by Z. Mrazec and R. Lucas [55]. The evolution of the microscopic structure of PVC grains is presented in Figure 2.17, where x represents the conversion of VC [56].

2.2.6 Properties

PVC properties, like thermal stability, are strongly related to the macromolecular structure of this polymer. Microstructure studies proved that its

Table 2.6 Consumptions for 1 metric ton of suspension PVC

Monomer	1.05–1.08 tons
Suspension agent	1–25 kg
Initiator	1–3 kg
Demineralized water	5–6 m^3
Nitrogen	30–40 m^3
Cooling water (+5°C)	40–60 m^3
Steam	15–20 tons
Electrical power	250–350 kWh
Compressed air	300–350 Nm^3

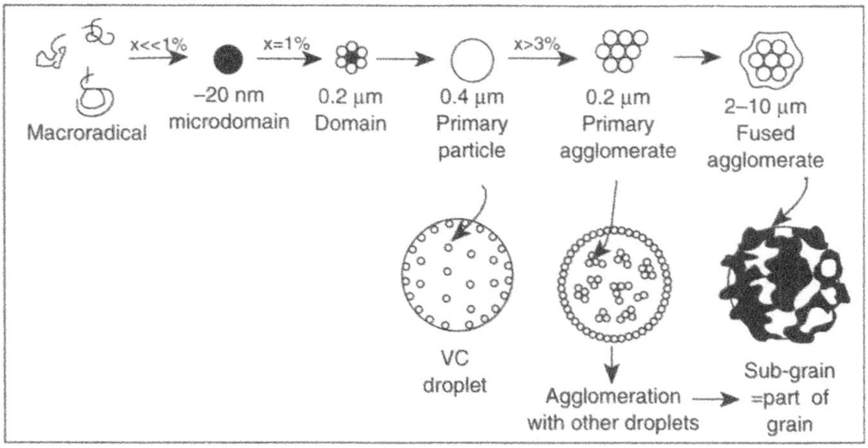

Figure 2.17 The evolution of the microscopic structure of PVC grains. (Reprinted by permission of Carl Hanser Verlag from *Polymer Reaction Engineering* by R.H. Reichert and W. Geisler, 1983. Carl Hanswer Verlag Munich–Cincinnati.)

units are bonded 'head to tail'. Industrial techniques where polymerization takes place through the radical mechanism lead to a polymer with a certain degree of branching. This characteristic depends on a series of factors like polymerization procedure, conversion and temperature. Generally, emulsion polymerization provides a PVC with a degree of branching which is three times higher than that of suspension PVC. Accounting for this effect is the solubility of VC in PVC which is only 30%.

A low polymerization temperature at industrial scale permits the formation of a linear polymer; from the energetic consideration point of view, a lower temperature is favorable to the formation of a perfect linear chain because of the fact that the chain propagation reaction needs a lower activation energy compared with that of chain transfer. The reaction rate of chain transfer is much higher in emulsion polymerization.

PVC branched microstructure looks like:

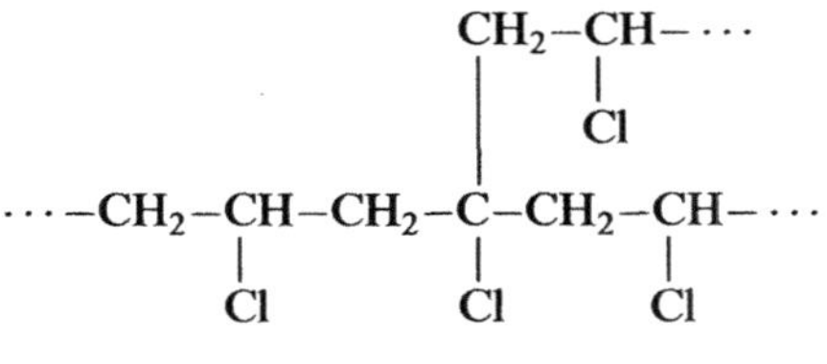

PVC obtained at industrial scale is generally atactic with short sequences of syndiotactic structures. Experimentally, it was proved that a lower temperature polymerization leads to syndiotactic segments.

Ping Xu *et al.* [57] have discovered the molecular structural defects of PVC obtained through suspension polymerization. PVC MW is one of the characteristics which determines the physical–mechanical properties of the end product's compatibility with plasticizers, and the rheological behavior of the melt. MW distribution can be established through the well known fractionation techniques.

With respect to processing and service life:

1. the PVC microstructure characteristics like MW, macromolecule geometry, degree of branching and polydispersity have the same importance as:
2. its supramolecular features such as particle shape, size, degree of crystallinity, porosity.

PVC for different applications has a MW in the range of 60 000 to 220 000. PVC with MW between 60 000 and 90 000 (*K* value 50–63) is used for blow molding, injection molding, injection blow molding and calendering; for extrusion of pipes, sidings, profiles and sheets higher MW PVC is used, up to approximately 220 000 (*K* value 79.5).

PVC is an amorphous polymer with 5% to 10% crystallinity [58–60]. It is one of the most adaptable synthetic polymers. Starting with a base polymer that is thermal and light sensitive and relatively brittle, technologies can produce articles ranging from soft to rigid. This tremendously wide spectrum of products is made possible through the addition of stabilizers, plasticizers, modifiers, etc. The microstructure of this polymer, which has a chlorine atom situated on each alternate carbon atom of the polymer chain, leads to a very rigid and relatively tough plastic in the unplasticized state. In the presence of a plasticizer, however, the dipole bonding polymer chains is much reduced, leading to increased freedom of chain movement and thus to a flexible material. By suitable choice of plasticizer type and amount, a whole range of flexible PVC can be produced.

The presence of tertiary carbon affects PVC's thermal stability. PVC's service life is also negatively influenced in respect to photo and thermal stability by the traces of initiator, emulsifier, etc. The thermal stability of PVC decreases with increase in polymerization temperature [61]. The degradation of PVC is always associated with hydrogen chloride evolution [61–67]; the earlier attempts to improve PVC thermal stability involved the addition of mild alkalis or organic bases to neutralize the acid formed. It has been assumed the PVC's low stability (dehydrochlorination starts at about 100°C) is caused by the irregularities in its chains.

The main irregularities in the PVC chain, which are in fact the causes of its instability, are [68–71]:

- head to head bonds;
- total and internal double bonds;
- chloroallyl and carbonyl-chloroallyl groups;
- oxygen-containing groups;
- branches;
- labile chlorine atoms;
- configuration and conformation of the polymer chain.

Both free radical and molecular mechanisms have been proposed for the thermal dehydrochlorination of PVC [71]. To avoid such thermal degradation one or two stabilizers are used. In relation to the synergism due to costabilization, some researchers show that the efficiency of a secondary stabilizer arises essentially from blocking of the polaron dehydrochlorination to favor allylic Cl substitution [72]. Initiator residues may also affect PVC's thermal stability. The recent progress in knowledge of the nature and abundance of irregularities in this polymer structure gives a far better possibility of elucidating the relationship between defects and thermal stability.

Irrespective of the nature of the initiation site, HCl generation is mainly a matter of polyene propagation. One important feature of this propagation is the restricted length of the resulting polyene sequences. The maximum number of conjugated double bonds is about 30, while mean values in the range 6–14 have been reported [73].

During PVC processing, the polymer is exposed to temperatures which cause thermal dehydrochlorination. More than this, the presence of oxygen leads to thermo-oxidative degradation, and the mechanical stresses may cause macromolecular scission. Stabilizers and lubricants diminish the degradation reactions.

When exposed to natural weathering, PVC deteriorates, becoming increasingly colored and brittle, with a steady drop in the tensile strength, elasticity and impact resistance. Owing to its large outdoor application, PVC has to be protected against photodegradation by the addition of adequate photostabilizers. The effects of UV light on the behavior of PVC were thoroughly investigated in order to obtain better information and control of its photodegradation [74].

It was pointed out that, in order to provide light stability, provision should be made for processing to be carried out with a good thermal stabilizer. The role of this additive is to quench the thermo-oxidative component of weathering as well as preventing degradation of PVC on processing. The photostability of the thermal stabilizer is obviously important for weathering applications [75,76].

Even with thermal stabilizers, PVC does not perform well outdoors unless steps have been taken to protect it from the damaging effects of sunlight.

Loss of impact strength is generally accompanied by a darkening of color although bleaching is sometimes observed.

Additives used for PVC or other polymers are generally classified according to their specific function as follows:

1. anti-aging additives (antioxidants, UV stabilizers, fungicides);
2. additives which assist processing (thermal stabilizers, lubricants, thixotropic agents);
3. modifiers of mechanical properties (plasticizers, fillers, toughening agents);

4. surface property modifiers (antistatics, slip additives, antiwear additives, antiblock additives, adhesion promoters);
5. optical property modifiers (pigments and dyes, nucleating agents);
6. additives to reduce formulation costs (fillers, dilutants);
7. others (blowing agents, flame retardants).

Rigid PVC is suitable to be processed by injection molding, compression molding, extrusion, calendering, thermoforming and laminating. Mold shrinkage is 0.7–1%. Plasticized PVC is suitable for injection molding, compression molding, blow molding, extrusion, rotational molding, calendering and laminating [77,78]. Its mold shrinkage is 1–5% depending on the plasticizer. Plastisols are made of finely divided unplasticized PVC dispersed in a plasticizer to form a creamy paste. Regarding the processing techniques, 61% of PVC are extrusions (pipe and tubing, siding and accessories, wire and cable covering, film and sheet), 4% coatings, 6% molded items, 11% calendered items and others (18%) which include paste processed items (except coatings), resellers and compounders [79]. PVC has a high ignition temperature, low flammability with little total heat released. Plasticized PVC products are obviously more flammable than the rigid ones [80,81].

Some physical–mechanical properties of PVC are shown in Table 2.7.

PVC is immiscible with PVAc [82], but polyblends with other polymers were obtained [83–89]. Besides VC copolymers with other vinylic monomers, copolymers with propylene, 1-butylene and 1-pentene can be obtained [90].

2.2.7 Applications

Rigid PVC is used for pipes, bars, tubes, cladding, window and door frames, ventilation ducts, gutters, roofing and linking for chemical tanks. PVC foams are used for thermal insulation and as separators for batteries [40, 91–93]. Flexible (plasticized) PVC is used for hoses, ducting, curtaining, raincoats, cable insulation, footwear, wall covering, flooring, etc. Radio frequency is used for welding PVC products [94]. Plastisols and organosols are used to

Table 2.7 Some physical–mechanical properties of PVC

Property	*Value*
T_g	80°C depending on MW
Specific gravity	1.38–1.50
Specific heat	0.85–0.9 kJ/kg·K
Water absorption	0.4% volume
Elastic modulus for rigid PVC	2800 kPa
Tensile strength	50 MPa

obtain heavy duty industrial clothing, handbags, shoes, gloves, etc. [95,96]. Good performance is shown by PVC composites containing mica and surface treated wood fibers.

2.2.8 Chlorinated PVC

The chlorination of PVC leads to an increase of chlorine content from 56.5 to 63–68%. Two types of polymer are produced commercially, those containing 63–64% and 67–68% chlorine. Although there are two known chlorination procedures which work at low (50°C) or high (100°C) temperature, the first one is preferred due to the better properties of the final product. The most common commercial process in use involves dispersing the PVC in water and swelling the particles with a chlorinated hydrocarbon. At 50°C chlorine is added and the system is irradiated with UV light to produce chlorine radicals. Further chlorine is added to maintain the pressure. Washing with an alkaline water and drying are the last operations [97]. Under the conditions of low temperature, chlorination substitution occurs mainly at the unsubstituted methylene groups so that the microstructure of the new polymer is:

$$\cdots-CHCl-CHCl-CHCl-CH_2-CHCl-CHCl-CHCl-\cdots$$

The higher chlorine content increases the T_g, affects the thermal stability and increases the melt viscosity; nevertheless, the so-modified PVC can be processed through extrusion and injection molding. The higher softening point permits a maximum service temperature of about 100°C compared to about 65°C for unplasticized PVC. As a result chlorinated PVC is used in plumbing applications, particularly hot and cold water distribution and hot corrosive effluent handling [46].

R. Newe *et al.* [98] have obtained a patent on continuous production of radiochemically chlorinated PVC prepared in suspension.

2.2.9 Vinyl chloride–vinyl acetate copolymers

About 90% of the PVC produced is used in the form of a homopolymer, the other 10% as copolymers and terpolymers. The most important commercial copolymers are those with vinyl acetate or with vinylidene chloride [99].

The introduction of vinyl acetate (2–10 mol%) in the PVC macromolecules leads to the increase of solubility and improvement of molding behavior by lowering the temperature at which this copolymer can be processed and by increasing melt flow. Copolymerization takes place generally by suspension or solution techniques. These copolymers are used for flooring compositions, fibers and for surface coatings. The last ones have very good resistance to water, acids, alkalis, aliphatic hydrocarbons, and oil and fats: they are flexible and non-flammable. The main application of coatings based on vinyl chloride–vinyl acetate copolymers are for industrial plants exposed to corrosive agents and marine equipment [100].

2.2.10 Vinyl chloride–vinylidene chloride copolymers

Vinyl chloride–vinylidene chloride copolymers are obtained especially through the emulsion technique. In some applications, the latex, after addition of stabilizers, is used directly as a coating; in other applications the copolymer is recovered through well known techniques in any form. The amount of vinylidene chloride in the copolymer varies in the range of 5–12% and the properties of the copolymer are similar to those of vinyl chloride–vinyl acetate copolymers. There are also commercial products containing higher amounts of vinylidene chloride. Such copolymers are often further modified by the incorporation of plasticizers. The combination of copolymerization and plasticization permits the design of copolymers ranging from soft rubbery materials and filaments to rigid plastics. The major applications of such copolymers lie in the field of packaging where its low permeability, good optical characteristics, excellent chemical resistance, non-flammability and low cost are of value [100,101].

2.3 POLY(VINYL ACETATE) (PVAc) AND ITS DERIVATIVES

2.3.1 Introduction

Vinyl acetate (VAc), a colorless liquid, has a boiling point of 72.5°C, is completely miscible with organic liquids and is soluble in water with a ratio of 2.5 g/100 g. It polymerizes to poly(vinyl acetate) in a solvent which aids in temperature control or in bulk polymerization. For high MW PVAc, suspension or emulsion polymerization are used. In commercial practice, the latter is predominant and the produced latices are directly used in water based paints and adhesives. For PVAc in solid form, the former procedure is used. The worldwide consumption of PVAc in 1979 was *c.* 800 000 tons [99].

Vinyl acetate polymerization occurs through free radicalic process; its induction period and kinetics are strongly affected by the monomer impurities which must be removed.

2.3.2 Bulk polymerization

The main factors controlling the bulk polymerization process are the monomer purity, the type and concentration of the initiator, and the nature of the chain transfer modifier. This technique is applied in continuous plants using organic peroxides as initiators and aldehydes as chain modifiers.

VAc bulk polymerization has limited applications because it leads to a low molecular weight polymer with a high degree of polydispersity. In this procedure, the MW is controlled more through the amounts of initiator and chain transfer agent and less through the polymerization temperature. An increase of the amount of initiator provides a low MW. For the synthesis of a low viscosity PVAc, acetic or propionic aldehyde are used as chain transfer agents. The latter is preferred because of its boiling temperature which is close to that of vinyl acetate. Impurities such as acetic acid, acetic anhydride, water and ethylidene diacetate, affect the polymerization rate.

The continuous plant for bulk polymerization of VAc consists of two vessels for the preparation of the reaction mixture, a polymerization tower, conveyor belt and a cooling device (Figure 2.18).

Bulk polymerization is carried out at 70–80°C in a nitrogen atmosphere and requires 8 hours; oxygen is a strong inhibitor of the process.

2.3.3 Solution polymerization

Organic solvents such as toluene, benzene, ethyl acetate, ethanol and sometimes methanol are used for solution polymerization. Toluene is an active chain transfer agent and leads to low MW; benzene has a lower activity from this point of view and is preferred. During solution polymerization in benzene, the viscosity of the solution increases until the process is complete, but the concentration of the system is too low to exhibit auto-acceleration due to the gel effect. The solution may be used as obtained, the solvent may be stripped off, or the polymer may be precipitated with a poor solvent

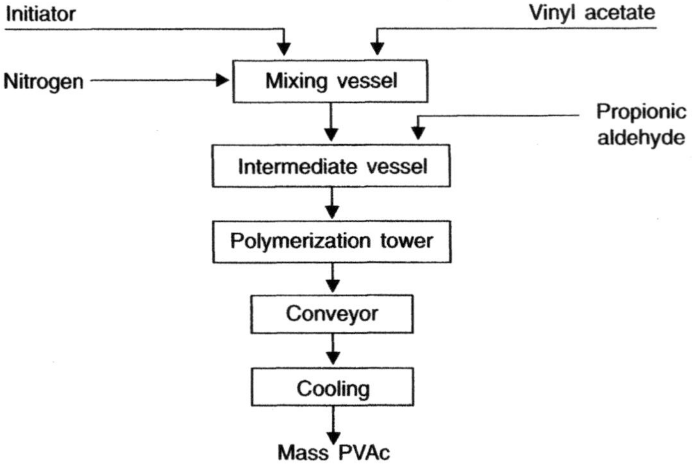

Figure 2.18 Vinyl acetate bulk polymerization.

for PVAc such as ethanol [3]. Aldehydes have the same influence as in the bulk polymerization process.

In some commercial plants, ethyl acetate is used as solvent and the MW is controlled by the initiator and the transfer agent concentrations.

To carry out the process, propionic aldehyde is dissolved in the monomer and the initiator is dissolved in ethyl acetate. The continuous solution polymerization takes place in plants with two columns (reactors) which can be completely separated or connected by a U-shaped pipe, at the boiling temperature of the medium, in 5–12 hours. The obtained PVAc solution can be used as is (a lacquer); from it PVAc can be separated, by running the reaction mixture to a still where the solvent and unchanged VAc are removed by steam distillation.

A mixed 'bulk solution' procedure is sometimes used in order to combine the advantages of the bulk and the solution polymerizations in one continuous procedure. The first step avoids the introduction of impurities in the medium with different components which may affect the MW; in the meantime, the duration of the process is reduced. For the second step, to (a) avoid the difficulty of the removal of the polymer from the polymerization reactor, (b) improve the heat transfer and (c) obtain a polymer solution usable directly in a continuous system of hydrolysis and acetylation, methanol, methyl acetate or their azeotropic mixture is added as a solvent; until its completion, the polymerization occurs in solution. It takes place with peroxide initiators, without chain modifiers and produces a less branched PVAc.

Solution polymerization of vinyl acetate has a series of drawbacks, such as low reaction rate, low MW (14 000–32 000) and high polydispersity; due to secondary reactions and impurities, PVAc is branched.

2.3.4 Emulsion polymerization

Since PVAc is often used in a latex form, it is primarily manufactured by free radical initiated emulsion polymerization. Emulsion polymerization of vinyl acetate is carried out to 50–55% solids in the presence of 0.05–1.5% surfactants, generally anionic, non-ionic or mixtures of these two types. In certain instances protective colloids such as poly(vinyl alcohol) and cellulose derivatives are used. PVAc emulsions can be made with surfactants alone, with protective colloids alone or with combinations of the two. Normally up to 3% emulsifier can be included in the recipe. The larger the quantity of surfactant, the smaller the particle size of the emulsion [105]. In some adhesive applications, the amount of poly(vinyl alcohol) may be more than 1% of the total amount of VAc. Potassium persulfate is often used as an initiator, and a pH of 4–6 is obtained with a small quantity of sodium bicarbonate as a buffer. If the pH is outside this range, polymer instability can result due to hydrolysis after polymerization is complete or even due to partial hydrolysis of VAc during the polymerization process. Unlike the polymerization of VAc

Table 2.8 Typical formulation for PVAc by emulsion polymerization

	Parts by weight
Vinyl acetate (inhibitor free)	100
Water	100
Hydroxyethylcellulose	2.5 (protective colloid)
Poly(ethylene glycol), ether of lauryl alcohol	2.5 (surfactant)
Sodium dodecylbenzenesulfonate	0.1 (surfactant)
Sodium bicarbonate	0.5 (buffer)
Potassium persulfate	0.5 (initiator)

with persulfate, during polymerization with the Fe^{2+}–H_2O_2 redox system, hydroxyl radicals fail to stabilize the precipitated polymer which undergoes massive coalescence [102]. In general there is no MW increase in the PVAc due to the change transfer of macroradicals to VAc [103].

A typical formulation might be as shown in Table 2.8 [46].

The most commonly used technique for producing PVAc latices is the delayed addition procedure in which 5–10% of the initial vinyl acetate is added and polymerized to form a seed. During the ensuing 2–4 hour period, the balance of the monomer is introduced gradually into the polymerization system. With this technique, fine particles in the 0.1–0.2 μm range can be formed at a relatively low surfactant concentration. Particle number and size data from a series of seeded emulsifier-free VAc emulsion polymerization experiments have been analyzed with the aid of polymerization and particle growth models [104]. The pH of most commercial emulsions is 4–6, solids content is 48–55% and viscosity is 200–4500 cP [3]. Nucleation of particles in seeded emulsion polymerization of VAc with poly(vinyl alcohol) as emulsifier was reported [105].

C.K. Lee and T.H. Forsyth [106] have investigated the feasibility of using a tubular reactor for the seeded emulsion polymerization of VAc and have studied the effects of process variables on conversion rates and latex properties. This study of the continuous, tubular, seeded emulsion polymerization of VAc led to the following conclusions:

- complete monomer conversion is possible with high concentrations of initiator and mixed emulsifier;
- the use of seed improves the latex stability, but phase separations still occur at low monomer conversions.

New completely automatic plants working 'in cascade' have three to five reactors; such continuous procedures permit the production of latices with constant quality.

D.M. Lange *et al.* [104] reported a kinetic analysis of seeded emulsion polymerization of VAc. The results of the analysis were used to evaluate the magnitude of water phase terminations, to estimate radical capture coefficients and to evaluate competitive growth. A newly developed PVAc emulsion latex was prepared by the semicontinuous process using an octyl aldehyde–sodium bisulfate–potassium persulfate redox pair initiator system [107].

2.3.5 Suspension polymerization

This procedure is used mainly when high molecular weight PVAc is needed for conversion to poly(vinyl alcohol). Polymerization takes place in water using a monomer soluble initiator such as benzoyl peroxide, a dispersing agent such as 0.1% poly(vinyl alcohol), some cellulose derivatives, or some polyelectrolytes such as the sodium salt of styrene–maleic acid anhydride copolymer. Dicyclohexylperoxidicarbonate can also be used for the polymerization [44]. Bead size is very large and depends upon monomer : water ratio, the nature and amount of dispersing agent, agitation rate, temperature and the presence of small amounts of water insoluble salts (sulfates or carbonates).

The stability of the dispersion can be much improved by adding small amounts of water insoluble salts. An example of a formulation is shown in Table 2.9.

The suspension polymerization process of vinyl acetate is presented in Figure 2.19. Polymerization takes place in a nitrogen atmosphere at 70–90°C, for 2–6 hours while using a stirring rate of 120 rpm. Generalities on suspension polymerization of vinyl monomers are discussed in H.G. Yuan *et al.* [108].

2.3.6 Properties

PVAc is a solid polymer which, depending on its MW, varies from a gum-like material to a leather-like one. It is too brittle and shows too much cold flow to be useful in bulk flow for rigid applications. It is transparent and colorless, its specific gravity is 1.191 and the refractive index is 1.4665. It swells a little in water and is sensitive to acids or concentrated alkalis. The softening temperature varies with the MW. The glass transition temperature has been reported to be 28–29°C and PVAc tends to degrade above 130°C;

Table 2.9 Typical formulation for PVAc by suspension polymerization

	Parts by weight
Vinyl acetate (inhibitor free)	100
Poly(vinyl alcohol) 5% solution	100
Magnesium carbonate dispersed in 10 parts H_2O	0.1
Benzoyl peroxide (dissolved in VAc)	0.5

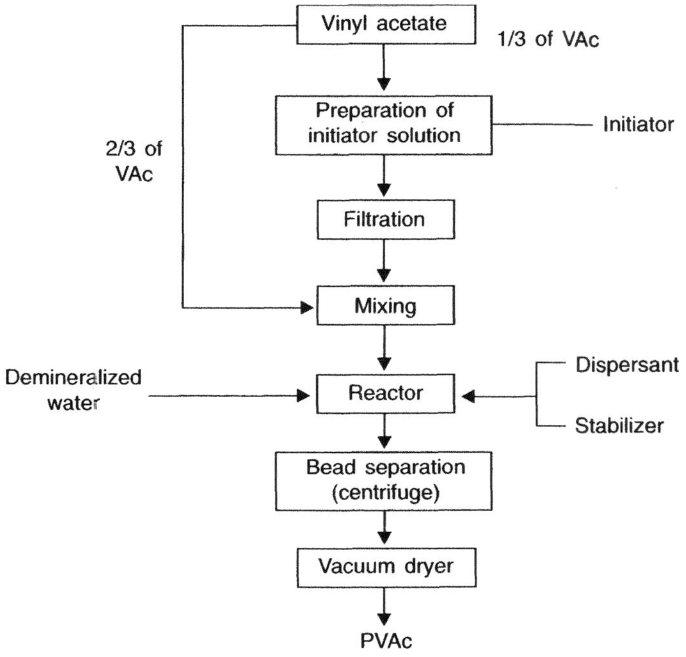

Figure 2.19 Vinyl acetate suspension polymerization.

when heated above 150°C, PVAc easily depolymerizes with the release of acetic acid. PVAc is soluble in acetic acid, benzyl alcohol, 95% ethyl alcohol, methanol, chloroform, dioxane, benzene and toluene; it is stable with gasoline, kerosene, mineral oils and greases. PVAc can be plasticized by conventional esters such as dibutyl phthalate, tritolyl phosphate, tributyrin and diphenyl. Internal plasticization is done by copolymerization with comonomers like vinyl propionate or vinyl caprate. X-ray diffraction analysis shows that PVAc is amorphous. PVAc has good adhesion to glass, textiles, paper, leather and wood. Adhesive properties are enhanced by adding 10–15% plasticizer in PVAc latex [18,46].

2.3.7 Applications

PVAc is used primarily for the production of poly(vinyl alcohol) and poly-acetals. All its applications depend on the MW (K value); some examples are:

1. PVAc, $K = 80$–100 (in solution) for poly(vinyl alcohol);
2. PVAc, $K = 70$–80 (mixed with a filler) for pressed items;
3. PVAc, $K = 50$–70 (mixed with a filler) for foundry;
4. PVAc, $K = 45$–55 (in solution) for poly(vinyl butyral);

5. PVAc, $K = 30$–45 (in solution) for lacquers, solutions for thermo-adhesive papers;
6. PVAc, $K = 30 \pm 2$ (in solution) for poly(vinyl formal).

PVAc's excellent adhesion properties are due to the nature of its functional groups. Its properties do not deteriorate at low temperatures or in the sunlight and are unaffected by bacteria, fungi or insects. This polymeric adhesive will not sustain prolonged stress, particularly at elevated temperatures. It swells slightly and becomes opaque on long immersion in water, but recovers upon drying; more recently synthesized appropriate copolymers show greater water resistance [18].

Some researchers underline the variation among PVAc adhesives in their resistance to moisture and to weathering; however, all proved vulnerable to some degree to continuous stressing on the glued joints [109]. Developments in the formulation of PVAc adhesives have greatly reduced its sensitivity to water and involve some degree of crosslinking of the polymer chains. There are two known types of PVAc wood adhesives, a conventional one and a crosslinking type one or two part system with improved water resistance. The crosslinking is done with metallic salts such as chromium (III) or aluminum nitrate [110]. PVAc adhesive is quick-setting, a property which shortens press and assembly times in manufacturing processes. The latex, which is obtained through emulsion polymerization, is free of odor, non-flammable, mold resistant and has a uniform viscosity and a good shelf life. It may be used with or without a plasticizer.

For some applications, such as for making sealable coatings, for grease proofing paper, or for the manufacture of laminating plastics, PVAc resins are sold as a solution in organic solvents [40].

PVAc is also used as latex paints, a binder for pigmented paper and paperboard coatings, for cotton textile sizing, as a binder for non-woven fabrics and an antishrink agent for glass fiber reinforced polyester molding resins.

2.3.8 Vinyl acetate copolymers

Because plasticizers added in PVAc products may migrate or evaporate, copolymerization as an internal plasticization with comonomers like butyl acrylate, 2-ethylhexyl acrylate, diethyl fumarate, diethyl maleate, and vinyl esters of fatty acids is recommended. The copolymerization takes place through the emulsion procedure and the amount of comonomer is up to 20%.

The aim of synthesizing VC–VAc copolymers is to increase solubility and to improve molding characteristics. Ethylene–vinyl acetate (EVA) copolymers with a VAc content in the range 5–20% by weight are widely used. Compared to LDPE, these copolymers show improved clarity, low temperature flexibility and impact strength. Such copolymers with increasing amounts of VAc find applications such as films for food packaging, medical tubing, squeeze bulbs, gaskets, coatings and hot melt adhesives [46].

A more recent development in VAc technology is the advent of ethylene copolymers involving a higher amount of VAc. These copolymers, obtained through the emulsion copolymerization procedure are recommended for various adhesive and paint applications. Ethylene adds flexibility to copolymer films, analogously to acrylates, with an economic advantage over the acrylates.

2.3.9 Poly(vinyl alcohol) (PVA)

Although the direct polymerization of vinyl alcohol (VA) was recently reported [111], this polymer is obtained by the hydrolysis of PVAc, since the monomer is the enolic non-stable tautomer of acetaldehyde and cannot be produced without reverting to the aldehyde form. Thus PVA is not synthesized by a polymerization process and its degree of polymerization depends primarily on the size of PVAc chains.

The transformation of PVAc to PVA is obtained by the base catalyzed alcoholysis or by the acid initiated hydrolysis of PVAc. In both cases, the acidic traces of sodium acetate formed in the base catalyzed alcoholysis affect the properties of PVA, primarily the stability. Such difficulties are avoided if PVA is produced using a small amount of base as catalyst. In methanol medium in the presence of sodium methoxide, the formation of PVA takes place as follows:

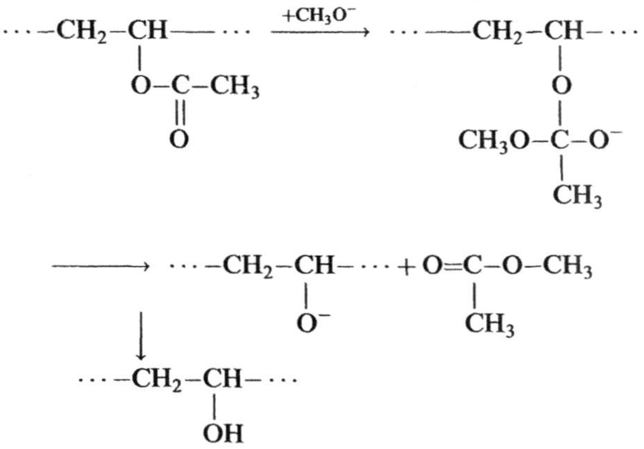

(PVA)

With sodium methoxide, the alcoholysis is produced in 6–24 hours at a temperature in the range of 25–35°C. If the process of alcoholysis is done without agitation, PVA is obtained in a gel form; alcoholysis with agitation leads to a polymer in powder form or pellets. Instead of sodium methoxide, alkaline hydroxides can be used, especially KOH. Saponification (hydrolysis degree) depends on the amount of hydroxide and the amount of water in the alcohol.

PVA produced through catalytic alcoholysis presents some advantages in comparison with that obtained through acidic hydrolysis; its stability is better and the polymer purification is easier. A pure PVA is obtained by precipitation with acetone (in excess) from an aqueous solution.

In the case of acidic saponification, the acid's nature influences the main characteristics of PVA; acids such as HCl, $HClO_4$, H_3PO_4 and $ClCH_2COOH$ can be used as saponifying agents. Hydrochloric acid is not recommended due to the fact that it leads to a colored PVA. Often sulfuric acid is used, because that agent causes a light colored powder which can be easily washed and dried; the formation of unstable esters constitutes the drawback of using sulfuric acid. Heating those esters causes a release of acid which degrades the polymer. The other three acids mentioned above do not have this drawback and thus they increase the stability of PVA. The pH and the temperature influence the rate of saponification. A high ratio polymer/liquid medium and an intense agitation lead to an increase of the fineness of the PVA powder.

(a) Discontinuous hydrolysis of PVAc

In the case of acidic hydrolysis, the saponification takes place at the boiling temperature of the solvent or at a lower temperature in 8–24 hours. The hydrolysis with alkalis occurs at 30–35°C, and at the end of the process the system is brought to boiling point. The gel produced during the initial stage disintegrates after new solutions of alkaline catalyst are added. Saponification continues until the desired content of hydroxylic groups is obtained.

(b) Continuous hydrolysis of PVAc

The flow chart of this technology is shown in Figure 2.20.

The process in the hydrolyzer needs 5–30 minutes at 50–60°C and leads to a PVA with 77–95% degree of hydrolysis which can be increased during maturation. The supplementary hydrolysis is necessary when a PVA with high degree of hydrolysis (99.5% hydroxylic groups) is required.

(c) Properties

The presence of secondary alcoholic groups allows the polymer to participate in reactions which are specific for secondary alcohols, such as oxidation, which leads to ketonic groups, formation of polyalcoholates with metallic sodium which can be used for the production of vinylic ethers. It also forms esters and the reaction with aldehydes leads to the formation of acetal; with ketones it forms poly(vinyl ketal) [112]. Since it is an alcohol-like cellulose, it also participates in polymer analogous reactions.

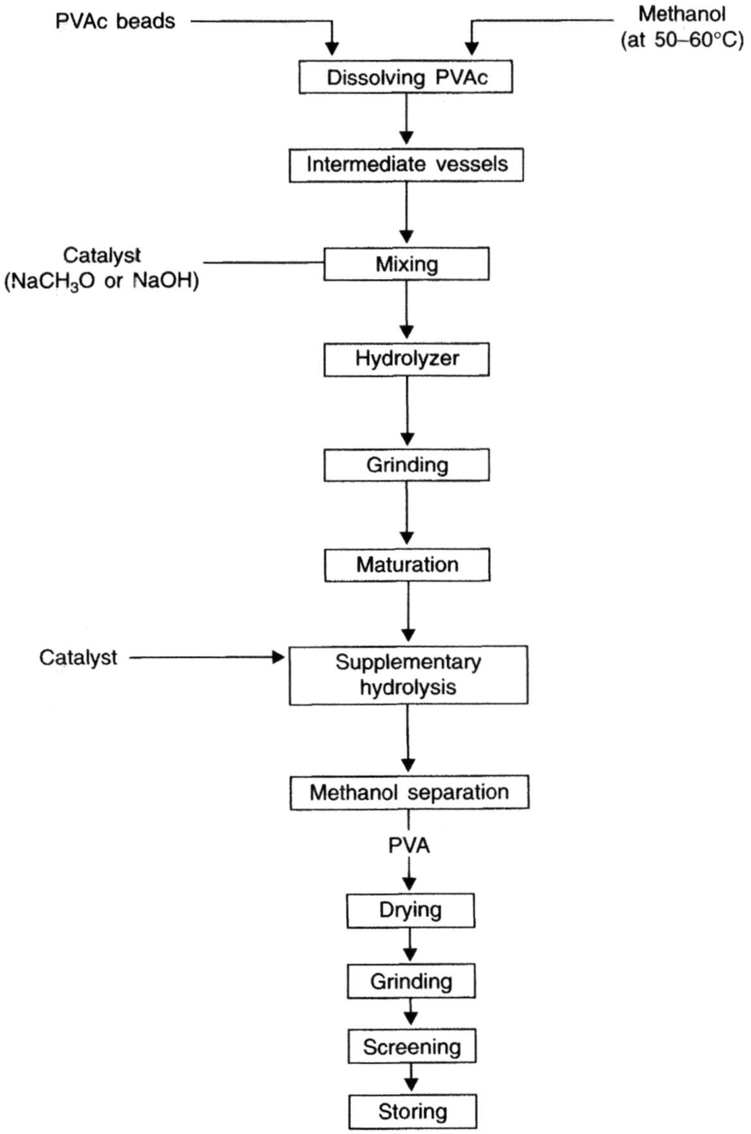

Figure 2.20 Poly(vinyl alcohol) production through PVAc hydrolysis.

Above 100°C dehydration occurs, this process being accelerated by the acid or alkali traces existing in the polymer. In the meantime, physical–chemical properties are changed, e.g. the water solubility decreases, the brittleness and the glass transition temperature increases, double bonds form, and ring and oxygen bridges form as a result of intermolecular reactions. In principle, the degree of polymerization does not vary during hydrolysis; however, if the initial PVAc has a very high degree of polymerization and the hydrolysis temperature is too high, PVA degradation occurs.

The modification of esteric groups in hydroxyls leads to an increase of the polarity of the polymer and, implicitly, of the intermolecular forces. Due to its high number of alcoholic groups, PVA is one of the most polar and hydrophilic synthetic polymers; it also form hydrogels [113].

PVA is water soluble, soluble in acids including phosphoric acid, in high polar solvents such as dimethyl formamide (DMF), dimethylsulfoxide (DMSO), phenol, formamide, pyridine, ethanolamine, dioxan, diethylene triamine and their aqueous solutions. It is soluble in concentrated aqueous solutions of these solvents; it is soluble in concentrated aqueous solutions of polyalcohols like glycols, glycerin, diethyleneglycol, etc. and insoluble in simple alcohols, ketones, ethers and esters.

In general, PVA properties such as water solubility, adhesion, tensile strength of films, solvent resistance, etc. depend upon the PVAc MW and the degree of hydrolysis.

The temperature at which complete aqueous solution occurs varies with the grade, depending mainly on the degree of hydrolysis. PVA with 87–89% hydrolysis shows maximum solubility in that it dissolves readily in both cold and hot water. Above 89–90% hydrolysis, heating is required for complete solution [114].

Solutions of fully hydrolyzed PVA show no apparent change in properties when heated for several days at temperatures up to 100°C.

Dehydration of PVA takes place (above 100°C) with the formation of double bonds which may lead to peroxide bonds, esteric cycles, other cycles or even crosslinks:

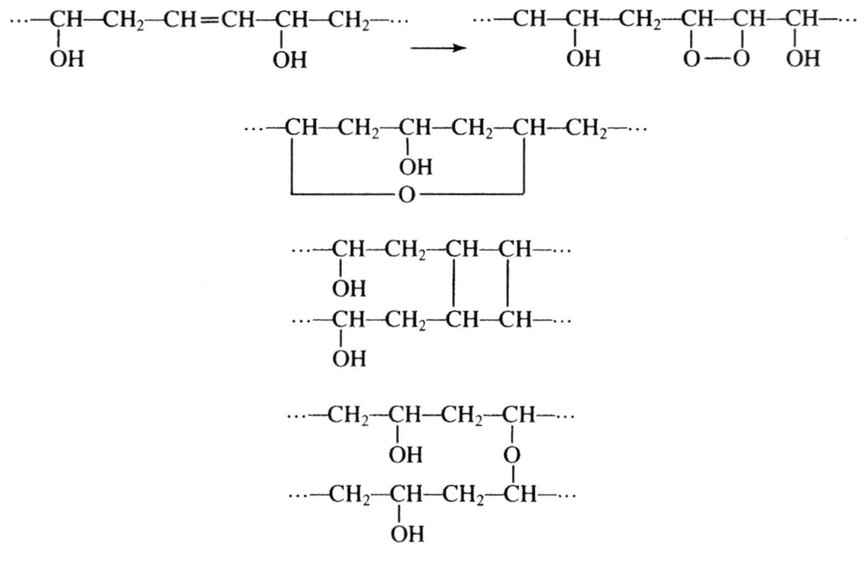

The intermolecular hydrogen bonding results in a high crystalline melting point (about 230°C for completely hydrolyzed grades) and appreciable degradation occurs before this temperature is reached. Thus PVA can be processed only in solution (as in solution spinning of fiber) or if plasticized (when compression molding or extrusion are possible) [46].

PVA is a white powder with a specific gravity in the range of 1.2–1.3 and a glass transition temperature of around 80°C. PVA with 50–80 mol% acetyl groups is water stable and has better thermal stability and mechanical properties than PVAc. By drawing, the degree of crystallinity increases; orientation is necessary in the case of films and PVA fibers.

An isotactic PVA can be prepared by acid cleavage of poly(vinyl *t*-butyl ether) made by cationic polymerization of its monomer at −78°C [19].

(d) Applications

PVA is used for the production of fibers, artificial leather, tubing, gaskets with good stability to gasoline and other oil derivatives, rubber-like items, transportation belts, surgical fibers, emulsifiers, adhesives for paper and paperboard, and in general purpose adhesives for bonding paper, textiles, leather and porous ceramic surfaces.

PVA is an efficient binder for solid particles, pigment, ceramic products, plaster, cement, fibers, non-woven fabrics, catalyst pellets, cork compositions, cement building products, etc. It is a good protective colloid for aqueous emulsions and is employed for this purpose in a large variety of emulsion and suspension systems. In the manufacture of textile fibers, the most important operation is to transform it into a product insoluble in water. This is obtained by treating PVA fiber with a concentrated aqueous solution of sodium sulfate containing certain amounts of formaldehyde and sulfuric acid. Such a treatment will produce formal groups:

$$\cdots-CH_2-\underset{\displaystyle OH}{\underset{|}{CH}}-CH_2-\underset{\displaystyle OH}{\underset{|}{CH}}-\cdots + CH_2O \cdots -\underset{|}{CH}-CH_2-\underset{|}{CH}-CH_2-\cdots$$
$$O-CH_2-O$$

Hydroxyl groups of neighboring PVA macromolecules may also participate in the formation of formal groups. After such treatment, PVA fiber becomes insoluble. In comparison with other synthetic fibers, PVA fibers show a higher rate of water absorption. Due to such characteristics, PVA fibers are used as substitutes for cotton fibers. These synthetic fibers have very good dimensional stability and tenacity [115]. High strength and high modulus PVA fibers have been obtained by crosslinking, wet spinning and multistep drawing [116].

Light polarized films can be manufactured by casting a PVA solution on a surface, and drying and peeling the resulting film [117].

The useful relative humidity range and mechanical properties of starch were significantly extended by blending with PVA. The degree of hydrolysis and MW of PVA significantly affects the water resistance and mechanical properties in blends with high amylose thermoplastic starch [118].

2.3.10 Poly(vinyl acetal)s

The poly(vinyl acetal)s are prepared by acidic catalyzed acetalization of PVA. The predominant 1,3-glycol structure of PVA strongly favors the formation of the 1,3-dioxan cycle [102].

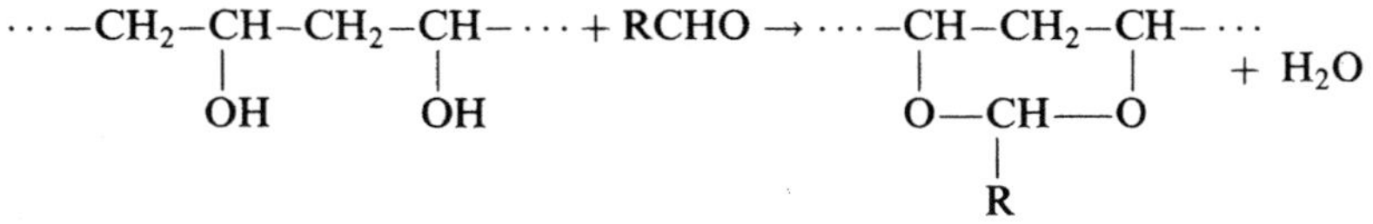

Such reaction occurs at random, and as a result, some hydroxyl groups remain unreacted. This explains why such polymers contain, besides the acetal groups, isolated hydroxyl groups. The microstructure of poly(vinyl acetal)s is more complex because there will also be residual acetate groups due to incomplete hydrolysis of PVAc. Thus the properties of poly(vinyl acetal)s will depend on the amounts of hydroxyl, acetal and acetate groups.

In industry, two procedures are used. In one of them, known as 'hydrolysis–acetalization', the hydrolysis of PVAc and the acetalization of PVA take place in a homogeneous medium in the same reactor. The other procedure, known as 'heterogeneous hydrolysis', uses aqueous solutions of PVA. In one variant, miscible solvents are used, such as alcohols. Another variant is based on the use of water miscible organic solvents such as benzene, butylacetate, etc. These two variants use two different reactors.

Poly(vinyl acetal) and poly(vinyl butyral) are produced in one reactor from an alcoholic solution of PVAc. To obtain poly(vinyl formal) with the same procedure, a solution of PVAc in acetic acid must be used.

In the case of hydrolysis–acetalization, an alcoholic solution of sulfuric acid and the aldehyde, 130–200% of the theoretic amount are added to the solution of PVAc. The hydrolysis needs 10–16 hours and takes place under continuous stirring at around 70°C. The PVA produced is transformed immediately, without precipitating, into poly(vinyl acetal) which is soluble in the same solvent as PVAc. Adding water to the resulting system leads to the precipitation of the poly(vinyl acetal).

Heterogeneous hydrolysis can be performed in an alcoholic medium where PVA is insoluble or in an aqueous solution of PVA in which the obtained acetal is insoluble.

The degree of acetalization depends on the nature and amount of the aldehyde, temperature and reaction time.

When the aldehyde is formaldehyde, the product is poly(vinyl formal). This polymer is, however, made directly from PVAc and formaldehyde without separating PVA [20].

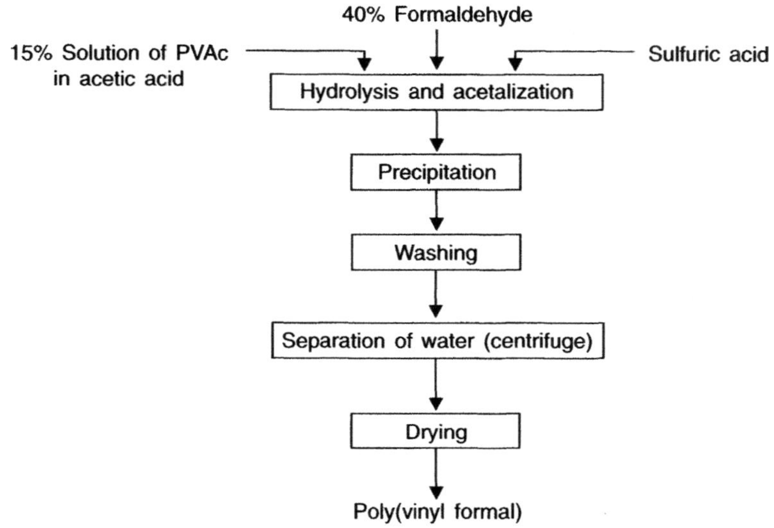

Figure 2.21 Production of poly(vinyl formal).

(a) Poly(vinyl formal)

This acetal is produced by dissolving a PVAc suspension in acetic acid and reacting it with formaldehyde in the presence of a catalytic amount of a mineral acid (sulfuric, hydrochloric or phosphoric). The mixture is heated to 70°C for up to 24 hours. Water is then added to the mixture with rapid agitation and the poly(vinyl formal) is precipitated as granules which are washed to neutralization and dried.

The hydrolysis–acetalization technology for the production of poly(vinyl formal) is presented in Figure 2.21.

The degree of acetalization of the commercial poly(vinyl formal) is 75–85%. Its solubility and other properties depend on this degree. Enamel formulations always contain this polymer and, more commonly, phenol–formaldehyde polymer. Poly(vinyl formal) is available in several MWs [119].

(b) Poly(vinyl acetal)

The acetal can be produced by both the homogeneous procedure in alcoholic medium or the heterogeneous one in aqueous medium. Acetaldehyde, with a low boiling point, and paraldehyde (2,4,6-trimethyl-1,3,5-trioxane) with a boiling point of 124.5°C are used for acetylation.

Poly(vinyl acetal) has a lower T_g and thermal stability than poly(vinyl formal) and is soluble in common organic solvents such as aromatic hydrocarbons, alcohols, esters and ketones.

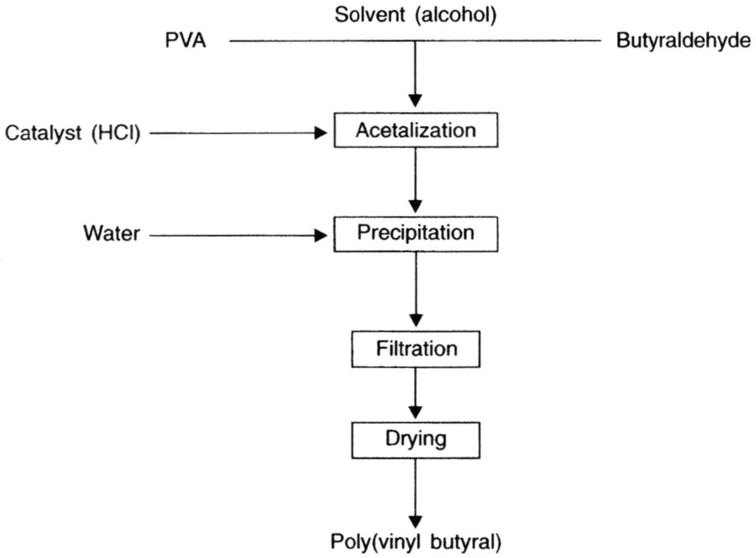

Figure 2.22 Production of poly(vinyl butyral).

(c) Poly(vinyl butyral)

When the aldehyde in the acetalization process is butyr-aldehyde, the end product is poly(vinyl butyral). More often for the synthesis of this acetal, a mixture of an aqueous solution of PVA and butyr-aldehyde is used; this mixture is stirred continuously and heated to 80°C in the catalytic presence of HCl for a few hours (Figure 2.22). The technical butyral separated by filtration usually contains 55–75% butyral groups. Its T_g is only 49–50°C compared to acetal with a T_g of 82°C and to formal with 116–160°C. Plasticized butyral is soft, flexible and shock resistant down to −20°C.

(d) Applications of poly(vinyl acetal)s

As has been previously mentioned, the properties of poly(vinyl acetal)s depend on the amount of acetal, acetyl and hydroxyl groups and the MW. The specific gravity of poly(vinyl acetal)s is in the range of 1.11–1.22, increasing from butyral to formal. The commercial applicability of poly(vinyl acetal)s depends upon good electrical properties, degree of crosslinking, adhesion to different materials, resistance to organic solvents and toughness.

Poly(vinyl formal) finds wide use as a wire enamel insulator. For this application, resolic, phenolic, cresolic or epoxy resins are used as crosslinking agents. Such resins are able to react (during heating) with the acetal functional groups and as a result improve its thermal stability, and increase insolubility and wear resistance. The use of poly(vinyl formal) as electrical wire insulation permits the decrease of the volume and weight of the electrical motors, and the increase of their output.

Poly(vinyl acetal) is slightly soluble in alcohols, ketones, aromatic hydrocarbons, esters, etc. It is soluble in ethanol, butanol, acetic acid, cyclohexanone, mixtures of benzene and alcohols and in other organic solvents. It is used as a component in lacquers and for the fabrication of some pressed or injected items. Plasticizers such as dibutyl-sebacate, dioctyl-sebacate or dioctyl phthalate in amounts of 15–35%, lower the T_g and provide poly(vinyl butyral) films with a good flexibility and low temperature stability. Poly(vinyl butyral) forms transparent, colorless films that are used in the fabrication of safety glass. It has good adhesion properties to glass and metals. It is flexible and possesses good light stability. This acetal can be processed by extrusion and injection. The adhesives made of this polymer are used in the construction industries for metals, plastics, wood, concrete, etc.

2.4 FLUOROPOLYMERS

2.4.1 Introduction

The microstructures of monomers and polymers belonging to this group are comparable with hydrocarbons where all or almost all of the hydrogen atoms are replaced with fluorine or chlorine. The most important polymers of this group are the homopolymers of tetrafluoroethylene, trifluorochloroethylene, vinyl fluoride, vinylidene fluoride and various copolymers based on these and other monomers. The polymer of trifluorochloroethylene is not a true fluorocarbon since it contains one chlorine atom for every three fluorine atoms, but its properties are close to those of poly(tetrafluoroethylene) (PTFE).

Major commercial fluoropolymers are PTFE, poly(vinyl fluoride) (PVF), poly(vinylidene fluoride) (PVDF), poly(chlorotrifluoroethylene) (PCTFE), poly(tetrafluoroethylene-co-hexafluoropropylene) (FEP), poly(tetrafluoroethylene-co-perfluoropropylvinylether) (PFA), poly(tetrafluoroethylene-co-ethylene) (ETFE) and poly(chlorotrifluoroethylene-co-ethylene) (ECTFE). Of these, PTFE has the largest commercial production [120].

The worldwide production capacity of fluoropolymers is 72 000 tons per year, of which 76% (55 000 tons) represents PTFE. Consumption has increased in the USA from 13 500 tons in 1990 to 14 000 tons in 1992, while in Japan it has decreased from 16 000 tons in 1990 to 14 000 tons in 1992 and it has remained constant in Western Europe at 17 000 tons during the same period [121,122].

2.4.2 Poly(tetrafluoroethylene) (PTFE)

Tetrafluoroethylene spontaneously polymerizes violently through a radicalic mechanism in the presence of small amounts of oxygen or common initiators at temperatures of 60–140°C and at a pressure of around 7 MPa. Initiation is usually done with ammonium persulfate, other persulfates, benzoyl peroxide, hydrogen peroxide, etc. The process can be carried out in emulsion, suspension or solution; however, polymerization in aqueous medium is preferred. The polymerization is strongly exothermic (≈25 kcal/mol) and has to be thoroughly controlled because of the low thermodynamic stability of the monomer and the high polymerization rate which suddenly increases with the increasing pressure.

Tetrafluoroethylene emulsion polymerization usually takes place under pressure at 70–80°C, under continuous stirring in the presence of an emulsifier and one of the above mentioned initiators, using a ratio of water to monomer of 100:6 up to 100:9. The MW of PTFE obtained through emulsion or suspension techniques is high, in the range of 4×10^5–9×10^6. It has to be this high otherwise the polymer is unsuitable for applications.

In the latex form, with a solids content of up to 60%, PTFE is used as a coating, in film casting and in the impregnation of fabrics. The latex particles (0.1 μm) form, through coagulation, larger particles with a diameter of about 0.5 mm. The beads obtained during suspension polymerization are about the same size as these aggregates.

PTFE is produced in granular, fine powder or aqueous dispersion (latex) form, depending upon the intensity of agitation and the amount of emulsifier. The process, known as aqueous dispersion polymerization, is based on mild agitation with a dispersing agent and leads to the formation of a latex with minute colloidal particles. Through gentle agitation in the presence of an emulsifier, coagulation is avoided and a fine PTFE powder is obtained. The granular PTFE is produced when little or no emulsifier is used and a vigorous agitation is employed [123]. Figures 2.23 and 2.24 show the formation of different types of PTFE.

The homogenous solution polymerization of highly fluorinated acrylic monomers has been recently achieved in supercritical carbon dioxide through a radical mechanism [124].

(a) Properties

PTFE is linear with practically no branching, and has a highly regular structure with a helix configuration. This is because branching requires the breakage of the strong C–F bonds (116 kcal) which is quite unlikely under the polymerization conditions. PTFE has a high degree of crystallinity (about 95%), a T_g of 126°C and a T_m of 327°C [28]. It has a high thermal and thermo-oxidative stability and it is completely solvent resistant; only certain fluorinated solvents dissolve it at temperatures near its T_m.

The mechanical strength of PTFE remains practically unchanged over a wide range of temperature from −100°C to +350°C [125]. It is a tough, relatively flexible material, with a tensile strength of about 20 MPa and an

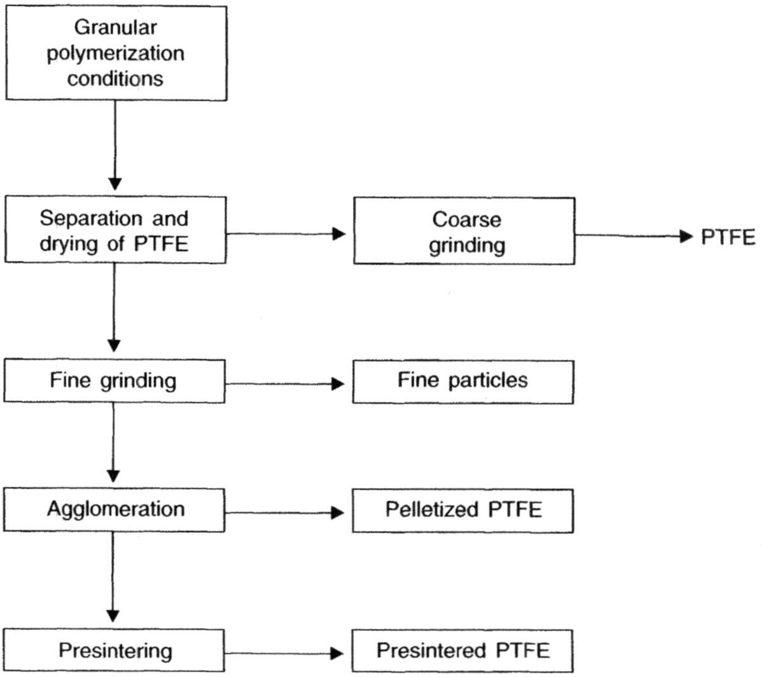

Figure 2.23 Granular polymerization. (Reprinted with the permission of John Wiley & Sons Inc. from *Encyclopedia of Polymer Science and Engineering* by H. Mark, N.M. Bikales, C.G. Overberger and G. Menges (eds), Vol. 16, 1989.)

elongation at break of 200–300%. Its high T_m and melt viscosity (10^{10}–10^{11} P at 350°C) also make PTFE impossible to process by conventional means and thus special techniques are required [46,126,127]. Moldings are made from granular polymers by a technique similar to that used in powder metallurgy. Products are made by sintering or special extrusion techniques.

Ram extrusion (a pressure sintering process for the production of high MW polymers) is predominantly used for processing fluoropolymers [128]. PTFE has outstandingly good electrical properties such as a low dielectric constant, low coefficient of friction, low flammability, low moisture absorption, excellent weatherability and low surface energy.

Improvements such as increased purity, better processability, etc. as a result of a broader range of free flowing types and a higher bulk density have constituted the trend during the past few years. To insure improvement of cold flow and hardness, fillers (5–40%) stable at the sintering temperature of PTFE are added. Examples of such additives are glass fiber, carbon black, bronze, graphite, carbon fibers and polymers such as polyimides, polyphenylene sulfide and polyetherketone [122].

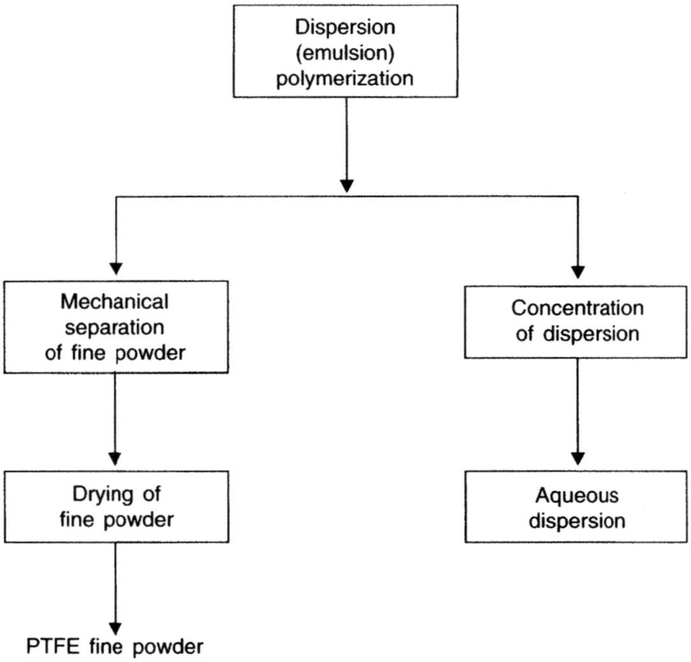

Figure 2.24 PTFE fine powder technology. (Reprinted with the permission of John Wiley & Sons Inc. from *Encyclopedia of Polymer Science and Engineering* by H. Mark, N.M. Bikales, C.G. Overberger and G. Menges (eds), Vol. 16, 1989.)

(b) Applications

The excellent electrical insulation properties and chemical inertness over a wide range of temperature account for its main uses, namely electrical and chemical applications. Electrical applications include wire insulation, whilst the chemical ones are gaskets, pump parts, laboratory equipment, seals and valves. Others include low friction linings for hoppers, non-stick coverings for kitchen utensils, and diaphragms for low temperature equipment, protective gloves, coatings, laminates and lubricants [129–133].

2.4.3 Poly(trifluorochloroethylene) (PCTFE)

The monomer trifluorochloroethylene is prepared from hydrogen fluoride and hexachloroethane, which yields trifluorotrichloroethane. Through dehalogenation with zinc as a catalyst, trifluorotrichloroethane gives trifluorochloroethylene [100]:

$$CCl_3{-}CCl_3 + HF \longrightarrow CClF_2{-}CCl_2F + Cl_2$$

$$CClF_2{-}CCl_2F \xrightarrow[-Cl_2]{} CClF{=}CF_2$$

Purification of the monomer is very important in order to obtain a polymer with acceptable properties. This monomer is the most toxic of the fluorine-containing ethylene monomers. Trifluorochloroethylene explosion surges, like those of tetrafluoroethylene, move at such a high speed that normal relief devices cannot handle them. This monomer has to be inhibited for shipping.

Trifluorochloroethylene is a gas with a boiling point of −27°C and a freezing point of −147°C. If it is stored without an inhibitor in its gas or liquid states, it is able to autopolymerize at room temperature.

At the commercial scale, the polymerization is done under pressure at 45–50°C with benzoyl peroxide as initiator, in an organic solvent such as carbon tetrachloride or chloroform. Poly(trifluorochloroethylene), being insoluble in these organic solvents, precipitates as fine particles or flakes.

A suspension process was developed for the synthesis of a high MW PCTFE with good physical properties such as melt flow and thermal stability. Perfluorocarbonperoxide was used as the initiator [134].

PCTFE is a linear polymer, like PTFE, but its degree of crystallinity is a little lower. Its chemical stability is similar to that of PTFE, but it is swollen at room temperature by ethers, halogen derivatives, toluene and some esters. At higher temperatures, it is dissolved by these compounds. Molten alkali metals, some metal halides and fluorine will attack PCTFE. This polymer may be melt processed by extrusion and injection molding, but careful temperature control is required since degradation occurs at the processing temperatures (230–290°C).

Typical applications of PCTFE include gaskets, diaphragms, valve seats, observation windows for chemical equipment, tubing and electrical components such as coils and forms for electrical equipment. Dispersions are used for films and coatings.

2.4.4 Poly(vinyl fluoride) (PVF) and poly(vinylidene fluoride) (PVDF)

Vinyl fluoride can be polymerized with PB as the initiator under high pressure (100 MPa) or in an acetone solution. PVF has superior mechanical properties compared with those of PVC, which it resembles from the microstructure point of view. This superiority is provided by its higher degree of crystallinity. Above 100°C, it is soluble in highly polar solvents such as DMF or cyclohexanone. Like PVC, it is stable to concentrated acids and alkalis.

PVF is generally processed as sheets and films which have better weathering properties compared with those of similar products made of PVC or PE. Cold drawn films can reach a tensile strength of 20 kgf/mm^2. PVF is used as a glazing film for outdoor use. It has a T_g of −20°C and a T_m of 200°C.

PVDF is produced by high pressure radical polymerization in aqueous systems. It has moderate mechanical properties and good chemical and electrical properties (piezoelectric and pyroelectric). It shows good heat resistance up to 150°C and its T_g is −35°C and its T_m is 171°C. The two grades available on the market have average MWs of 300 000 and 600 000.

2.4.5 Fluorocarbon copolymers

Copolymers based on fluorinated monomers have properties similar to the fluoropolymers discussed above. The most interesting are [46,135].

- tetrafluoroethylene–hexafluoropropylene copolymer;
- tetrafluoroethylene–ethylene copolymer;
- tetrafluoroethylene–perfluoroalkoxy copolymer;
- chlorotrifluoroethylene–ethylene copolymer;
- vinylidene fluoride–hexafluoropropylene copolymer;
- vinylidene fluoride–chlorotrifluoroethylene copolymer;
- hydropentafluoropropylene copolymer;
- hexafluoroisobutene copolymer.

Other fluoropolymers produced on a commercial scale are polyfluoroprene, polyfluoroacrylate and polyfluorosilicone.

The copolymerization of tetrafluoroethylene with hexafluoropropylene is carried out in an aqueous medium using difluoroacylperoxide as the initiator, wherein a fluoroalkylcarboxylic acid is fed to the reaction system. The same comonomers can be polymerized in the presence of ω-hydroperfluoroheptanoic acid using di(ω-hydrododecafluoroheptanoil) peroxide as the initiator, and leading to a copolymer with a melting point of 266°C and a low metal ion content [136].

Fluoroalkyl sulfinates are used as a source of fluoroalkyl radicals in aqueous emulsion polymerization of fluorine-containing monomers to give polymers having a high level of perfluoroalkyl end groups. These disulfinates act as monomers yielding fluoropolymers with specific microstructural fragments derived from the parent perfluorodisulfinate. Hexafluoropropylene–vinylidene fluoride copolymer prepared by using ammonium persulfate as initiator in the presence of $C_8H_{17}SO_2Na$ contains 0.46% C_8H_{17} end groups [137].

Major uses of fluoroelastomers based on monomers such as vinylidene fluoride, chlorotrifluoroethylene, tetrafluoroethylene, hexafluoropropylene and perfluoromethyl vinyl are in the aerospace industry (20%), industrial equipment, wire and cable jacketing and other insulation applications [20]. Vulcanization of these elastomers is carried out with diamines, peroxides and bisphenol [138].

The recently synthesized amorphous fluoroplastics combine the outstanding chemical, thermal and electrical properties of crystalline perfluoropolymers with high optical clarity, better mechanical properties and solubility in selected fluorocarbon solvents.

Some fluoropolymers can be obtained through a polycondensation process; they combine the mechanical and thermal properties of engineering polymers with the low moisture sensitivity and excellent electrical properties of crystalline fluoropolymers. They are aimed at electronics, aerospace or composite applications [121].

2.5 POLYACRYLICS

2.5.1 Introduction

High polymers known as acrylics or polyacrylates are obtained from derivatives of acrylic and methacrylic acids; the group contains also their copolymers with monomers like vinylic, allylic, etc.

Monomers commonly used in the production of these polymers are acrylonitrile, acrylic acid, methyl, ethyl and butyl acrylates, methacrylonitrile, methacrylamide, methacrylic acid, methyl, ethyl or butyl methacrylate, acrolein, methacrolein, acrylamide, 2-ethyl-hexyl acrylate, cetyl methacrylate, etc. Acrolein can be polymerized by 1,2 addition at the vinyl group or at the carbonyl one, or as a conjugated diene by 1,4 addition. Its polymers that have been studied were obtained by radical or anionic polymerization.

Acrylic and methacrylic monomers can be easily polymerized through well known techniques to obtain a wide range of products with properties determined by the type of monomer or monomers used for their synthesis, by the applied polymerization procedure and the polymerization conditions. Polymerizations of acrylates and methacrylates are accompanied by the release of high amounts of heat and substantial decrease in volume. Both of these factors influence most manufacturing processes.

Bulk polymerization is used when the end product has to be a sheet (organic glass), bar, block or tube. Solution polymerization provides mainly adhesives, for use in concrete, paper or wood, and coatings.

Emulsion polymerization is used for latices with various applications or for the production of powder-like polymers or copolymers for processing through pressing and injection molding.

For extrusion processing or for dentistry applications, acrylic polymers are obtained through suspension polymerization. These polymers have a high MW, a high degree of purity and good optical properties.

2.5.2 Poly(methyl methacrylate)

(a) Bulk polymerization

This process occurs through a radicalic mechanism, using peroxides (benzoyl peroxide) or azo (azobisisobutyronitrile) initiators. In some cases, 5–70% plasticizer and a dye are added, and these have to be well homogenized with

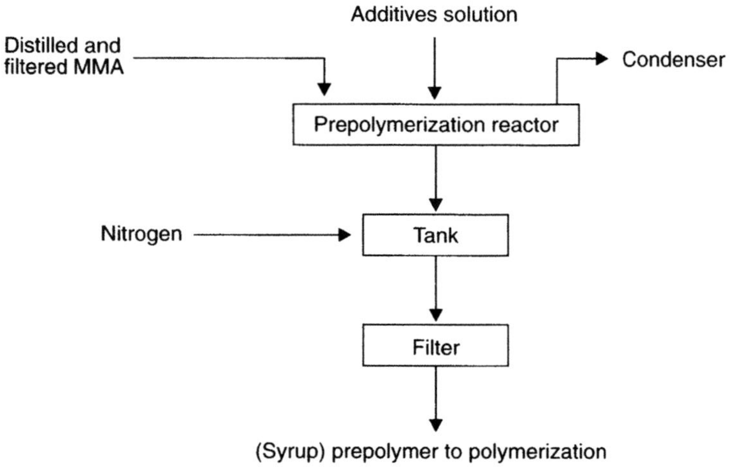

Figure 2.25 Production of MMA prepolymer.

the peroxide solution in the acrylic monomer(s). The reaction rate and the MW are especially determined by temperature and initiator concentration.

The procedure for obtaining poly(methyl methacrylate) (PMMA) is based on the following operation: dosing the monomer and the auxiliary components such as initiator, plasticizer, dye; dissolution of the additives in the monomer; prepolymerization, prepolymer filtration, washing and mold preparation; prepolymer casting in the mold and polymerization; and finally sheet removal.

The scheme of prepolymer preparation is presented in Figure 2.25. A common formulation contains, besides MMA, 0.1 part initiator, 5 parts plasticizer and 0.05 part dye. Prepolymerization occurs at 100°C and the duration of the process depends on the prepolymer viscosity. For thin sheets, a viscosity of 50 cP is necessary, but for the thick sheets, 200 cP is recommended. After cooling at room temperature, the prepolymer is filtered under nitrogen pressure: prepolymer coloring may be done at this stage by adding a filtered dye solution in MMA.

In MMA bulk polymerization, an autoacceleration – known as the Trommsdorff effect – is observed beginning at 20–50% conversion. This autoacceleration which continues up to high conversion is attributed to the increase of the viscosity to such an extent that the diffusion rate, and therefore the termination reaction, is reduced. In other techniques, the prepolymer can be obtained by dissolving a certain amount of ground PMMA in the monomer.

The polymerization molds are made of silicate glass; prepolymer-filled molds are put on carts with shelves which are placed in thermal rooms or water baths for the polymerization phase. Since the polymerization process is exothermal the continuous release of the reaction heat is necessary for the production of a high quality item. Due to the low thermal conductivity of PMMA some difficulties may be encountered while attempting to keep the polymerization temperature constant.

The heat transfer (q) between the polymerization medium and the air circulation around the molds can be expressed by the following equation:

$$q = \frac{T_1 - T_2}{\frac{1}{\alpha} + \frac{d_1}{K_1} + \frac{d_2}{K_2}} \tag{2.1}$$

where q = heat transfer (W); T_1 = prepolymer temperature (°C); T_2 = circulating air temperature (°C); α = heat transfer coefficient; K_1 = silicate glass thermal conductivity (W/m · °C); K_2 = PMMA thermal conductivity (W/m · °C); d_1 = silicate glass thickness (mm); and d_2 = PMMA sheet thickness (mm).

The value of α depends on the circulating air speed, which has to be greater than 7 m/s.

The polymerization temperature is a function of the initiator concentration, sheet thickness, and the type of prepolymer; it varies generally between 30 and 130°C. In the case of a continuous polymerization, the following thermal conditions are followed: 1st thermal room: 55°C (8 hours), 2nd thermal room: 50–62°C (8–10 hours), 3rd room: 125°C (8 hours). For more details, see Reference [26].

(b) Solution polymerization

Solution polymerization of methacrylic esters is less frequently used than bulk polymerization, and is used mainly for the production of lacquers. The lacquer viscosity depends on the polymer MW, which in turn depends on the monomer and initiator concentration, and temperature. As the monomer concentration increases, its MW increases. Thus, for instance, at 5% monomer, MW is 26 000, at 10%, 33 000 and at 20% 52 000. Solution polymerization is suitable for the synthesis of polymers having MW up to 200 000.

The production of a dry PMMA follows the flowchart presented in Figure 2.26. Solution polymerization is carried out at between 75 and 110°C in an aromatic solvent such as benzene or toluene, a ketone such as methyl-ethyl-ketone, or a water–methanol mixture in the presence of around 40% MMA and 0.2–1% initiator. As a result of this process, a 40–60% solution of PMMA is obtained which is used as it is, or processed according to Figure 2.26.

(c) Emulsion polymerization

This process is not applied extensively in commercial operations. However, this technique can be used in non-aqueous media for the production of

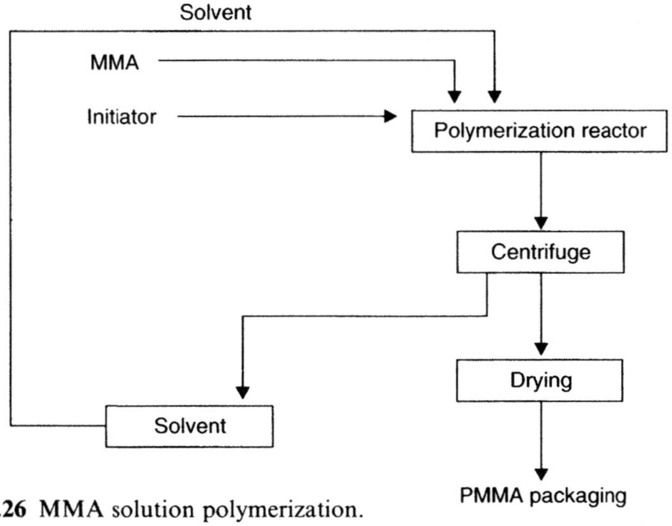

Figure 2.26 MMA solution polymerization.

certain surface coatings. The polymerization of ethyl acrylate is carried out most often in emulsion.

Emulsion polymerizations of acrylic monomers are usually carried out by batch processes in kettles designed to withstand an internal pressure of at least 450 kPa. The temperature is controlled by circulating steam or cold water through the kettle's jacket. Figure 2.27 shows the details of this procedure.

Polymerization starts at 85°C and after the addition of monomer emulsion is complete the temperature in the reactor is raised to 95°C to complete the conversion.

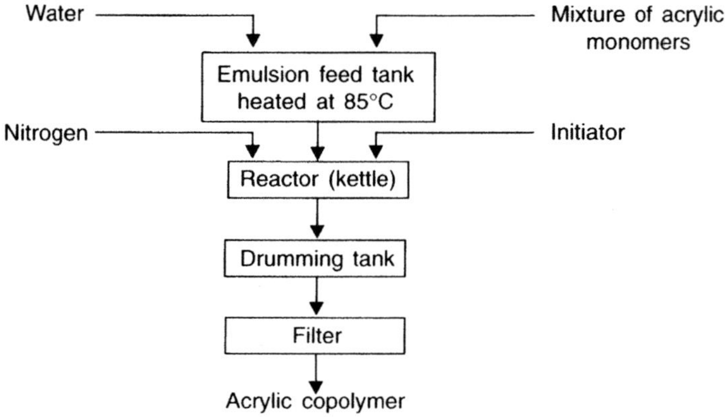

Figure 2.27 Emulsion copolymerization of acrylic monomers.

Emulsion polymerization of butylacrylate is faster with an aqueous solution of ammonium peroxidedisulfate than with an oil solution of dibenzoyl peroxide. The overall activation energy is 50–60 kJ/mol, which is nearly half of that of the bulk polymerization [139].

An emulsion of a polymer comprising a recurring unit derived from methacrylate and a vinylic monomer containing an $-SO_3R$ group forms a protective film on concrete, metals and mineral materials. Such an emulsion also has an affinity for acrylic polymers and is used as a dental adhesive [140].

Aqueous dispersions of PMMA are produced for the paint, paper, textile, floor polish and leather industries where they are used as coatings or binders. The most common are copolymers of MMA with either ethyl or butyl acrylates.

(d) Suspension polymerization

Suspension polymerization of acrylic monomers follows the same rules as the bulk polymerization with the difference that the process occurs in the monomer drops dispersed in water. Peroxidic initiators such as benzoyl peroxide or acetyl peroxide most commonly are used. AIBN or other initiators of its family are used for colored items because they do not affect the integrity of the dye. Peroxides are used in amounts of 0.5–1%. The formulation includes a suspending agent such as sodium polyacrylate and buffers. MMA suspension polymerization takes only about 1–4 hours at 75–110°C and *c.* 0.3 MPa [4]. The process is controlled by the periodic testing of the unreacted monomer. The equipment for acrylic and methacrylic polymerization processes in general is similar to that used for emulsion polymerization. The flow chart is shown in Figure 2.28.

The clear or colored beads are then sent to the pelletizer.

(e) Copolymers

The copolymers of acrylic monomers can be grouped as follows:

- acrylic esters copolymers [141];
- acrylic esters–vinyl monomers copolymers [142–145];
- acrylic esters–tetrafunctional components copolymers.

Copolymers of the first group are elastic, soft at normal temperature and have adhesive properties. They may be processed as transparent sheets, which are useable for special synthetic glasses, or as powders for injection molding. Latices based on acrylic copolymers are used for leather treatment and for textiles and paper impregnation.

The second group refers to copolymers with vinyl chloride, vinyl acetate, styrene, vinyl pyridine and methyl methacrylate. An ethyl acrylate–chloroethyl vinyl ether copolymer has been marketed as an oil resistant, high thermal stability elastomer.

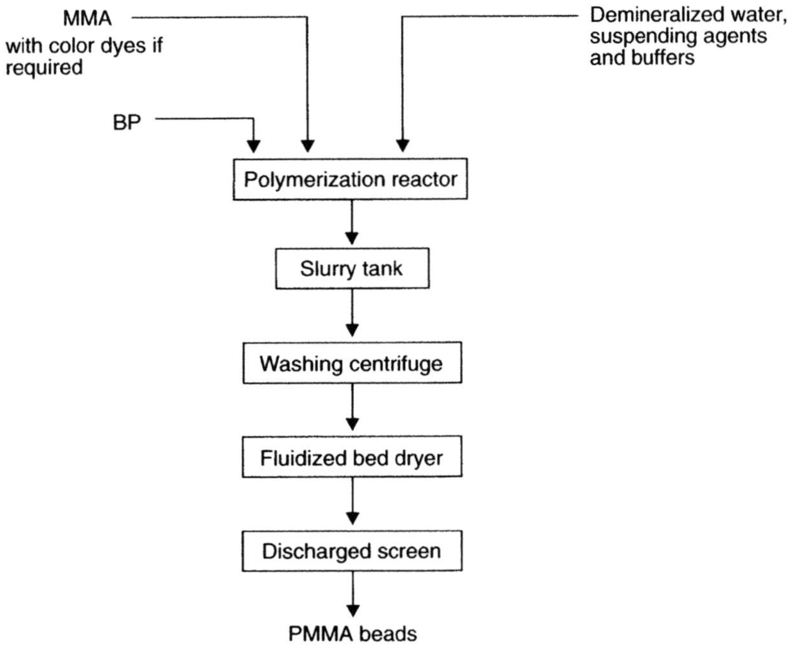

Figure 2.28 MMA suspension polymerization.

The emulsion copolymerization of *n*-butylacrylate and sulfonated styrene was developed by J.E. McGrath *et al.* [146] in order to produce ionomers. They demonstrated that emulsion copolymerization is a useful and potentially important technique in the synthesis of sulfonate ion-containing polymers by a reaction between the ionogenic and covalent monomers, provided that a water soluble initiator system is used.

Vulcanizable rubbers made of polymers of ethyl or butyl acrylates or their copolymers with acrylonitrile, styrene or dienes are prepared by emulsion technology. To make the vulcanization of minor amounts of chlorine-containing monomers such as chloroethyl vinyl ether more convenient, the monomers are usually introduced into the acrylic polymer. Many acrylic copolymers are thermoplastic; minor amounts of acrylic or methacrylic acids are able to improve the adhesion characteristics of the copolymers. In latices, such small amounts also impart mechanical storage and freeze–thaw stability.

(f) Properties

The polymers belonging to this group are amorphous. PMMA microstructure has recently been discussed by R.H.G. Brinkhuis and A.J. Shouten [147,148].

Polymethacrylic esters have a higher T_g and a better thermal stability than the corresponding polyacrylic esters. The higher T_g is due to the lower mobility of the quaternary carbon determined by non-symmetrical substitution. As the length of the alcoholic radical increases, the value of T_g decreases; this characteristic is valid for both polyacrylates and polymethacrylates.

The solubility and the viscosity of this group of polymers also depend on the alcoholic group. The polymers of inferior esters are soluble in esters, aromatic and chlorinated hydrocarbons. Polymers of superior esters are also soluble in paraffinic hydrocarbons.

Polymethacrylates are superior to polyacrylates in many respects, e.g. chemical stability, thermal stability, transparency to light (90%), especially the UV spectrum, and stability to water.

PMMA depolymerizes at 160°C with the formation of its monomer; polyacrylates decompose in similar conditions forming decomposition products, but not their monomers.

When heated, polyacrylates, with the exception of poly(*n*-dodecyl acrylate), soften before the corresponding polymethacrylates; this is also reflected at the brittle temperature (see Table 2.10).

(g) Applications

PMMA has numerous applications in different technical fields and in domestic uses. It is used as organic glass for windows for transportation vehicles, as molding powder for technical items and consumer goods, medical and laboratory equipment, dentures, films, etc. Biaxial stretching improves both the mechanical properties and surface smoothness of PMMA films [149].

Through polymerization of acrylic and methacrylic monomers bearing phosphonic esters, phosphonic acids and semisalts, B. Boutevin *et al.* [150] obtained additives for adhesive products.

Low molecular weight PMMA and poly(methyl methacrylate)-co-(*n*-butyl methacrylate) emulsion based polymers were developed to be used as binders

Table 2.10 Brittle temperature of polyacrylates and polymethacrylates (°C) [18]

Alkyl group	*Polyacrylates*	*Polymethacrylates*
Methyl	6	105
Ethyl	−24	65
n-Butyl	−55	20
n-Octyl	−65	−20
n-Dodecyl	−3	−65
iso-Butyl	−24	53

of high temperature powders for the rapid prototyping method known as 'selective laser sintering'.

Poly(butyl methacrylate), which belongs to the groups of soft polymethacrylates, possesses good adhesive properties and good solubility in aromatic hydrocarbons. As a film it is elastic and used as a core layer in special glasses; it is also used in print compositions, leather treatment, and in the food and pharmaceutical industries.

Polyacrylates are also used for extruded flexible tubes, as cable covers; they have important advantages such as a lack of volatile plasticizers, good stability to oils, ozone and other atmospheric agents.

Polyacrylamide is water soluble, but being non-ionic the viscosity of its solutions is less affected by alkali. This polymer softens at 210°C. Some of the applications of its water solutions are flocculation of ores, textile sizes and retention of paper fillers.

2.5.3 Polyacrylonitrile (PAN)

(a) Acrylonitrile polymerization and copolymerization

The high reactivity of acrylonitrile is explained by the two following features:

1. the presence of a high polar substituent;
2. the double and triple bond conjunction.

Although in the pure state it is stable when heated, it is able to polymerize easily under the influence of X-ray, gamma or UV radiation, and ultrasounds or high energy electrons. Organic or inorganic initiators or redox systems are often used. Bulk and solution polymerization (organic solvents or concentrated mineral salts solution) or aqueous medium polymerization (emulsion, suspension) can be applied. Solution polymerization is generally used for obtaining acrylic polymers (PAN and AN copolymers) suitable for direct wet or dry spinning fiber manufacture. Common solvents are dimethylformamide (DMF), dimethylacetamide, dimethylsulfoxide (DMSO), ethylene carbonate (EC) or propylene carbonate (PC), and concentrated aqueous solutions of NaSCN, HNO_3, H_2SO_4 and $ZnCl_2$.

From the commercial point of view, the most important technologies are:

- acrylonitrile polymerization in aqueous medium;
- acrylonitrile polymerization in organic solvents.

Both are used with the aim of producing PAN fibers.

(b) Polymerization in aqueous solutions

This procedure permits an easy thermal regulation and leads to a polymer with homogeneous MW and low cost. Because PAN homopolymer has a limited solubility and colorability, it is used commercially with low amounts (5–10%) of other comonomers able to increase PAN solubility and the diffusion of colorants in the fiber, although a third comonomer able to increase colorability much more is preferred in many cases. This is why at present, PAN homopolymer for synthetic fibers has been completely replaced with co- or terpolymers with 5–15% comonomer(s). Comonomers like vinyl acetate, itaconic acid, acrylic acid, methyl acrylate, methacrylonitrile, vinyl-methyl-ketone, morfoline, vinyl-pyridine and α-methyl styrene are often used in this respect.

(c) Copolymerization in aqueous medium

This technology is known as solution–suspension polymerization, due to the fact that it starts from a solution of the monomer in water, and ends as a suspension in the same medium. The process has two phases:

- homogeneous polymerization of AN in water solution with the formation of a dispersion of PAN particles in water;
- heterogeneous polymerization of unpolymerized AN in the previous phase.

Because of a lower activation energy, 11.1 kcal/mol (4.65×10^4 J/mol), the redox initiation system based on potassium persulfate–sodium thiosulfate ($K_2S_2O_8$–$Na_2S_2O_3$) is preferred. The rate of the bimolecular reaction persulfate–reduction agent increases in the presence of metallic ions with variable valencies (Cu^{2+}, Fe^{2+}). The rational use of such cationic promoters (0.0001%) permits a good control of the rate of polymerization and of the MW.

The conversion and MW of the produced PAN depend on the following factors: initiator type and amount, pH (2–2.5), temperature, time and reactants' throughput [4].

To avoid crosslinkings and branchings which may produce difficulties during polymer dissolution and high viscosity spinning solutions, polymerization temperatures not higher than 60°C are recommended. Conversions above 80% lead to chain transfer reactions and branched polymers unsuitable for spinning. For dry spinning, the MW has to be 35 000–50 000 and for wet spinning, 60 000–80 000.

Continuous polymerization has the advantages of a high productivity, low volume installation, reproducibility and uniformity of the product.

A terpolymerization process for the production of a fiber-forming PAN consists of the following main operations:

- preparation of the comonomer mixture;
- preparation of the redox initiator system and promoter;
- terpolymerization;
- monomer recuperation;
- filtration and drying.

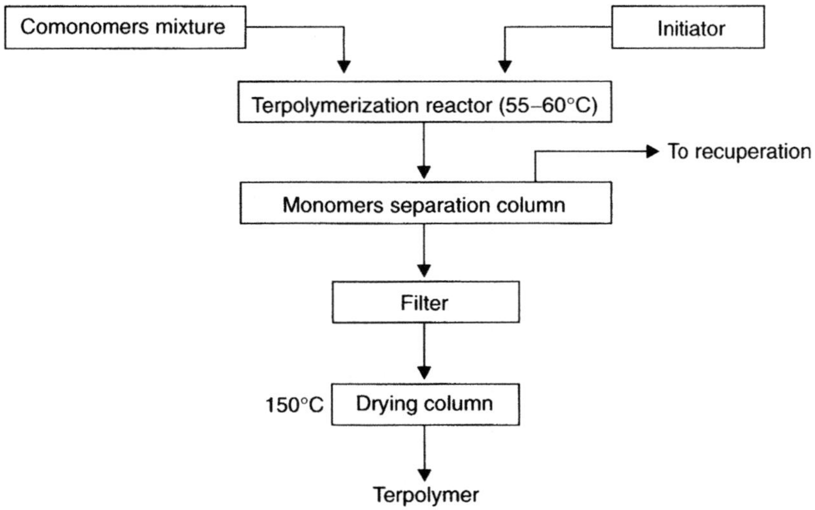

Figure 2.29 Continuous terpolymerization of AN with other vinyl comonomers.

The flow chart is presented in Figure 2.29.

(d) Copolymerization in organic solvents

AN copolymerization in media able to dissolve both the initial comonomers mixture and the terpolymer permits the formation of solutions which may be used directly in the spinning process. This advanced procedure simplifies the terpolymer synthesis technology excluding operations like separation, washing and drying of the terpolymer. Also, the phase of spinning solution preparation, dissolution, filtration and homogenization are no longer necessary.

Commercial plants use organic solvents like DMF, EC and DMSO. BP and AIBN are used as initiators, the process taking place at 50–60°C. In this procedure, AN, comonomers, the organic solvent and the initiator as a solution are introduced continuously in a certain ratio into the reactor. Due to the fact that with the increase of conversion the viscosity of the system increases, the stirrer is built to be able to insure the same stirring effects during the copolymerization. The conversion is limited to 60–80% because of the medium viscosity and the desired properties of the polymer. A schematic of this procedure is shown in Figure 2.30.

(e) Copolymerization in aqueous concentrated solution of mineral salts

Nowadays, this procedure is almost completely replaced by that based on organic solvents. Among the salts used in this technology are calcium

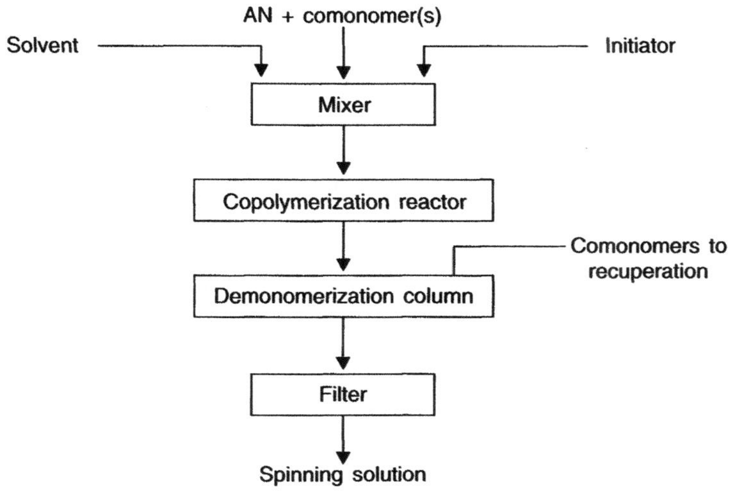

Figure 2.30 AN copolymerization in organic solvents.

chloride, zinc chloride, sodium or potassium thiocyanate, perchlorates, etc. The initiators are water soluble and when UV rays are used as the energy source, the process can be initiated by iron chloride–hydrochloric acid system. By stopping the copolymerization at 75% conversion, a PAN solution with a viscosity of 2000 cP and a MW of 75 000–80 000 can be obtained.

The monomer mixture usually contains, besides AN, methyl acrylate and itaconic acid. Their water solution is mixed with the copolymerization medium made of a water solution of 50% potassium thiocyanate. In the case of a thermal process, copolymerization takes place at 80°C until a conversion of 70–75% is reached. From a technical point of view, there are no significant differences between the equipment necessary for this procedure and the copolymerization in organic solvents. At the end of the process, a 12–13% PAN spinning solution is obtained.

(f) PAN properties

PAN and its copolymers, with AN in a predominant amount, are white powders up to a temperature of 250°C when they become darker because of the beginning of degradation. Having relatively high values of T_g (90 and 140°C), they have a low thermal plasticity and cannot be used as a plastic material. Their high crystalline melting point (317°C), limited solubility in certain solvents and the superior mechanical properties of the fibers are due to the intermolecular forces between the polymer chains. Appreciable electrostatic forces occur between the dipoles of adjacent $-C{\equiv}N$ groups on the same macromolecule. This intramolecular interaction restricts bond rotation and leads to a stiff chain [46]. Roentgenographically, the stretched PAN fiber presents a high degree of crystallinity.

As mentioned previously, PAN is soluble in a limited number of polar liquids like DMF, DMSO, dimethylacetamide, dimethylsulfone, tetramethylsulfide and aqueous solutions of ethylene carbonate or of some mineral salts. In DMF, concentrations up to 25% and above can be obtained; EC permits formation of solutions with up to 14% PAN.

(g) Applications

Staple acrylic fibers, being soft and resilient, are used as a substitute or diluent for wool, and fabrics made from them show good crease resistance and crease retention (e.g. in permanent pleats); they are also made into resilient bonded batting.

The rot and light resistant properties suggest numerous outdoor applications for acrylic fibers in heavy duty and finer fabrics, and in netting and filter fabrics [18].

PAN is one of the important raw materials for the production of carbon fibers. When PAN is heated in vacuum or air to over 200°C, a number of reactions take place, resulting in the formation of some by-products which are liberated as volatiles. The carbon ring structure produced during the carbonization of PAN fibers under tension has a crystalline form [151]. The dehydrogenation and denitrogenation steps are presented in Figure 2.31. A complete flow diagram for the preparation of carbon fibers from PAN is given in Figure 2.32.

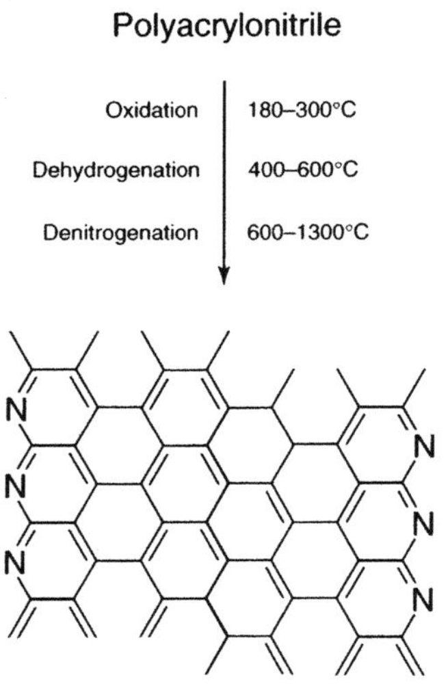

Figure 2.31 PAN transformation in carbon fiber.

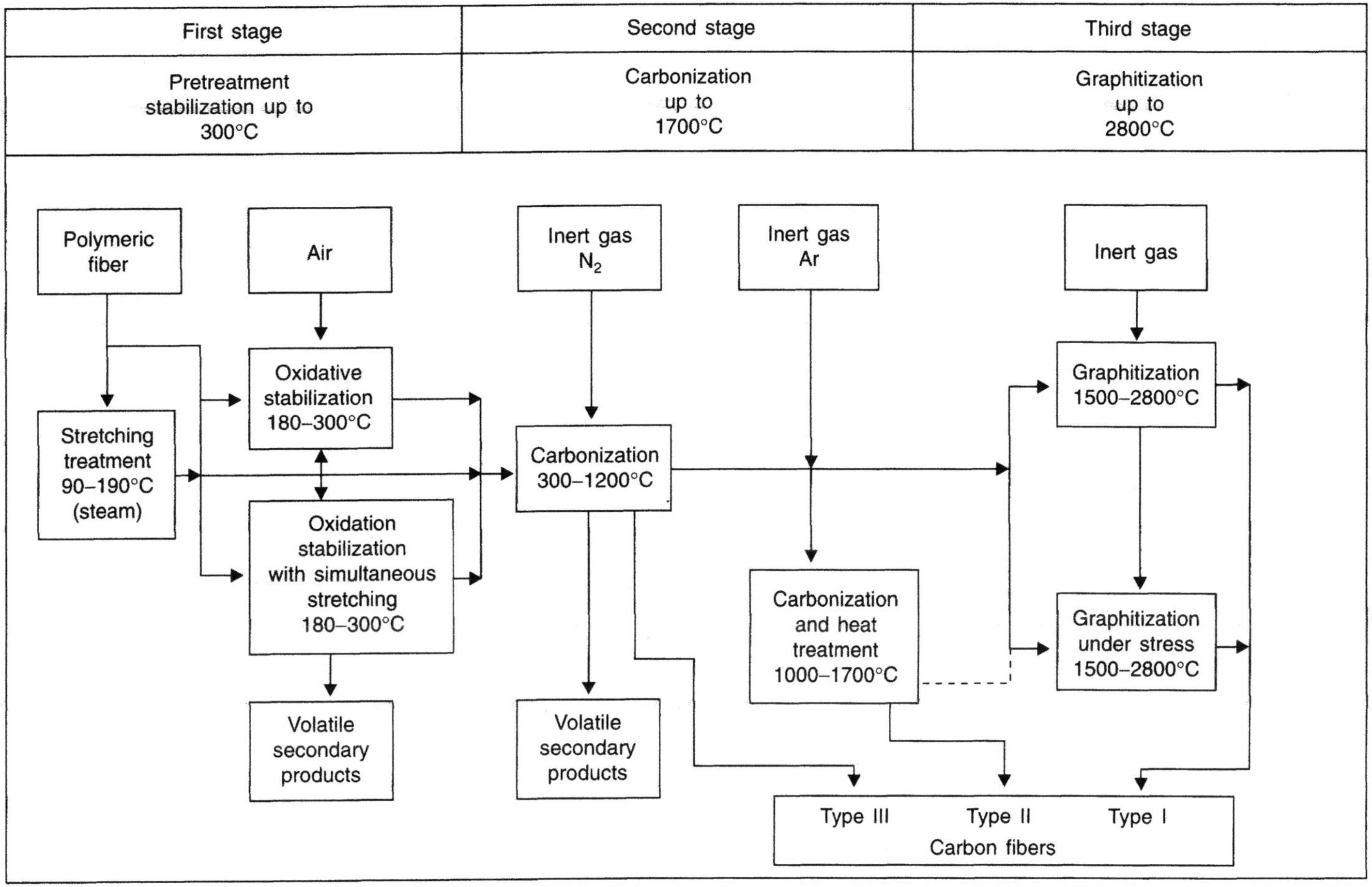

Figure 2.32 Schematic representation of carbon fiber production from PAN.

Table 2.11 Relative mechanical properties of composites made with E-glass and carbon fiber [152]. (Reprinted with permission from *Carbon Fibres in Composite Materials* by R.M. Gill, 1972. Copyright The Plastics and Rubber Institute)

Materials	*Tensile modulus (MN/cm^2) ($10^6\,lbf/in^2$)*	*Specific gravity*	*Specific tensile modulus (MN/cm^2) ($10^6\,lbf/in^2$)*
E-glass fiber composite (glass 50% by volume)	1–4 (2–0)	1–66	0.8 (1–2)
Carbon fiber tows with parallel alignment (high strength fiber 50% by volume)	12–6 (18–0)	1–53	8–1 (11–8)

Table 2.11 indicates the relative mechanical properties of a glass fiber composite in the form of a mat or rovings, compared with the high strength form of PAN based, carbon fiber aligned tows. More data on the use of PAN–carbon fiber may be found in the technical literature [153,154]. PAN fibers are also being used as reinforcement for cementitious composites [155].

2.5.4 Poly(acrylic acid) (PAA)

All acrylic polymers can be represented by the following microstructure:

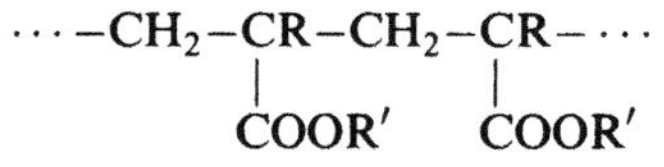

where R = H or CH_3 and R′ = H, alkyl or a metal cation.

Both poly(acrylic acid) and poly(methacrylic acid) are obtained by polymerization of acrylic or methacrylic acid. Polymerization in aqueous solution at less than 20–30% concentration is readily controlled. However, the high heat of polymerization makes it difficult to control the process of highly concentrated solutions, and an uncontrolled exotherm and crosslinked polymer may result. In the case of acrylic acid (AA) a hazard exists if the system is in a closed or poorly vented vessel. Polymerization of AA or methacrylic acid (MAA) can be carried out in bulk, aqueous solution, non-aqueous solution (benzene), inverse phase emulsion and suspension using the free radical initiation systems common for other vinyl monomers. Besides water, a conventional recipe involves an initiating system (potassium persulfate and sodium thiosulfate) and a chain transfer agent. The process may be carried out at 50–100°C or by radiopolymerization by exposing the system to ^{60}Co radiation. In aqueous media, salts of AA and MAA may also be polymerized.

In addition to the thermal or radiopolymerization of the appropriate monomer, the hydrolysis of a suitable polymer constitutes another convenient technological procedure for the preparation of both the polymeric acid and salt.

AA has been polymerized in supercritical CO_2 in the presence of AIBN as the initiator and an ethylmercaptan as a chain transfer agent under pressure (11–34 MPa) with no applicable effect of pressure on MW, polydispersity, particle size and morphology [156].

Graft polymerization of AA on glass beads can improve their compatibility and adhesion with a polyester matrix [157].

The hydrolysis of poly(isopropyl acrylate) leads to a water soluble poly(acrylic acid) if the first one is syndiotactic and to a water insoluble one if poly(isopropyl acrylate) is isotactic. The hydrolysis takes place in an acetic acid aqueous solution; after the solution has been completed, *p*-toluenesulfonic acid is added and the system is heated to 110°C. The alkyl acetate by-product is removed by passing a steam of nitrogen through the solution. Poly(acrylic acid) is isolated by precipitation. Such polymeric acids can be transformed into useful macromolecular derivatives through different reactions, such as neutralization, esterification, complex formation, dehydration, etc.

Polyacrylic acids and their monovalent cation salts are hard, brittle and water soluble. The pH has a significant effect on the viscosities of the aqueous solutions due to the uncoiling. The maximum viscosity is reached at neutral pH or mildly alkaline conditions.

(a) Applications

The applications of poly(acrylic acid) and other water soluble acrylic polymers are the result of their properties and chemical reactions already mentioned. The most important are the uses as adhesives, thickeners, coatings, flocculants, dispersants, fluidizers, in cosmetics, pharmaceuticals, ion exchange processes, etc. [18]. PAA and PMAA are important warp sizes in adhesives, dispersants and textile processing. These acids and their sodium and ammonium salts are used for thickening aqueous latices, textile sizes and in drilling muds.

REFERENCES

1. Mark, H., Bikales, N.M., Overberger, C.G. *et al.* (1985) *Encyclopedia of Polymer Science and Engineering*, 2nd edn, Vol. 1, J. Wiley & Sons, New York.
2. Anon. (1994) *Chem. Eng. News*, **72**(27), 30.
3. Seymour, R.B. and Carraher Jr, C.E. (1992) *Polymer Chemistry*, 3rd edn, Marcel Dekker, New York.

4. Feldman, D. (1974) *Macromolecular Compounds Technology*, Ed. Technica, Bucharest.
5. Nguyen-Khac Tin, Flashel, E. and Renken, A. (1983) in *Polymer Reaction Engineering* (eds K.H. Reichert and W. Geisler), Hanser Publishers, Munich, p. 175.
6. Chen, G.C. (1994) *Polym. Plast. Technol. Eng.*, **33**(1), 55.
7. Kawaguchi, H., Sugi, Y. and Ohtsuka, Y. (1981) in *Emulsion Polymers and Emulsion Polymerization* (eds D.R. Basset and A.E. Hamielec), ACS Symposium Series 165, American Chemical Society, Washington, p. 145.
8. Said, Z.F.M., Hassan, S.A. and Dunn, A.S. (1981) in *Emulsion Polymers and Emulsion Polymerization* (eds D.R. Basset and A.E. Hamielec), ACS Symposium Series 165, American Chemical Society, Washington, p. 471.
9. Nemirovski, N. and Narkis, M. (1994) *Polym. Eng. Sci.*, **34**(9), 750.
10. Sundberg, D.C., Hsieh, J.Y., Soh, S.K. and Baldus, R.F. (1981) in *Emulsion Polymers and Emulsion Polymerization* (eds D.R. Basset and A.E. Hamielec), ACS Symposium Series 165, American Chemical Society, Washington, p. 327.
11. Vivaldo-Lima, E. and Saldivar-Guerra, E. (1992) *50th Annu. Tech. Conf. Soc. Plast. Eng.*, Vol. 1, p. 329.
12. Ranby, B. (1993) *J. Macromol. Sci. Pure Appl. Chem.*, **A30**(9,10), 583.
13. Wang, J.L. and Favstritsky, N.A. (1994) *Polym. Preprints*, **35**(2), 701.
14. Akitaya, S. and Tukazawa, T. (01 May 1993) Jpn. Kokai Tokkyo Koho, JP 05,310,997 [93,310,997] (Cl C08K3/32), 22 Nov. 1993, Appl. 92/139,940.
15. Aerdts, A.M., Detan, J.W. and German, A.L. (1991) *Macromolecules*, **24**(71), 1473.
16. Kirk–Othmer (1985) *Concise Encyclopedia of Chemical Technology*, J. Wiley & Sons, New York.
17. Billmeyer, Jr, F.W. (1984) *Textbook of Polymer Science*, 3rd edn, J. Wiley & Sons, New York.
18. Roff, W.J. and Scott, J.R. (1971) *Fibres, Films, Plastics and Rubbers. Handbook of Common Polymers*, Butterworth, London.
19. Stevens, M.P. (1990) *Polymer Chemistry*, 2nd edn, Oxford University Press, New York.
20. Chanda, M. and Roy, S.K. (1987) *Plastics Technology Handbook*, Marcel Dekker, New York.
21. Alger, M.S.M. (1989) *Polymer Science Dictionary*, Elsevier Applied Science, London.
22. Erbyl, H.Y. and Baysal, B.M. (1987) in *Synthetic Polymeric Membranes* (eds B. Sedlacek and J. Kahovec), Walter de Gruyter, Berlin.
23. Steiberg, M. (1975) in *Copolymers, Polyblends and Composites* (ed. N.A.J. Platzer), Advances in Chemistry Series 142, American Chemical Society, Washington.
24. Savenije, H.B. (1993) *Kunststoffe*, **83**(10) 763.
25. Albright, L.G. (1985) *Process for Major Addition-type Plastics and their Monomers*, R.E. Krieger Publ. Co., Malabar, Fl.
26. Rodriguez, F. (1982) *Principles of Polymer Systems*, 2nd edn, McGraw-Hill Book Co., New York.

27. Rubin, I.J. (ed.) (1990) *Handbook of Plastic Materials and Technology*. J. Wiley & Sons, New York.
28. Jouan, X. and Gardette, J.L. (1991) *J. Polym. Sci. Part A*, **29**(5), 685.
29. Deanin, R.D., Hanoochehri, A. and Kanakia, C.J. (1987) in *Proceedings of Engineering Polymers and Composites*, 23–24 Sep. 1987, the Society of Plastics Engineers, Rosemont, Illinois, USA, p. 183.
30. Reiss, G. and Gaillard, P. (1983) in *Polymer Reaction Engineering* (eds K.H. Reichert and W. Geisler), Hanser Publishers, Munich, p. 24.
31. Ray, W.H. (1983) *Chemical Engineering Reactions*, ACS Symposium Series No. 226. American Chemical Society, Washington, p. 101.
32. Meyer, M.H. (1993) *Kunststoffe*, **83**(10), 764.
33. Zilberman, E.N. (1992) *J. Macromol. Sci. Reviews in Macromol. Chem. Physics*, **32**(2), 235.
34. Danusso, F. (1955) *Ric. Sci. Suppl. 25*, p. 46.
35. Sidiropulou, E. and Kiparissides, C. (1990) *J. Macromol. Sci. Chemistry*, **A27**(3), 257.
36. Poehlein, G.W. and Schork, F.J. (1988) *Polymer News*, **31**, 231.
37. Roe, C.P. (1968) *Industrial and Engineering Chemistry*, September, p. 20.
38. Blackley, D.C. (1975) *Emulsion Polymerization*, J. Wiley & Sons, New York.
39. Dunn, A.S. (1991), *J.O.C.C.A.*, **2**, 50.
40. Feldman, D. (1989) *Polymeric Building Materials*, Elsevier Applied Science, Barking.
41. Shelton, L.G., Hamilton, D.E. and Fipackerly, R.H. (1971) in *Vinyl and Diene Monomers. Part III* (ed. E.D. Leonard), Wiley-Interscience, New York, p. 1274.
42. Tauer, K., Reinisch, G., Sajewski, H. and Mueller, T. (1991) *J. Macromol. Sci., Chem.* **A28**(3–4), 431.
43. Kelsall, D.G. and Maitland, G.L. (1983) in *Polymer Reaction Engineering* (eds K.H. Reicher and W. Geisler), Hanser Publishers, Munich.
44. Reichert, K.H., Moritz, H.U., Gabel, Ch. and Deiringer, G. (1983) in *Polymer Reaction Engineering* (eds K.H. Reicher and W. Geisler), Hanser Publishers, Munich.
45. Pinto, J.C. (1990) *J. Vinyl Technol.*, **12**(1), 7.
46. Saunders, K.J. (1988) *Organic Polymer Chemistry*, 2nd edn, Chapman & Hall, London.
47. Gottesman (1975) in *Applied Polymer Science* (eds J.K. Craven and R.W. Tess), American Chemical Society, Washington, pp. 349–383.
48. Kaneko, J., Fujiwara, M. and Fujimoto, T. (15 Mar. 1990) US, US 5,294,684 (Cl 526–62; C08F2/18), 15 Mar. 1990, JP Appl. 90/132,246: 22 May 1990; *CA*, **120**, 324445b.
49. Fabini, M., Bobula, S., Rusina, N. *et al.* (1994) *Polymer*, **35**(10), 2201.
50. Smirnov, Y.A., Evdokimov, Y.P. and Kozlov, V.F. (1991) *Chemical and Petroleum Engineering* (English translation of *Khimiceskoe i Naftganoe Mashinostroenie*), **27**(3–4), 187.

51. Xie, T.Y., Hamielec, A.E., Wood, P.E. and Woods, D.R. (1991) *Polymer*, **32**(6), 1098.
52. Xie, T.Y., Hamielec, A.E., Wood, P.E. and Woods, D.R. (1991) *J. Appl. Polym. Sci.*, **43**(7), 1259.
53. Matthews, G. (1972) *Vinyl and Allied Polymers*, Vol. 2, Iliffe Books, London, p. 36.
54. Okawa, M., Oochi, J., Wakamori, H. *et al.* (01 Feb. 1994) Jpn Kokai Tokkyo Koho, JP 06, 25,313 [94,25,313] (Cl. C08F2/20), Appl. 91/150,239, 21 Jun. 1991; *CA*, **120**, 324453c.
55. Mrazek, Z. and Lucas, R. (1991) *Polym. Eng. Sci.*, **31**(5), 313.
56. Kelsall, D.G. and Maitland, G.C. (1983) in *Polymer Reaction Engineering* (eds K.H. Reichert and W. Geisler), Hanser Publishers, Munich.
57. Ping, Xu, Zhou, D. and Zhao, D. (1989) *Eur. Polym. J.*, **25**(6), 575–579.
58. Kulshreshtha, A.K. (1993) *Polym. Plast. Technol. Eng.*, **32**(6), 551.
59. Gilbert, M. (1994) *J.B.S. Macromol. Chem. Phys.*, **C34**(1), 77.
60. Rabinovitch, E.B. and Summers, J.W. (1980) *J. Vinyl Technol.*, **2**(3), 165.
61. Xie, T.Y., Hamielec, A.E., Wood, P.E. *et al.* (1991) *Polymer*, **32**(9), 1696.
62. Leplyanin, G.V., Salimgareeva, V.N., Sannikova, N.S. *et al.* (1990) *Polymer Science USSR* (English translation of *Vysokomolekulyarnye Soyedineniya Series A*), **32**(6), 1223.
63. Gardette, J.L. and Lemaire, J. (1993) *J. Vinyl Technol.*, **15**(2), 113.
64. Summers, J.V. (1983) *J. Vinyl Technol.*, **5**(2), 43.
65. Capocci, G. (1989) *J. Vinyl Technol.*, **11**(4), 195.
66. Behnisch, J. and Zimmermann, H. (1992) *Int. J. Polymeric Mat.*, **16,** 143.
67. Patel, K., Velazquez, A., Calderon, H.S. and Brown, G.R. (1992) *J. Appl. Polym. Sci.*, **46,** 179.
68. Wypych, J. (1986) *Poly(vinyl chloride) Stabilization*. Elsevier Science Publ., Amsterdam, pp. 1–35.
69. Caraculacu, A.A. (1987) *Pure Appl. Chem.*, **53,** 385.
70. Ravek, J.F., Skoronski, T.A. and Rimley, B. (1980) *Polymer*, **21,** 226.
71. Grassie, N. and Scott, G. (1985) *Polymer Degradation and Stabilization*. Cambridge University Press, Cambridge.
72. Tran, V.H., Nguyen T.P. and Molinie, P. (1994) *Polym. Degrad. Stab.*, **44**(2), 151.
73. Hjertberg, T. and Sörvik, E.M. (1984) in *Degradation and Stabilization of PVC* (ed. E.D. Owen), Elsevier Applied Science, London, pp. 21–79.
74. Decker, C. (1984) in *Degradation and Stabilization of PVC* (ed. E.D. Owen), Elsevier Applied Science, London, p. 81.
75. Davis, A. and Sims, D. (1983) *Weathering of Polymers*, Applied Science Publ. London and New York, p. 192.
76. Capocci, G. (1989) *J. Vinyl Technol.*, **11**(4), 195.
77. Gomez, J.L. (ed.) (1984) *Engineering With Rigid PVC*, Marcel Dekker, New York.
78. Briston, J.H. and Katan, L.L. (1989) *Plastic Films*, 3rd edn, Longman Scientific and Technical.

79. Greek, B.F. (1991) *Chem. Eng. News*, **69**(23), 39.
80. Galloway, F.M. and Hirschler, M.M. (1990) in *Fire and Polymers* (ed. G.L. Nelson), ACS Symposium Series 425, American Chemical Society, Washington, p. 592.
81. Montando, G. and Puglisi, C. (1991) *Polymer Degradation and Stability*, **33**(2), 229.
82. De Jager, J. and TenBrinke, G. (1991) *Macromolecules*, **24**(11), 3454.
83. Wolfe, Jr, J.R. (1987) in *Thermoplastic Elastomers* (eds N.R. Legge, G. Holden and H.E. Schroeder), Hanser Publishers, Munich, p. 118.
84. Nadkarni, V.M. and Jog, J.P. (1991) in *Two Phase Polymer Systems* (ed. L.A. Utracki), Hanser Publishers, Munich, p. 213.
85. Kijima, Y., Kawaki, T. and Inone, T. (1991) in *Two Phase Polymer Systems* (ed. L.A. Utracki), Hanser Publishers, Munich, p. 246.
86. Feldman, D. and Rusu, M. (1971) *Rev. Gen. Caoutch. Plast.* **48**, 687.
87. Titow, W.V. (1984) *PVC Technology*, 4th edn, Elsevier Applied Science, London.
88. Lath, D. and Lathova, E. (1993) *Makromol. Chem.*, **194**, 3087.
89. Akovali, G. and Aslan, S. (1993) *J. Appl. Polym. Sci.*, **50**, 1747.
90. Osawa, Z., Sunakami, T. and Fukuda, Y. (1994) *Polymer Degradation and Stability*, **43**, 61.
91. Mrazek, Z., Lucas, R. and Pradova, O. (1991). *J. Appl. Polym. Sci.*, **43**(5), 977.
92. Feldman, D., Banu, D. and El-Raghi, S. (1994) *J. Macromol. Sci. Pure Appl. Chem.*, **A31**(5), 555.
93. Ainsworth, S.J. (1992) *Eng. Chem. News*. **70**(28), 15.
94. Paroli, R.M., Dutt, O., Delgado, A.H. and Stenman, J. (1993) *J. Mat. Civil Eng.*, **5**(1), 83.
95. Leichton, J. and Szabo, E. (1993) *J. Vinyl Technol.*, **15**(3), 188.
96. Rabinovitch, E.B. (1989) *Makromol. Chem. Makromol. Symp.*, **29**, 241–251.
97. Mark, H., Bikales, N.M., Overberger, C.G. *et al.* (1989) *Encyclopedia of Polymer Science and Technology*, 2nd edn, Vol. 17, J. Wiley & Sons, New York, p. 327.
98. Newe, R., Schmidt, P., Grensing, R. *et al.* Ger(East) DD283,285. *CA*, **114**, 2296942.
99. Ulrich, H. (1982) *Introduction to Industrial Polymers*, Hanser Publishers, Munich.
100. Brydson, J.A. (1989) *Plastics Materials*, 5th edn, Butterworths, London.
101. Wessling, R.A. (1977) *Polyvinylidene Chloride*, Gordon and Breach, New York.
102. Mark, H., Bikales, N.M., Overberger, C.G. and Menges, G. (1989) *Encyclopedia of Polymer Science and Engineering*, 2nd edn, Vol. 17, J. Wiley & Sons, New York.
103. Litt, M.H. and Chang, K.H.S. (1981) in *Emulsion Polymers and Emulsion Polymerization* (eds D.R. Bassett and A.E. Hamielec), ACS Symposium Series 165, American Chemical Society, Washington, DC, p. 455.
104. Lange, D.M., Poehlein, G.W., Stayashi, S. *et al.* (1991) *J. Polym. Sci. Part A: Polymer Chemistry*, **29**, 785.

105. Lepizzera, S.M. and Hamielec, A.E. (1994) *Macromol. Chem. Phys.*, **195**(9), 3103.
106. Lee, C.K. and Forsyth, T.H. (1981) in *Emulsion Polymers and Emulsion Polymerization* (eds D.R. Basset and A.E. Hamielec), ACS Symposium Series 165, American Chemical Society, Washington, p. 567.
107. Mohsen, R.M., Ramadan, A.M., Shaffer, K.A. and Badran, A.S. (1993) *Pigm. Resin Technol.*, **22**(4), 4.
108. Yuan, H.G., Kalfas, G. and Ray, W.H. (1991) *J. Macromol. Sci. Rev. Macromol. Phys.*, **C31**(2,3), 215.
109. Beech, J.C. (1979) *J. Inst. Wood Sci.*, **7**(6), 7.
110. Woodley, B.A., Pritchard, J. and Armstrong, A.A. (1982) in *Polymer Latices and their Applications* (ed. K.O. Calvert), Applied Science Publ., London.
111. Anna, K. and Novak, B.M. (1994) *J. Amer. Chem. Soc.*, **116**(9), 4073.
112. Nakamura, N. (1991) *J. Appl. Polym. Sci.*, **42**(5), 1341.
113. Hyon, S.H. (1990) *Kobunski Karo.*, **39**(6), 326; *CA*, **114**, 144305x.
114. Modi, T.W. (1980) in *Handbook of Water Soluble Gums and Resins*, McGraw-Hill Book Co., New York.
115. Gowariker, V.R., Viswanathan, N.V. and Sreedhar, J. (1986) *Polymer Science*, Halsted Press, New York.
116. Hwang, K.S., Lin, C.A. and Liu, C.H. (1994) *J. Appl. Polym. Sci.*, **52**(9), 1181.
117. Mukai, Y., Fujita, S., Kubotsu, A. and Nakao, S. (1992) Jpn Kokai Tokkyo Koho JP 05,337,967 [93,337,967] (Cl. B29 C41/12), 21 Dec. 1993, Appl. 92/179,334, 11 Jun. 1992.
118. Park, E.H., George, E.R., Muldoon, M.A. and Flammino, A. (1994) *Polymer News*, **19,** 230.
119. Lupinsky, J.H. (1985) in *Applied Polymer Science* (eds R.W. Tess and G.W. Pochlein), ACS Symposium Series 285, American Chemical Society, Washington, DC, p. 22.
120. Korinek, P.M. (1990) *Kunststoffe*, **80,** 44.
121. Feiring, A.E., Imbalzano, J.F. and Kerbow, D.L. (1994) *Trends in Polymer Science*, **2**(1), 26.
122. Korinek, P.M. and Homburg, B. (1993) *Kunststoffe*, **3,** 782.
123. Mark, H., Bikales, N.M., Overberger, C.J. and Menges, G. (1989) *Encyclopedia of Polymer Science and Engineering*, 2nd edn, Vol. 16, J. Wiley & Sons, New York.
124. DeSimone, J.M., Guan, Z. and Elsbernd, C.S. (1992) *Science*, **256**(5072), 945.
125. Gowariker, V.R., Viswanathan, N.V. and Sreedhar, J. (1986) *Polymer Science*, J. Wiley & Sons, New York, p. 240.
126. Alger, M.S.M. (1989) *Polymer Science Dictionary*, Elsevier Applied Science, London.
127. Wright, W.W. and Lee, W.A. (1968) in *Progress in High Polymers* (eds J.C. Robb and F.W. Peaker), Hippe Books, London, p. 200.
128. Stamprech, P. (1988) in *Plastics Extrusion Technology* (ed. F. Hensen), Hanser Publishers, Munich, p. 614.
129. Carpenter, M.S., Chillons, S.E. and Will, R.R. (1991) *Polym. Preprints*, **32**(1) 252.

130. Anon. (1991) *Plastverarbeiter*, **42**(11), 146.
131. Kobayashi, S and Narui, S. (1992) Jpn Kokai Tokkyo Koho JP 05,339,537 [93,339,537] (Cl. C09 D127/12) 21 Dec. 1993, Appl. 92/172,015, 05 Jun. 1992; *CA*, **120**, 301296q.
132. Knox, J.B., Delaney, W.E. and Connelly, J.M. (1993) Fr. Demande FR 2,694,940 (Cl. C08 J9/16) 25 Feb. 1994, Appl. 93/9,620, 04 Aug. 1993.
133. Simkin, B. (1992) Eur. Pat. Appl. EP 572,764 (Cl. C08 L27/16), 08 Dec. 1993, US Appl. 892,457, 02 Jun. 1992; *CA*, **120**, 325873b.
134. Andrews, Jr, M.H., Olsen, R.J., Eian, G.L. and Allen, R.C. (1993) *Polym. Prepr. Amer. Chem. Soc., Div. Polym. Chem.*, **34**(1), 409.
135. Bonardelli, P., Moggi, G. and Russo, S. (1985) in *Emulsion Copolymerization* (eds J. Guillot and C. Pichot), Hüthig and Wepf Verlag, Basel, p. 11.
136. Noda, T., Honda, H. and Okuno, C. (1992) PCT Int. Appl. WO 9402,522 (Cl. C08 F214/26), 03 Feb. 1994, JP Appl. 92/195,147, 22 Jul. 1992; *CA*, *121*, 10234k.
137. Groataert, W.M. (1993) US, US 5,285,002 (Cl. S26–222; C08 F2/00), 08 Feb. 1994, Appl. 36,073, 23 Mar. 1993; *CA*, *121*, 10237p.
138. Lee, K.S. (1990) in *Engineering Polymers* (ed. R.W. Dyson), Blackie, New York.
139. Potisk, P. (1994) *Angew. Macromol. Chem.*, **218,** 53.
140. Nakabayashi, N., Ishitara, K. and Saima, Y. (1992) Eur. Pat. Appl. EP 572,227 (Cl. C08 F2/26), 01 Dec. 1993, JP Appl. 92/158,945, 27 May 1992; *CA*, **120**, 300651q.
141. Emilie, B., Pichot, C. and Guillot, J. (1985) in *Emulsion Polymerization* (eds J. Guillot and C. Pichot), Hüthig and Wepf Verlag, Basel, p. 25.
142. Nomura, M. and Fujita K. (1985) in *Emulsion Polymerization* (eds J. Guillot and C. Pichot), Hüthig and Wepf Verlag, Basel, p. 25.
143. Morgan, L.W. and Jensen, D.P. (1985) in *Emulsion Polymerization* (eds J. Guillot and C. Pichot), Hüthig and Wepf Verlag, Basel, p. 59.
144. Guillaume, J.L., Pichot, C. and Revillon, A. (1985) in *Emulsion Polymerization* (eds J. Guillot and C. Pichot), Hüthig and Wepf Verlag, Basel, p. 69.
145. Cruz, M.A., Palacrios, J.A., Garcia-Rejon, A.G. and Rios, L.G. (1985) in *Emulsion Polymerization* (eds J. Guillot and C. Pichot), Hüthig and Wepf Verlag, Basel, p. 87.
146. McGrath, J.E., Wilkes, G.L., Ward, T.C. *et al.* (1988) *New Elastomer Synthesis for High Performance Applications*, Noyes Data Co., Park Ridge, NJ.
147. Brinkhuis, R.H.G. and Shouten, A.J. (1991) *Macromolecules*, **24**(7), 1487.
148. Brinkhuis, R.H.G. and Shouten, A.J. (1991) *Macromolecules*, **24**(7), 1496.
149. Wong, Y.D. and Cakmak, M. (1993) *Int. Polym. Process*, **8**(2), 143.
150. Boutevin, B., Hamoui, B. and Parisi, J.P. (1994) *J. Appl. Polym. Sci.*, **52**(3), p. 449.
151. Donnet, J.B. and Bansal, R.C. (1984) *Carbon Fibers*, Marcel Dekker, New York.
152. Gill, R.M. (1972) *Carbon Fibres in Composites Materials*, Hippe Books, London.

153. Katz, H.S. (1978) in *Handbook of Fillers and Reinforcements for Plastics* (eds H.S. Katz and J.V. Milewski), Van Nostrand Reinhold Co., New York, p. 562.
154. Bentur, A. and Mindess, S. (1990) *Fibre Reinforced Cementitious Composites*, Elsevier Applied Science, London.
155. Hahne, H., Karl, S. and Worner, J. (1987) in *Fiber Reinforced Concrete. Properties and Applications* (eds S.P. Shak and G.B. Batson), SP-105, American Concrete Institute, Detroit, p. 211.
156. Romack, T.J., Maury, E.E. and De Simone, J.M. (1994) *Polym. Preprints*, **35**(2), p. 741.
157. Yin, R., Ottenbrite, R.M. and Siddiqui, J.A. (1994) *Polym. Preprints*, **35**(2), 705.

Diene polymers

3

3.1 POLYBUTADIENE (PB)

3.1.1 Introduction

Although this elastomer was first produced in Europe in the early 1930s, it was virtually unknown in the USA prior to 1960. The discovery of organometal catalysts for butadiene polymerization in hydrocarbon solvents brought a remarkable change to the industry. Solution polybutadiene has found its way into tire and non-tire compounds as well. As a result, PB has taken over the number two position among synthetic rubbers, second only to emulsion styrene–butadiene copolymer. In only one decade (1969–1979) the consumption of PB in the USA increased from 260 000 tons to 380 000 tons [1,2] and reached 406 000 tons in 1990 [3]. In 1993, total sales of PB in the USA and Canada were 473 000 tons [4].

The vast majority of production is solution process; small but unknown quantities of emulsion process PB are made in various parts of the world [5]. Solution polymerized PB in straight as well as oil extended grades is manufactured by a large number of companies; there is also some emulsion PB available.

Butadiene, the monomer, is produced through the dehydrogenation of butylenes or butane, or by thermal cracking of petroleum fractions. It was not until the exploitation of alkyl-lithium and Ziegler–Natta catalyst systems in the 1950s that polybutadiene and other elastomeric homopolymers were produced with sufficient stereoregularity to enable them to be a marketable commodity. The discovery of stereospecific catalysts based on transitional metals has changed completely the commercial production of polydiolefins. With such a catalyst, homopolymers with 98% identical units (1,4-*trans* or 1,4-*cis*), head to tail linked and with a more regular microstructure than the synthetic polymers known before, were obtained.

The first catalyst systems used for the polymerization of conjugated diolefins were based on $TiCl_3$ or $TiCl_4$ and on organometallic components such as

Table 3.1 Effect of catalysts on polybutadiene composition [2]. (Reprinted by permission of C. Hanser Verlag, from *Introduction to Industrial Polymers* by H. Ulrich, 1982)

Catalyst	% *trans*-1,4	% *cis*-1,4	% 1,2
Et_3Al/VCl_3	97–98	–	2–3
$Et_3Al/TiCl_4$	49	49	2
$Et_3Al/Ti(OBu)_4$	0–10	–	90
$Et_3Al/Cr(acac)_3$	18	12	70
$Et_2AlCl/CoCl_2$	3	93–94	3–4

$Al(C_2H_5)_3$, $AlLi(C_2H_5)_4$, etc. Later, a great number of components with the transitional metals of the IV–VIII groups of the periodical system like halogenates, alcoholates, acetyl-acetonates, Co-complexes, amines, isonitriles, nitriles, etc. were used for the preparation of polymerization catalysts for conjugated diolefins. The most utilized are chlorides, iodides, the complexes of chlorides with Lewis bases, alcoholates and acetyl-acetonates.

The effect of the nature of a catalyst on the PB microstructure is presented in Table 3.1. 1,4, 1,2 and 1,2/1,4 polybutadienes are of commercial interest. *Cis*-1,4 forms are the most important polybutadienes, making up the butadiene rubbers. Four general types are produced [6,7].

1. High *cis*-1,4-PB (with approximately 92% *cis*-1,4 content) using Ziegler–Natta catalysts such as dialkyl-aluminum chloride–cobalt(II) halide combinations.
2. Medium *cis*-1,4-PB (with approximately 92% *cis*-1,4 content) using Ziegler–Natta catalysts such as dialkyl-aluminum chloride–titanium tetra-iodide combinations.
3. Low *cis*-1,4-PB (with approximately 40% *cis*-1,4 content) using alkyl-lithium catalysts.
4. High *trans*-PB by the Ba–Mg–Al catalyst system.

The application of Ziegler–Natta catalysts and of other stereospecific systems to diene monomers drew the attention of professionals on PB that has the advantage of being cheaper and more accessible than isoprene and other diolefins. A list of catalyst systems for B polymerization and the main characteristics of PB with the respective references is given by M. Morton [8].

3.1.2 Solution polymerization

In most cases, B polymerization takes place in solution and the main parameters of this process are the nature of catalytic system, monomer purity and the nature of the solvent which constitutes the polymerization medium.

As in all catalytic systems based on two components, the ratio between them (e.g. Al/Co) affects the conversion, polymerization rate and MW of

PB. At an excess of organometallic component, the yield of polymerization and the MW decrease, probably due to the chain transfer effects observed.

Syndiotactic 1,2-PB (with 90–93% 1,2 units) are obtained by using a $(alkyl)_3$ Al–Co(II) halide–ligand Ziegler–Natta system. 1,2–1,4-PB (with 1–65% 1,2 units) is prepared with an alkyl-Li–Lewis base catalyst system [9]. Recently, a syndiotactic 1,2-PB was obtained by suspension polymerization of 1,3-B [10]. This process was carried out in the presence of a catalyst made with Co–octoate and $(C_2H_5)_3Al$ in heptane, water, PVA, CS_2, CH_2Cl_2 and cyclohexanone containing the monomer [10].

PB produced by emulsion polymerization is not useful as an elastomer. However, the copolymers with styrene (SBR) and AN (NBR) are widely used as elastomers [4].

3.1.3 Properties

In practical vulcanization of PB, sulfur is used as a vulcanizing agent [11]. The resilience, heat build-up and abrasion resistance of vulcanized PB are superior to those of vulcanized natural rubber. However, because of the lower tensile strength, tear resistance and relatively poor road adhesion, PB is blended with natural rubber or styrene–butadiene rubber when it is used for tire production.

3.1.4 Applications

The most important uses of PB are for tires, hoses and belts. PB is also used in the manufacture of impact modified PS systems. Low volume specialty PB products include liquid *trans*-1,4-PB used in protective coatings inside metallic cans, low MW liquid 1,3-PB (60–80%, 1,2 units) used as potting product for transformers and submersible electric motors and pumps, and liquid hydroxy-terminated PB intended for the formulation of PU elastomers [2].

In 1990, the USA's consumption of 365 000 tons was distributed as follows:

- 78% for tires (45% for passenger cars and 28% for trucks and buses, 4% retread rubber, 1% other);
- 20% for plastics modifiers (including 2% mechanical goods) [5].

3.2 STYRENE–BUTADIENE RUBBER (SBR)

Copolymers of butadiene and styrene are designated as SBR and are by far the most important synthetic rubbers at the present time. In the past, SBR was produced in larger quantities in the USA than is produced now. In 1993 the USA consumed *c.* 3.8 billion tons, of which SBR represented 39% (including solid, latex and carboxylated forms) followed by natural rubber (27%), PB (13%), ethylene–propylene (6%), NBR (3%), polychloroprene (2%) and others (10%) [12].

Polystyrene homopolymer has a number of deficiencies that limit its utility: it is brittle and has poor impact strength due to very stiff chains. Copolymerization of 1,3-butadiene with styrene imparts sufficient flexibility to yield elastomeric products. SBR is a 25% styrene–75% butadiene copolymer used especially for automobile tires. A higher styrene content (50–75%) leads to a product used in latex paints. The higher styrene content provides better abrasion resistance.

Although the double bond in the butadiene unit is slightly less active than in isoprene, it still responds strongly to sulfur vulcanization and degradation by oxygen. Thus the temperature limits placed on the use of natural rubber products apply similarly to products based on SBR, and the same compounding principles apply.

While solution SBR technology allows much more control of the polymer structure than that produced in emulsion technology, the latter provides almost all of the necessary SBR. In solution copolymerization, the reactivity ratios are quite different than those observed with free radical polymerizations. In hydrocarbon solvents, it is noticed that r_1/r_2 (1 for B and 2 for S) is at least 50/1. As a consequence, there is a pronounced tendency to block polymerization with B polymerizing first followed by S.

The SBR latex can be obtained through copolymerization at 50°C or around 0°C (cold rubber) and extended with mineral oils, carbon black or with both. The emulsion copolymerization is produced continuously. Beside the specific peculiarities of the continuous processes, the emulsion copolymerization is characterized by the following:

- a lower specific tension than that of the latices provided discontinuously;
- latex particles have a higher size distribution as a result of the difference in the time required in the reaction zone.

The polymerization rate decreases and the latex particles size increases when the continuous technology is used; the emulsifier distribution in the micelles and the polymer–monomer particles formation starts practically in the first reactor when the process is conducted in a battery of reactors.

The cold SBR obtained by copolymerization at around 0°C possesses better technical characteristics than the SBR obtained at 50°C; the decrease of the polymerization temperature to under 0°C does not cause more desirable properties, and this explains why at the commercial scale the copolymerization temperature is usually +5°C [13].

Normally, the phases ratio is 1–2.5 parts of aqueous phase for 1 part monomers phase; the ratio is determined by the copolymerization conditions and by the latex uses. When the hot rubber has to be produced and a highly concentrated latex is necessary, ratios of 1.05:1 or even 1:1 are recommended; a lower amount of aqueous phase leads to difficulties, due

Table 3.2 Typical polymerization recipes for hot and cold emulsion polymerized SBR [14]. (Reprinted by permission of Elsevier Science Publ. Ltd. from *Rubbery Materials* by J.A. Brydson, 1988)

Component	*Hot SBR (SBR 1000)*	*Cold SBR (SBR 1500)*	*Comments*
Monomers			
Butadiene	75	72	
Styrene	25	28	
Water	180	180	
Emulsifier system			
Fatty acid soap	4.5	4.5	May include resin acid to improve tack.
KCl		0.3	As electrolyte helps to stabilize emulsion. The $K_2S_2O_8$ has this additional role in hot SBR formulation.
Sodium naphthalene sulfonate		0.3	Auxiliary surfactant
Initiator system			
$K_2S_2O_8$		0.3	
p-Methane-hydroperoxide		0.06	Redox oxidant
$FeSO_4 \cdot 7H_2O$		0.01	Reducing agent
Sodium formaldehyde sulfoxylate		0.05	Reducing agent
Sodium salt of EDTA		0.05	EDTA = ethylene diamine tetra-acetic acid
Regulator			
t-Dodecyl mercaptan	0.28	0.2	
Polymerization stopper			
Hydroquinone			
Sodium dimethyl dithiocarbonate	0.05	0.05	
Polymerization temp. (°C)	50	5	
Conversion (%)	72	60	

to the increase of the viscosity of the system and at the same time a worsening of the copolymer quality due to the intensification of chain transfer reactions.

Because of the increase of the viscosity when the copolymerization is carried out at +5°C, it is necessary to increase the amount of water phase. Although a 2:1 ratio is favorable to the process and to the SBR properties, it leads to an inefficient use of the plant. Typical recipes for styrene–butadiene copolymerization are presented in Table 3.2 [14].

Soaps of fatty or sulfonic acids, such as stearic, oleic, palmitic, synthetic fatty acids of C_{10}–C_{16} fraction and rosin acid are used as emulsifiers. Nekal, which is the sodium salt of dibutyl-naphthalene sulfonic acid, was used previously and was recently replaced because of its polluting effects.

The use of rosin acid soap has several advantages such as:

- during coagulation, rosin acid remains almost completely in the rubber increasing its output;

- residual waters do not contain soap and their purification is no longer necessary.
- technical characteristics of the SBR, e.g. adhesive properties of the raw compounds, tension, wear and tear strength, are improved.

To avoid the main drawback related to the use of abietic acid salts, namely the decrease of the copolymerization rate, rosin is initially modified through halogenation, catalytic hydrogenation or disproportionation. Electrolytes such as KCl or Na_3PO_4 are used to maintain the system's superficial tension. Often an auxiliary surfactant is used, such as sodium naphthalene sulfonate or a sodium or potassium salt of the condensation product of β-naphthalene sulfonic acid with formaldehyde.

It is customary in emulsion polymerization to generate free radicals in the aqueous phase using either the thermally induced decomposition of water soluble initiators (such as persulfate anions) or redox couples [15]. Water soluble persulfates are used as initiators in the copolymerization process which occurs at 50°C. However, the copolymerization at lower temperatures needs more efficient initiators and for such processes, redox systems are preferred, such as systems based on hydroperoxides (as oxidant agent) and ferrous sulfate as a reducing agent. An auxiliary reducing agent is added for the continuous transformation of Fe^{3+} in Fe^{2+}.

In an alkaline medium, iron ions cannot exist as ions because of the insoluble hydroxide formation. For this reason, the sodium salt of EDTA is used, which is able to form with $FeSO_4$ a water soluble complex:

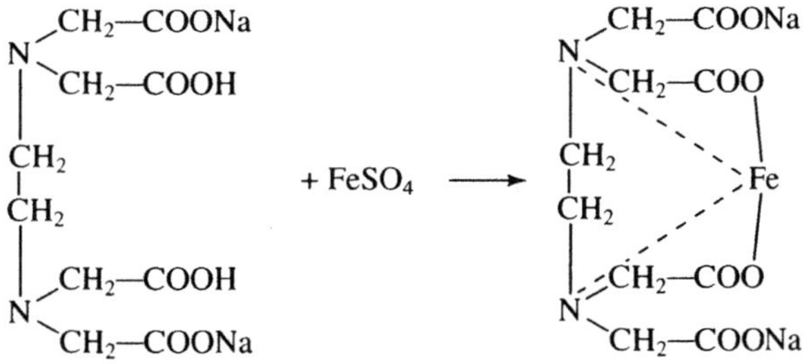

In water, this iron complex reacts with the hydroperoxide (initiator):

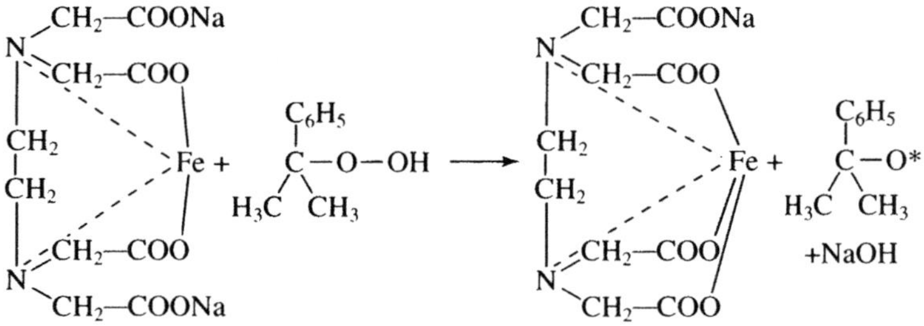

The polymerization rate will increase in such a system given an increase in the following parameters: initiator concentration, Fe^{2+} ion concentration and temperature [16].

Sodium formaldehyde sulfoxilate, which is an auxiliary reducing agent, reacts in the alkaline medium with the ferric complex leading to the remaking of Fe^{2+} ions.

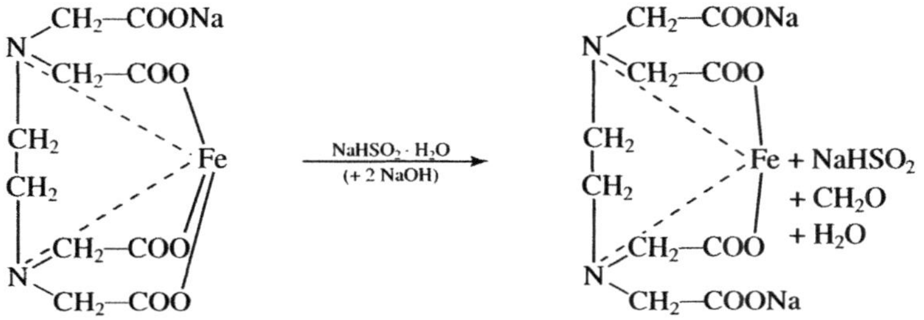

p-Methane hydroperoxide which is used very often is a mixture of hydroperoxides:

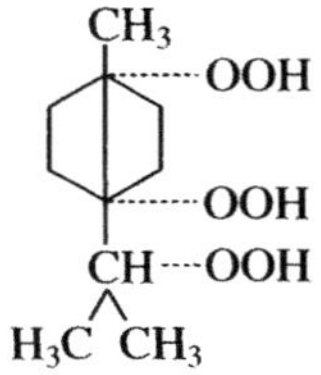

Its hydroperoxide groups replace hydrogen atoms at any of the positions shown. It has a half-life time of 25 h/120°C and 2 h/150°C and half-life temperatures of 135°C/10 h, 110°C/1 h, and 210°C/1 min. It is sometimes used to replace cumene hydroperoxide as a redox initiator component [17].

Depending on the pH, the secondary reaction between the hydroperoxide and sodium formaldehyde sulfoxilate takes place at a higher or lower rate. Na_3PO_4 is used as a buffer for maintaining the optimum pH.

Semicontinuous or continuous addition of initiator and chain transfer agent is employed as necessary to maintain a relatively constant initiation rate and MW. A surfactant is also often added through the emulsion polymerization. The rate and total amount of surfactant added can have a large effect on particle size and size distribution [18]. For MW control, aliphatic mercaptans with 10–16 carbon atoms are used; dodecylmercaptan is preferred in the case of hot SBR production.

When a 60% conversion is reached, the copolymerization is stopped because the increase of conversion leads to the increase of the role of chain transfer reactions which worsens the technical properties of the rubber. Hydroquinone, *p*-oxy-diphenylamine or sodium-dimethyl-dithiocarbamate are used as stoppers.

The average reactivity ratios for the emulsion copolymerization of butadiene with styrene are $r_1 = 1.6$ and $r_2 = 0.5$ [19]. The copolymerization heat of the butadiene–styrene system at 25°C is in the range of 17.1–17.7 kcal/mol [20].

In the continuous process, S, B, emulsifier, initiator and activator are introduced continuously from storage tanks through a series of agitated reactors at such a rate that the desired conversion (*c.* 60%) is reached in the last reactor [21].

For various commercial grades of SBR, butadiene polymers and latices, numbers are assigned such as:

- 1000, hot polymers;
- 1100, hot black master batch with 14 or less parts of oil per hundred parts SBR;
- 1500, cold polymers;
- 1600, cold black master batch with 14 or less parts per hundred parts SBR;
- 1700, cold oil master batch;
- 1800, cold oil-black master batch with more than 14 parts per hundred oil;
- 1900, miscellaneous dry polymer master batches;
- 2000, hot latices;
- 2100, cold latices.

These grade numbers were assigned by the International Institute of Synthetic Rubber Producers [22].

3.2.1 Butadiene–styrene emulsion copolymerization at low temperature

Hot copolymerized grades are characterized by highly branched macromolecules and contain amounts of microgel. Branching makes polymers more difficult to process, although in some respects processing is better than that of cold SBR. The vulcanizate properties of hot polymerized grades are poorer than those of the cold grades.

The linearity of the macromolecules obtained by cold copolymerization is much improved and gel is absent. These grades do not require premastication and are easy to process. The copolymerization at *c.* +5°C has some features which make it different from the copolymerization which leads to the production of the hot rubber; these differences are initiation with a highly efficient redox system; the ratio of phases is in the range 1.8:1 to 2.5:1; the use of MW regulators.

An important aspect consists in the reaction heat elimination, the thermal effect of butadiene–styrene copolymerization (70:30 in weight) is 250 kcal/kg. The amount of heat which must be eliminated in a time unit is determined by the thermal exchange surface, the total coefficient of heat transfer, and the temperature difference between the reaction medium and the cooling agent. If the reaction temperature is 50°C the temperature difference between the reaction medium and the cooling water is 30–35°C; to keep such a difference in the case of copolymerization at +5°C, brine with a temperature of −25 or −30°C is necessary. However, with such a brine, the control of the copolymerization becomes impossible because of the emulsion decomposition and the latex coagulation.

The production technology of the styrenic rubbers is characterized by two main steps:

1. latex production and comonomers' recycling;
2. copolymer separation from latex (coagulation).

(a) Latex production

The operations necessary to fulfil this stage are as follows [16]:

- preparation of comonomers mixture, aqueous phase, initiator solution, activator solution, regulator solution, stopper solution and antioxidant solution;
- copolymerization of butadiene with styrene;
- degassing the latex and recuperation of the unreacted monomers.

The comonomers must have a high purity, a minimum of 96% for butadiene and 99.5% for styrene. The pH of the aqueous phase has to be *c.* 10 and the ratio of the phases is 2.5:1. Copolymerization takes place in a battery with 8–12 reactors. Before mixing, the two phases are cooled to +5°C in coolers with coils in which brine is circulated. With regulated flow rates, the two phases are sent to a vessel for the preparation of comonomers emulsion and then to the copolymerization reactors (Figure 3.1).

After passing through the reactors, a final conversion of 58–63% is obtained. From the last reactor the latex of B–S elastomer is sent to a filter which removes the coagulated particles and from the filter the latex is sent to the recuperation of unreacted monomers. Before filtration, stopper is added. Unreacted monomers' recuperation is realized in vacuum; some plants circulate both the latex and vapor in the same direction while others use counter-current systems until residual styrene is less than 0.3%. The degassed latex is stocked under agitation in a reservoir where an antioxidant (2–3% phenyl-β-naphthylamine) is introduced (Figure 3.2).

(b) Butadiene–styrene copolymer separation

An average latex composition is copolymer, 29–33%; styrene less than 0.3%; free alkalis less than 0.1%; emulsifier less than 4%; and antioxidant 2.3–3%.

Latex is stocked in reservoirs with stirrers (40 rpm) having a few hundred cubic meter capacity. Coagulation can be done by using coagulating agents, or congelation.

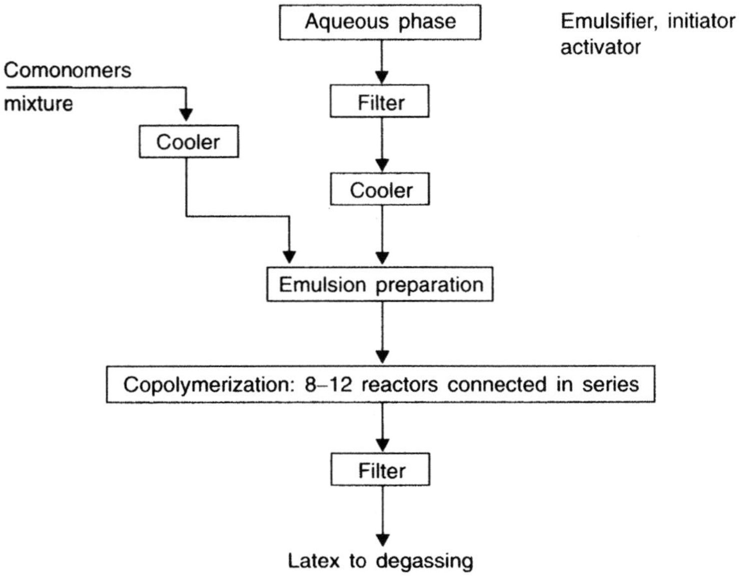

Figure 3.1 Butadiene–styrene emulsion copolymerization.

Latex stability [23] is due to the fact that the copolymer particles have the same electrical charge which causes their reciprocal rejection. R. Buscall and R.H. Ottewill [24] give a qualitative description of latex particles and underline that the origin of the basic interparticle forces has to be taken into account in understanding the stability behavior of dispersions in both aqueous and non-aqueous dispersion media. These can be summarized as follows.

1. Electrostatic effects, arising from ionized surface groups, usually repulsive, but particles having opposite charges will undergo electrostatic attraction which depends on the electrolyte concentration.
2. Steric effects, arising from the geometry and conformation of adsorbed or grafted molecules on the surface; these can also charge effects when the molecules are polyelectrolytes. These can be repulsive or attractive and are often temperature dependent and will also depend on electrolyte concentration.
3. Attractive effects, arising from the long-range dispersion forces when, essentially, the density and polarizability of the molecules constituting the particles are different from those of the medium. Generally these are electrolyte and temperature independent.
4. Solvation effects, arising from the organization of solvent molecules near an interface or a macromolecule in a different manner from that of the bulk phase.

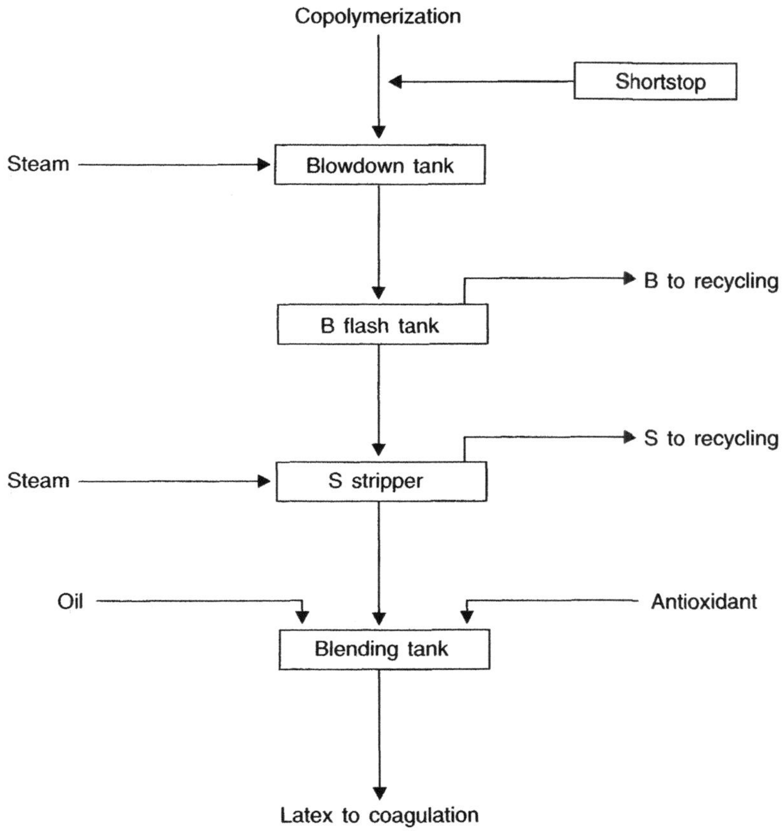

Figure 3.2 Recuperation of monomers.

It is considered that the latex destabilizing influences operate by chemical mechanisms. These influences, which are known generally as coagulants or coacervants, can be conveniently classified as follows.

1. Direct or contact coagulants.
2. Heat sensitive coagulants are substances which have relatively little effect upon the stability of a rubber latex at room temperature, but which cause the latex to destabilize rapidly if this temperature is raised above a certain value.
3. Delayed-action coagulants, which have relatively little effect upon the stability of a rubber latex when first added, but which cause the latex to destabilize rapidly after a certain time has elapsed [25].

Precipitating agents are acids, electrolytes or mixtures of acids and electrolytes.

The addition of aqueous solutions of alkaline metal salts followed by a diluted acid treatment of the latex leads to the decomposition of the soap-type emulsifier and to its transformation into a fatty acid; under such conditions the latex stability is affected and the elastomer coagulates in the form of porous solid particles.

The addition of the electrolyte solution during the first coagulation stage determines the agglomeration of the copolymer particles and the latex looks like a dense cream (creaming); introducing the diluted acid leads to the separation of the polymer into non-adhesive particles. The high specific surface of the copolymer particles makes washing the salts and acids and even the moisture released during drying easier.

Particle size depends on a series of parameters such as the medium's pH, temperature, storage duration of the latex prior to coagulation, type of latex, nature of coagulation agent and coagulation duration. The coagulation process depends on the latex state, its concentration, temperature and mixing procedure with acid. Common coagulation agents are solutions of electrolytes which contain Mg^{2+}, K^+, Na^+, Cl^-, SO_4^{2-}, etc. 'Cascade' coagulation techniques are used, based on stirred reservoirs fixed at different levels. The coagulation flow chart is presented in Figure 3.3.

When the rubber is produced in strips, $CaCl_2$ and acetic acid are used for the coagulation; washing and dewatering are done on equipment with plane screens. The MW of both hot and cold SBR is around 100 000 with a polydispersity of 7.5 for the former and 4–6 for the latter. Solution polymerization provides higher MWs and less polydispersed (1.5–2.0) SBR [26]. The possibility of producing a gel-free high MW SBR at reduced polymerization temperature enables the preparation of a high Mooney viscosity (*c.* 100) polymer which can be plasticized by low cost petroleum oils (oil extended rubber) and still retain its mechanical properties [22].

Dynamic modeling of the batch, semibatch and continuous production of styrene–butadiene copolymers by emulsion copolymerization has been reported [27].

3.2.2 Oil extended SBR

Generally, an easily processed elastomer with good vulcanizate properties should contain a high amount of linear macromolecules and have a wide polydispersity. Such an elastomer contains 15–20% macromolecules in the range up to MW = 30 000, which do not crosslink during vulcanization and act merely as a processing additive. It should also contain a high MW fraction in the order of 10^6, to obtain satisfactory end properties. To a great extent, mineral oils can take over the role of the low MW fraction in aiding processability and in realizing economic grade rubbers. Three types of mineral oil, which are incorporated as an emulsion in SBR latex prior to coagulation, are in general use:

- the highly aromatic, which are general purpose staining, and give excellent processability;

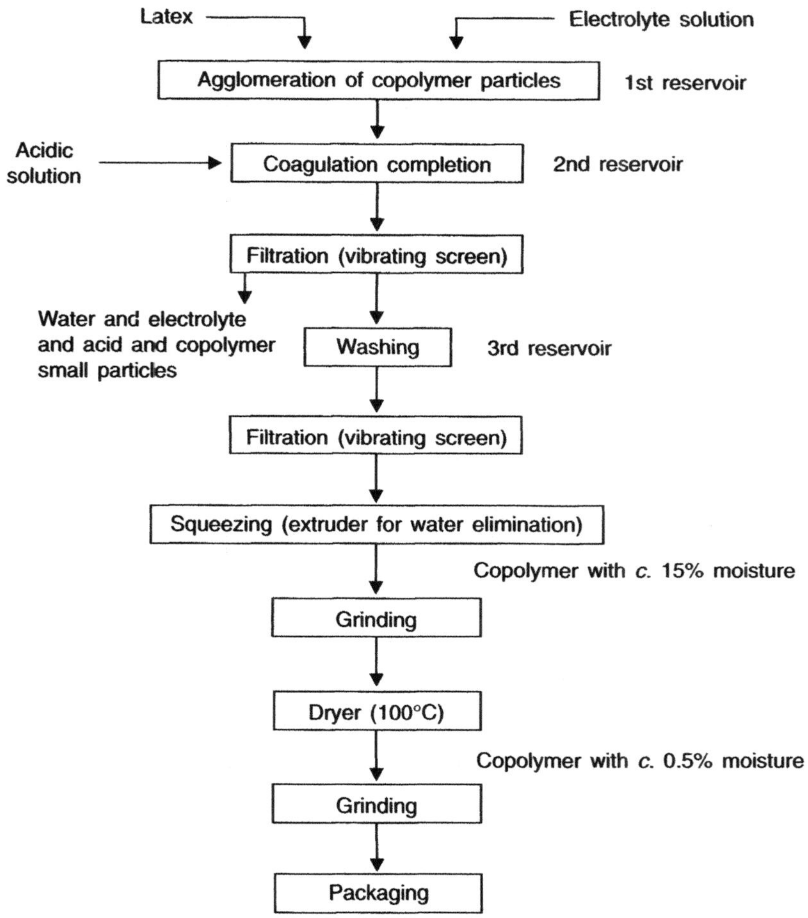

Figure 3.3 Latex coagulation.

- the aromatic, which are general purpose staining and give good processability;
- the naphthenic, which are for light colored products and non-staining application; the processability of these oils is satisfactory, provided that the paraffinic content is low, whereas the strength properties are generally lower than for the other types of mineral oil.

The affinity of the different types of SBR to diene elastomers, diminishes in the order aromatic > naphthenic > paraffinic. Paraffinic oils cannot be used with diene elastomers as they are incompatible; they 'sweat out', as do the paraffinic vaselines and paraffinic waxes, forming a 'bloom' [16,28].

In the case of oil extension, it is not necessary to use chain size regulators. Mineral oil is added in amounts of 20–50 weight per hundred (phr) of SBR weight which means 15–33% of oil extended SBR. The important decrease of their hardness when mineral oils are added makes plastification unnecessary during processing of oil extended SBR products. Tires made of oil extended rubbers have a service life of 15–20% higher than that of tires made of unextended SBR.

The technology of the manufacturing of oil extended SBR in a mixer is based on the homogenization of the SBR latex after degassing with the oil emulsion. For this, the necessary amount of oil is heated at 80°C and is introduced into an emulsification tank. Stearic acid is necessary for the preparation of a 3% emulsion with demineralized water. In the emulsification tank, the stearic acid emulsion is poured at the same time as the oil, and after 30 minutes' stirring, the system is cooled to 40°C; an aqueous 3% solution of the 3-ethanolamine is added. 3-Ethanolamine stearate is formed which constitutes the emulsifier of the mineral oil. After another 60 minutes' stirring, an emulsion with the following composition is obtained: mineral oil 40% (wt); stearic acid 3.7%; 3-ethanolamine 1.8%; water 54.5%.

The emulsified mineral oil is continuously sent for blending with SBR latex, in a mixing tank which has a strong stirrer and where a homogeneous distribution of the oil in latex is obtained in 7–8 minutes. Coagulation of this extended latex takes place in the same conditions as in the case of simple SBR latex.

Efficient antioxidants for oil extended SBR are diphenylenediamine, phenyl-isopropyl-phenylenediamine, *p*-oxy-diphenyl β-naphthylamine (oxineozon) and mercaptobenzimidazol. 0.5 parts of such additives are used with 1.5 parts phenyl-β-naphthylamine (Neozon D).

3.2.3 Carbon black and oil–carbon black SBR

Adding carbon black during processing SBR leads to a rubber with physical–mechanical properties almost identical with those of vulcanizates based on natural rubber. Blending of this ingredient with the rubber takes place during milling; reciprocal interpenetration of carbon black particles is difficult and needs a high energy supply.

However, while the rubber has a high viscosity, water is a low viscosity medium. For this reason, the conditions for dispersing carbon black in latex are more advantageous than those in the mill-mixed compounds. For such conditions a better blending of both components with a much lower energy consumption can be realized. Other advantages besides the improved SBR quality are easier processing in the rubber manufacturing plant, shorter mixing cycles, improved extrusion, uniformity of properties, cheaper and cleaner operation of the plant and a reduction in the number of weighing and mixing operations.

The production technology of carbon black filled SBR implies the following steps:

- production of SBR latex;
- preparation of carbon black suspension;
- latex and carbon black blending;
- carbon black filled SBR coagulation and drying.

Practically, there are two variants: one procedure based on the use of chemical dispersants and the other one without such additives. The following are used as dispersants: alkaline-lignin, sodium lignosulfonate and sodium-naphthyl-sulfonate, which are effective in alkaline medium. For this reason, the pH is regulated in the range 10–11 with alkalis. The dispersant amount depends on the type of carbon black; generally it is 5% of the carbon black weight.

For this procedure the carbon black has to be very well ground and after grinding it is mixed with water and dispersant. The presence of dispersant permits the formation of stable colloidal dispersions with an amount of up to 20% carbon black. Carbon black suspension is mixed with SBR latex in the proportion of 1 part carbon black to 2 parts rubber. For coagulation the electrolyte solution and the acid are introduced simultaneously; this is because the electrolyte alone would produce carbon black coagulation before that of the elastomer. Besides the many advantages already mentioned, this procedure has some drawbacks, for example, after coagulation, some carbon black deposits on the dryer walls, creating a fire danger. Figure 3.4 shows, in principle, how this procedure works. For carbon black dispersion in water, the procedure without dispersant uses strong machines with rotors which work at a speed of 3500–6000 rpm.

3.2.4 Solution SBR

Copolymerization of styrene with butadiene is carried out in a hydrocarbon solvent in an inert, moistureless atmosphere with alkyl-lithium catalyst. This catalyst is added to the thoroughly dry mixture of styrene, butadiene and solvent. Afterwards copolymerization stopper and stabilizer are added; copolymer MW is controlled through the amount of catalyst and the temperature. If oil extended SBR is desired, the mineral oil is introduced after the stabilizer but before the steam stripping operation.

Activation of the catalyst, e.g. by complexing with ethers, results in a more random microstructure. Such copolymers have a *cis* content of about 30% and a vinyl content of 25%. Increasing the polarity of the solvent gives a further increase in the vinyl configuration.

The solution process leads to copolymers with very low polydispersity since all the growing macromolecules have equal probability of propagation. The presence of impurities may produce a broadening of the MW distribution. In some instances, reagents may be added to introduce branching, or increase polydispersity to enhance SBR storage stability and processing behavior.

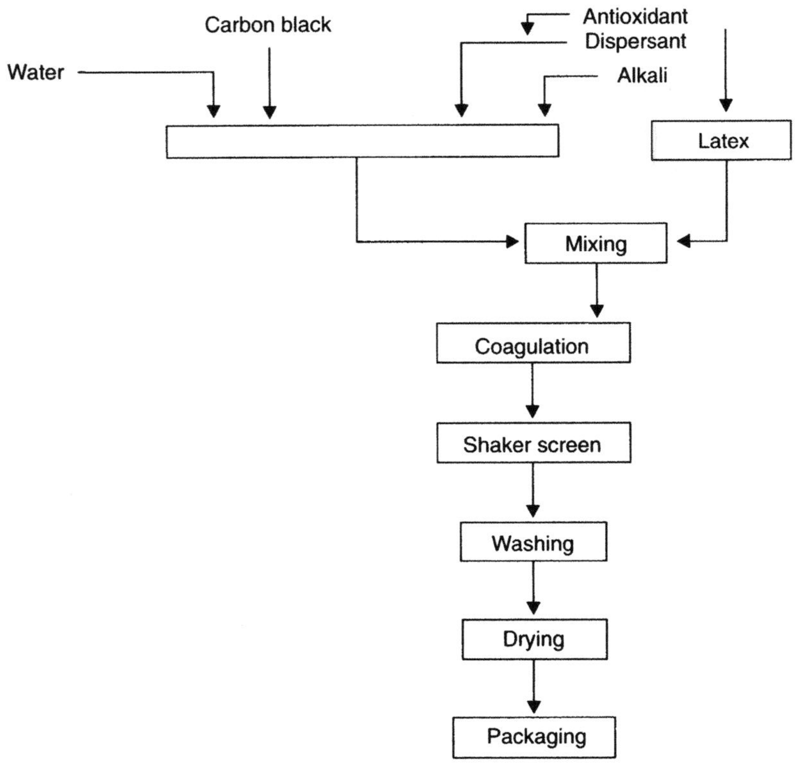

Figure 3.4 Production of carbon black filled SBR using a chemical dispersant.

With alkyl-lithium in a batch reactor, the reactivity ratios are such that butadiene units polymerize first and with total exclusion of any styrene present. Only when all the butadiene is consumed does the bulk of the styrene enter the macromolecule. Thus a block copolymer is produced. Since there is no termination step, triblock polymers may be made by adding in succession the monomers to the solution [29].

In polar media, the value of the r_1/r_2 ratio decreases and when using butyllithium catalyst at 30°C a more random and less blocky copolymer is produced [26].

The different possibilities of combining the two comonomers, either randomly arranged or in blocks, yield a variety of products with widely diverging characteristics, a survey of which is presented in Table 3.3 [30]. The blockiness provides a thermoplastic character to the rubber, leading to the improvement of some processability aspects such as extrusion and stability of milled sheets, and also in harder but less elastic vulcanizates [30].

The advantages of solution SBR over emulsion SBR [22] are easily controlled MW and narrow polydispersity, little or no branching, a higher

Table 3.3 Comparison of the features of solution and emulsion SBR [30]

Characteristics	*Solution*	*Emulsion*
Styrene configuration	Block or random	Random
Ash (% wt)	0.1	0.75
Organic acid (% wt)	0–0.5	6.75
Rubber hydrocarbon (% wt)	98+	92
Monomer incorporation	Controlled	Random
Polydispersity	Narrow	Broad
Long chain branching	Slight	More extensive
Color	White	Light to dark

concentration of pure elastomer since there is no emulsifier and more control over the structure of macromolecules. Some characteristics of solution SBR are presented in Table 3.4.

3.2.5 Carboxylated SBR

In the last decades, the production of latices based on carboxylated SBR has increased due to the demand of paper and textile carpet industries. The presence of carboxylic groups confers high adhesive strength and the possibility of using sulfurless crosslinking agents. The acidic comonomers used are unsaturated mono and dicarboxylic acids such as acrylic, methacrylic and itaconic, and their selection depends on the relative reactivity on terpolymerization, solubility in water and latex applications. These terpolymers contain up to 5% carboxylic monomers and 35–85% S providing a variation in the degree of stiffness.

The terpolymerization can be carried out by batch, semicontinuous or continuous technologies, the latter being preferred. The reaction temperature is controlled at between 60°C and 100°C so as to avoid high pressure, rapid depletion of radicals and the formation of very highly gelled products with very high MW. The process takes place at a pH between 2 and 4 in the presence of persulfates as initiators and the sodium salt of sulfonic acid as the emulsifier [16,31].

Commercially available carboxylated SBR latices have the following characteristics: total solid content 50–55%; average particle size 140–220 nm; surface tension above 38 mN/m (depending on end use); viscosity lower than 300 m Pa s; Mooney (ML1 + 4) viscosity 80–160 (depending on end use).

The main application of carboxylated SBR latices is in the carpet industry for precoat, secondary baking, backsizing, needlefelt and foam [31].

3.2.6 Properties

SBR is a general purpose rubber. Nearly 50% of the total tonnage of rubber consumed in the USA is SBR. It excels most grades of natural rubber in

Table 3.4 Characteristics of solution SBR [28]. (Reprinted with permission of the publishers, Butterworth-Heinemannn Ltd from *Rubber Technology and Manufacture* edited by C.M. Blow and C. Hepburn, 1982)

Elastomer type	*Conventional*			*Thermoplastic*
Vulcanization	With sulfur irreversible			Reversible via domain formation
Structure	Random		Certain degree of blockiness	Tailor-made blocks $S_xB_yS_x$
cis	40	30		
vinyl	<10	25		
Average MW	$(200-500)\times 10^3$			$(80-160)\times 10^3$
T_g (°C)	−40	−3	No clear transition point	Two clear T_g values +90 and −70
Bond styrene (% wt)	10–25		25–50	30–45
Bond styrene present in blocks (%)	0		approx. 60	100

uniformity and cleanliness and is normally preferred if it yields products of equal quality and cost. Its permeability, heat resistance, resistance to wear and aging is superior to natural rubber in many commercial products. It is inferior in workability, tensile strength, tear resistance, cut growth, tackiness and heat build-up due to hysteresis.

SBR requires more severe milling than natural rubber; however, over-milling causes a considerable decrease in the tensile strength of the product, thus mastication must be closely controlled. Larger amounts of plasticizer and softener are also required. Vulcanization does not require as much sulfur; however, greater amounts of accelerators must be used. Curing agents, fillers and activators are similar to those used for natural rubber.

SBR may be vulcanized with vulcanizing agents without sulfur like diazo derivatives, quinones etc. The vulcanization rate in the case of SBR products is lower than that for natural rubber; for this reason, more active accelerators are necessary. The SBR vulcanization rate depends more on the amount of accelerators than on that of sulfur, especially in the initial period of the process. At an increase of the vulcanization temperature up to 150–160°C, a decrease of the Young's modulus, relative elongation and even abrasion resistance has been observed. SBR resists ozone and light better, and has a lower water absorption and is a better electrical insulator than natural rubber. It also resists manganese, iron, and copper salts action much better, and its solubility in mineral oils is lower than that of natural rubber.

SBR is compatible with natural rubber and other general purpose synthetic rubbers. The reinforcing action of carbon black is much more pronounced, particularly when comparatively large amounts are used. The type of carbon black added for specific uses is also very important: special carbon blacks have been developed for reinforcing SBR. With respect to carbon black reinforcement, recent studies have revealed effects arising from the termination of *n*-butyl-lithium polymerized SBR using 4,4′-bis(diethylamino)benzophenone. The resulting chain end carries two tertiary amine sites and one hydroxyl site capable of interacting with the oxygenated surface of carbon black by acid–base and H bonding mechanisms, respectively [32]. Blends of SBR with PB or natural rubber show a heterogeneous morphology at microscopic scale and are compatible at macroscopic level at all ratios. Investigations have been done on curing characteristics, physical properties and extrusion behavior [33].

3.2.7 Applications

Besides its main use for tires and related products, SBR is also present in cables, mechanical goods, footwear and shoes, foamed items, belts, belting, hose, tubing, sponge and adhesive.

Since tire production consumes the major proportion of rubber products, the use of SBR in tires is very important. For relatively light-duty service, such as passenger or off-road tires, SBR is slightly better than natural rubber; treads usually show 10–25% longer wear.

The higher temperatures caused by greater hysteresis losses cause SBR to deteriorate much more rapidly than natural rubber, and for that reason, SBR is not widely used for heavy-duty high speed tires.

3.3 STYRENE–BUTADIENE COPOLYMERS (THERMOPLASTIC ELASTOMERS)

Thermoplastic elastomers copolymers of styrene and butadiene (B) or styrene and isoprene (I), with 50% or more styrene are of considerable commercial importance; the concept of thermosplastic elastomers originated with the commercial introduction of styrene–butadiene–styrene (SBS) triblock copolymers in 1965 [34]. The properties of thermoplastic elastomers are often intermediate between plastics and elastomers. Those with about 55–60% styrene are used in latex-type paints.

Latices are produced by batch emulsion polymerization similar to the technique of SBR synthesis. To improve their adhesive characteristics, polar comonomers like acrylic acid or acrylates are added.

Most of the polystyrene–polydiene block copolymers that are thermoplastic elastomers have the basic microstructure:

...–SSSSSSSSSS–BBBBBBBB–SSSSSSS–...   (S–B–S)

...–SSSSSSSS–IIIIIIIII–SSSSSSS–...   (S–I–S)

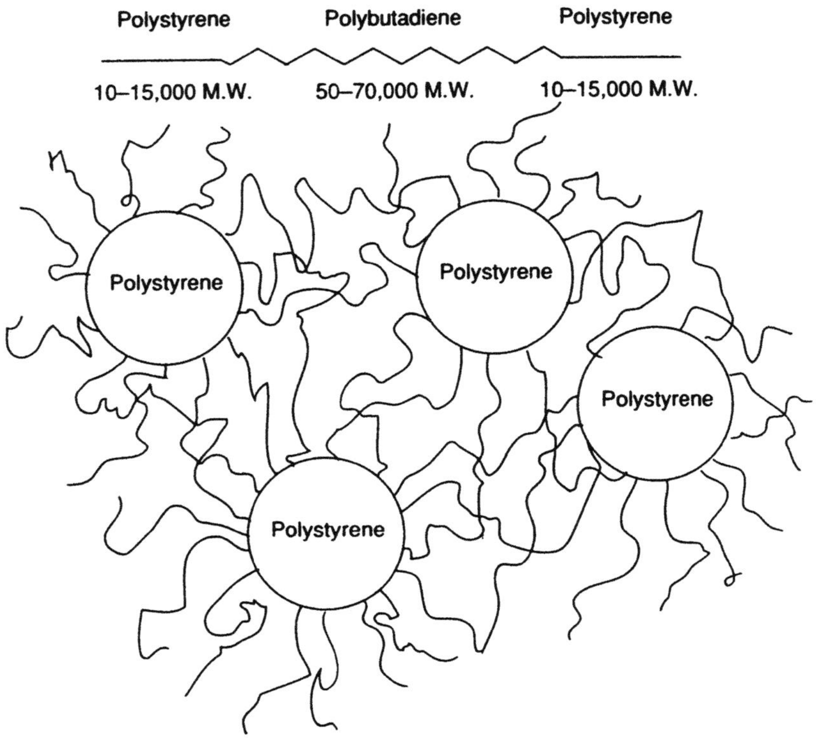

Figure 3.5 Schematic representation of S–B–S triblock copolymer.

The two separate phases (polystyrene and polydiene) retain the properties of the respective homopolymers and are characterized by their different T_g. This means that in S–B–S and S–I–S block copolymers, at room temperature, the PS phase is strong and rigid while the polydiene phase is soft and elastic.

Interesting data were obtained by V.A. Behrstein *et al.* [35], who studied B–S diblock copolymers with a varying content of the interphases (i.e. regions of segment mixing in superstructures formed by segregated PS and PB domains) by DSC, NMR and dynamic mechanical measurements.

S–B–S and S–I–S block copolymers are obtained by anionic copolymerization. The synthesis is carried out in an inert solvent (cyclohexane or toluene) and inert atmosphere without traces of oxygen or water, with organolithium (butyl-lithium) as the catalyst, although others can be utilized. Using this technology, it is possible to control the essential details of the chain, including tacticity, microstructure, MW and its distribution in each block sequence of the copolymer. Heterogeneous (Ziegler-type) catalytic copolymerization, and mechanochemistry can also be used to provide compositions which exhibit the properties of thermoplastic elastomers [36,37]. When batch copolymerization is used, the relative reactivities are such that butadiene polymerizes initially and only when it is almost consumed does styrene polymerize in blocks on the end of the copolymer chain.

S–B–S copolymers vary in properties from soft rubbery to hard semirigid products. The properties depend on the relative amounts of the comonomers and with the sizes of the blocks. More styrene leads to a more rigid and harder copolymer (Figure 3.5).

From the technical point of view, the most important of these block copolymers is their resemblance, at least at room temperature, to vulcanized natural rubber [38].

Styrene content in S–B–S block copolymers is in the range of 30–40% although for oil extended grades, this amount may reach 55%. While S–B–S block copolymers do not require vulcanization and can be quite adequate for many purposes without the need of fillers, there is some scope for compounding. The principal application for S–B–S is as an additive for asphalt, where it improves many properties, especially durability, and is used extensively both for road surfaces and for roofing [14]. Unlike asphalt, *cis*-1,4-polyisoprene is immiscible with S–B–S block copolymers [39].

3.4 NITRILE RUBBER (NBR)

3.4.1 Introduction

Nitrile rubber is a copolymer of acrylonitrile and butadiene. In 1990, the USA's production of the various nitrile elastomers was more than 56 200 tons. Exports exceeded imports slightly, with a domestic consumption of about 53 500 tons. Nitrile rubber production showed the largest decline in 1990, decreasing by 19% while others, including neoprene, butyl and polyisoprene, declined by 16% [3]. The peak year for nitrile rubber production was 1974 when the total of dry and latex forms was almost 88 000 tons. NBR consumption in the USA is expected to hold second place among non-specialty elastomers, averaging 3.7% annually [5]. In 1993 nitrile rubber represented only 3% of the total rubber consumption in the USA [12].

3.4.2 Emulsion copolymerization

Butadiene–acrylonitrile copolymerization takes place in aqueous emulsion by a radicalic mechanism and has many features similar to butadiene–styrene copolymerization. The preparation of hydrocarbon and aqueous phases does not have significant differences compared to that for styrenic rubber synthesis. For the copolymerization initiation, persulfates, organic peroxides or hydroperoxides are used. Many NBR-type copolymers are produced in different industrialized countries, the main differences between them being related to the amount of AN, plasticity and the type of antioxidant.

The ratio between AN and B may vary in a very broad range depending on the kind of NBR that is required. A whole series of NBRs are available, with AN contents varying from 18 to 48% (the preferred amount is in the range of 26 to 40%) [16].

Both hot and cold processes may be used but at present only the latter one (with temperatures in the range 5–30°C) are operated commercially. Comonomer units are randomly distributed in the copolymer chains. The butadiene content is made up of *trans*-1,4-, *cis*-1,4- and 1,2 units; as an example, for a copolymer containing 72% butadiene and obtained at 28°C the respective contributions are 77.5, 12.5 and 10%. NBR obtained by the cold process is less branched and has a lower polydispersity.

For the copolymerization at 50°C, the following reactivity ratios were found: r_1(butadiene) $= 0.35 \pm 0.08$; r_2(acrylonitrile) $= 0.0 \pm 0.04$. The possible negative value for r_2 is due to AN water solubility. The product of these reactivity ratios $r_1 r_2 \ll 0.016$. For the copolymerization at +5°C, $r_1 = 0.28$ and $r_2 = 0.02$ [40]. Anion active, cation active or non-ionic emulsifiers can be used, although the first group is preferred and the second is used only for the production of special latices. Sodium dibutyl-naphthalene-sulfonate is preferably used together with a non-ionic emulsifier. The type and amount of the initiator and the temperature are important parameters of this copolymerization process; both determine NBR properties. A lower temperature favors the synthesis of a more regular microstructure. For the same Mooney viscosity, cold NBR shows a better workability than that produced by the hot process. Additionally, it also possesses a lower thermal crosslinking tendency which is an advantageous characteristic when high temperature processing is necessary.

Both *tert*-dodecyl mercaptan and diisopropylxantogenate-disulfide are used together as chain regulators. Copolymers obtained without regulator contain branched or crosslinked macromolecules and as a result are difficult to process [16]. Hydroquinone is added as a stopper after *c*. 90% conversion, and the latex is stabilized with phenyl-β-naphthylamine.

A typical recipe is shown in Table 3.5 [21].

Table 3.5 A typical recipe for the production of NBR

	Parts by weight	
Butadiene	75	
Acrylonitrile	25	
Water	180	
Fatty acid soap (emulsifier)	4.5	
Tert-dodecyl mercaptan (modifier)	0.5	
Potassium chloride (stabilizer)	0.3	
Sodium pyrophosphate	0.1	(initiator system)
Ferric sulfate	0.02	(initiator system)
Hydrogen peroxide	0.35	(initiator system)

Very often potassium persulfate is also used as an initiator. Because so many modifications are marketed (for example, low, medium and high AN content) and the total volume is smaller than SBR, production is more often batchwise rather than continuous [41]. For the continuous polymerization a ratio of 2:1 between aqueous phase and hydrocarbon phase is recommended. The initiator is introduced periodically in the mixture of the phases.

Copolymerization requires 14 to 15 hours and takes place at 30°C during the circulation of the reaction system through a battery of reactors (usually eight). Produced latex is stripped for the recuperation of unreacted butadiene; before stripping, copolymerization is interrupted by adding 0.1% hydroquinone and at the same time an antioxidant is introduced.

Latex coagulation is done with a 25% sodium chloride solution and sulfuric acid in a plant identical with that used for SBR latices. The flow chart of NBR production is presented in Figure 3.6.

After coagulation the mixture is made alkaline, and the crumbs are filtered, washed, dried and sheeted. Crude NBR appears on the market in a number of forms such as sheets, crumbs, pellets, rolls or blocks, depending on the producer.

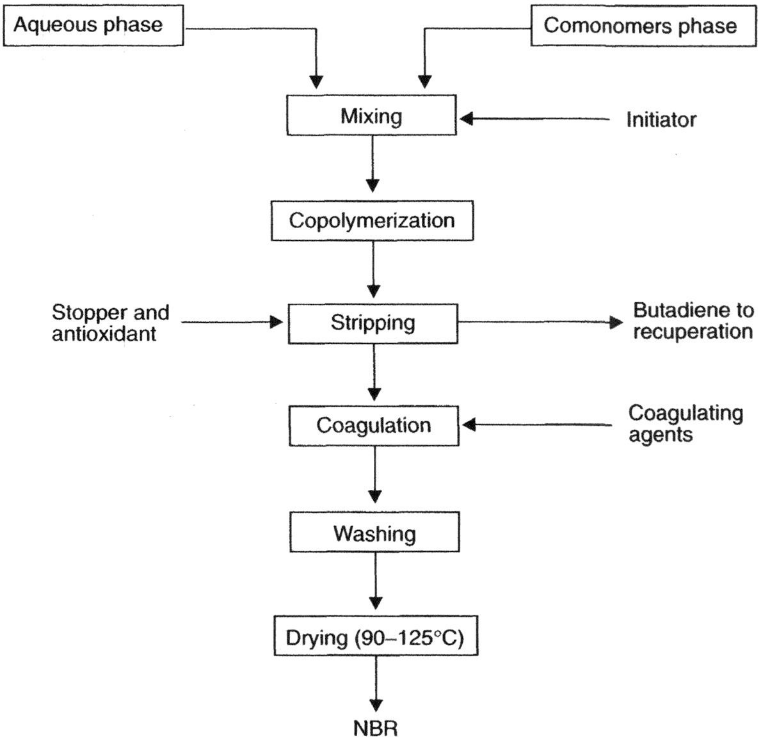

Figure 3.6 Butadiene–acrylonitrile continuous copolymerization.

3.4.3 Properties

The two comonomers are randomly distributed, with butadiene bonded especially in the 1,4 position, but 1,2 bonds are not excluded; thus the NBR microstructure can be represented as follows:

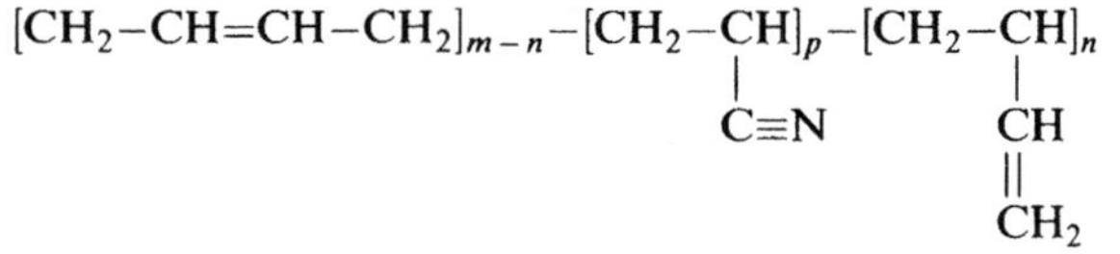

Vinylic groups can polymerize forming lateral chains.

In the case of a copolymer with 28% AN, only 10.5% of butadiene units are bonded in the 1,2 position.

The NBR technological properties depend on the following microstructural features: ratio between the amounts of units 1,4 or 1,2 bonded, branching and crosslinking degree, unsaturation degree, comonomers distribution in the copolymer chain, MW and polydispersity.

Regarding the dependence of T_g on AN amount, the literature provides different data. Some researchers [41] found that T_g varies between −25 and −55°C if the amount of AN in the copolymer varies from 20 to 40%. Other researchers [42] have found that T_g is −56°C for the copolymer 20:80% AN to B and −41°C for the copolymer with 30:70% AN to B. The specific gravity is 1.11.

The ingredients used for NBR vulcanization are generally the same as those used for natural rubber. Sulfur and accelerators are used and in special cases peroxides [11]. The amount of sulfur required is lower and the amount of accelerator necessary is higher than for natural rubber. A higher sulfur content increases the hardness and in very high proportions may affect the thermal resistance, tear resistance and elongation at failure. Copolymers with higher AN content need less sulfur for vulcanization [16]. NBR chemical stability is good when exposed to acids, fair for bases, variable for solvents, but excellent for oils and fats.

Nakajima *et al.* [43] have established that for a higher heat and chemical resistance, a high AN content is preferred. However, if a superior mechanical performance at low temperature is required, a low AN content must be chosen, even though the above characteristic resistances are somewhat sacrificed. Excessive amounts of gel and long chain branching cause relative difficulty in processing, although no significant milling problem occurred in this study. The higher MW gives a better product tensile strength. Both elongation at failure and the De Mattia Flex Test are adversely affected by the presence of gel and long branching. Therefore the higher MW with the lower gel content and the lower degree of branching is the preferred choice. This conclusion is affected by the processing conditions, compound recipes, product design and any other variables. A patent was obtained for nitrile rubber compositions resistant to heat and oils in the presence of metals [44].

NBR is compatible with PVC in all ratios and their polyblend combines well the chemical stability of PVC with the oil and fat resistance, and elasticity of NBR. At a high PVC amount, NBR acts as a plasticizer kept in PVC without any exudation. In NBR vulcanizates, PVC reinforces the rubber, and improves tension, tear, abrasion, solvent resistance and aging in different conditions. Commercial NBR:PVC polyblends contain 30–40 parts by weight PVC to 100 parts by weight rubber. Also used are the polyblends with styrene–acrylonitrile copolymer or with ABS.

Polyblends NBR–phenoplaste have a wide range of properties and are being used for a variety of molded products ranging from golf club heads to gears and housings. They can be very hard and yet possess high impact strength, resilience, and good chemical and water resistance.

3.4.4 Applications

Due to its properties, NBR is extremely useful in the manufacturing of items which come in contact with oil and fat such as carburetor diaphragms, fuel tanks, hoses and gaskets, gasoline hoses for service station pumps, large-size hoses for handling oil, and for cables that must resist oils and abrasion. It is used for manufacturing clothing resistant to oil, molded articles, especially mountings for machinery and shoe soles for service where oils are present. Printing rolls, blankets and plates are additional applications.

3.5 POLYISOPRENE (PI)

3.5.1 Introduction

The common form of this elastomer is *cis*-1,4-polyisoprene which occurs in the latex of many trees and plants, as natural rubber. *Trans*-1,4-polyisoprene can be isolated from some plants such as gutta percha or balata. Both of these forms may also be synthesized by the use of a stereospecific catalyst in anionic polymerization.

3.5.2 Natural rubber

Natural rubber still has a considerable technical and commercial importance. The worldwide production amounted to 5.5 million tons in 1992, representing 34% of the total rubber production (15 million tons), synthetics included [45]. The global consumption of natural rubber will reach 6 million tons in 1995 [46]. The rate of increase of natural rubber production is slightly higher than that of synthetic rubber. There are around 20 natural rubber growing countries. The following three account for almost 75%: Thailand,

Table 3.6 Composition of fresh latex and dry rubber

	Latex %	*Dry rubber %*
Rubber hydrocarbon	36	93.7
Protein	1.4	2.2
Carbohydrates	1.6	0.4

1.4 million tons in 1992; Indonesia, 1.38 million tons; and Malaysia, 1.21 million tons in the same year [45,47].

The earliest contributions to the knowledge of this natural polymer belong to Faraday, Williams, Tilden and Bouchardat [48].

Despite the competition of highly developed synthetic polymers, natural rubber has retained a leading place among commodity and engineering elastomers and production increases year by year.

Commercial raw natural rubber has a small, but highly important, number of non-elastomer components. These may comprise as much as 5–8% of the total composition and include proteins, sugars and fatty acids. Some of these act as natural antioxidants or curing activators. Trace elements, usually in amounts of 2–3 ppm, are important as catalysts to promote the oxidation of vulcanized natural rubber.

Rubber cultivation consists of the production of a stereoregular hydrocarbon polymer as a crop, by biosynthesis, using solar energy, CO_2, water and the elaborate metabolic processes of plant tissue [5].

Natural rubber latex is a dispersion of rubber particles (0.1–1.6 μm) in water. Its preservation is discussed in detail by D.C. Blackley [49]. The latex collected from trees is treated (for instance with ammonia, *c.* 0.05%) to prevent premature coagulation. In the factory it is either concentrated by partially removing the water or coagulated and processed into solid dry rubber. The main components of a typical mixture of fresh latex and dry rubber are illustrated in Table 3.6 [50]. The other components are lipids (neutral, glycolipids and phospholipids) and minerals.

Unvulcanized raw elastomer obtained by coagulation and by drying the latex has a very high MW (20×10^4–50×10^4 and more) and before compounding with the common additives, the rubber is masticated. Mastication, which consists of mechanically working the natural elastomer, reduces the chain length by a factor of about 10 and converts the raw gum to a plastic mass.

X-ray diffraction of raw elastomer in an unstretched state reveals a pattern of two concentric broad bands, indicating an amorphous state at room temperature. However, by stretching to at least 700% elongation, well defined interference spots (crystallites) appear.

Modified natural rubber products have attained some commercial applications as follows.

1. Depolymerized rubber is a thermally (140°C) degraded natural rubber with a very low MW used for potting compounds, binders for abrasive wheels and casting molds.
2. Cyclized rubber; thermoprene type, obtained with acidic catalysts, is used as adhesive for rubber–metal bonds, coatings for protecting metallic surfaces. Amphoteric halide type is based on a complex with heavy metal chlorides; it is used as a molding material or as a vehicle for coating compounds.
3. Chlorinated rubber; the fully chlorinated type contains 65% chlorine, has a specific gravity of 1.63–1.66 and is soluble in chlorinated and aromatic hydrocarbon solvents. It is used as an adhesive for metal to rubber bonding and for chemically resistant paints.
4. Rubber hydrochloride is the product that results after treatment with hydrochloric acid. It contains chlorine (28–30%) and has a specific gravity of 1.533. It is used for packaging due to its low permeability to water vapor and good transparency [48].
5. Of the recent chemical modifications of natural rubber, epoxidized natural rubber is the most attractive from both a technical and economic point of view [51].

3.5.3 Synthetic PI

The only PI which has commercial value is that with a high *cis*-1,4 content. Of the many catalysts that are capable of polymerizing isoprene in order to obtain PI only four have attained industrial importance, as follows [52]:

1. a coordination catalyst containing trialkylaluminum and titanium tetrachloride (Al-Ti) (Ziegler–Natta catalyst);
2. lithium and lithium alkyl;
3. a coordination catalyst of alane (aluminum-hydride derivative) and $TiCl_4$;
4. a coordination catalyst of aluminum alkyl and a vanadium salt for making *trans*-1,4-PI.

There appears to be no appreciable difference in either the processing or product performance of natural rubber and synthetic *cis*-polyisoprene except in those cases where the proteins, fatty acids, ash, etc. present in natural rubber affect its performance. In order to make the synthetic products exactly similar in performance, it is sometimes necessary to add materials of this type. Based on the above-mentioned procedures, the current large scale production of *cis*-1,4-polyisoprene is running at about 20% that of natural rubber (Table 3.7).

(a) Polymerization with Ziegler–Natta catalysts

Solution polymerization permits a good reaction control and the formation of a homogeneous polymer. The catalyst is prepared in aliphatic or aromatic

Table 3.7 Estimated 1986 world production of elastomers [6]. (Reproduced with permission from K.J. Saunders, *Organic Polymer Chemistry*, 2nd edn; published by Chapman & Hall, 1988)

Elastomer	*Production (million tons)*
Natural rubber	4.43
Styrene–butadiene	5.02
Others	2.87
Total	13.84

hydrocarbon solvents. Isoprene purity is an important factor because isoprene dimers influence polymer microstructure, while the presence of isobutene or cyclopentadiene affects the polymerization rate. The last one at 0.1% of the monomer amount determines the complete poisoning of the catalyst. Acetylenic or diene hydrocarbons, oxygenated compounds or sulfur components increase the induction period and decrease the reaction rate.

As polymerization medium, hydrocarbons such as butane, octane, cyclohexane, isopentane, isooctane, etc. are used. Polyisoprene, being insoluble in butane, octane, and cyclohexane gradually covers the surface of the catalyst, a complex able to enhance its activity. Polar solvents, especially H_4-furan can drastically alter the PI microstructure from the 70–90% *cis*-1,4, obtained in hydrocarbon media to mainly 3,4 (and some 1,2). These drastic changes generally occur even when very low concentrations of such polar cosolvents as H_4-furan are added to the hydrocarbon media [53]. In isohydrocarbons, polyisoprene MW decreases when the conversions increase; e.g. in isooctane, while the conversion passes from 20 to 50%, the average MW decreases from 50×10^6 to 28×10^4; in isopentane, the decrease is less significant. The important dissolving capacity of polyisoprene, pentane and isopentane favors a high isoprene conversion. Compared with other high boiling point hydrocarbons, it presents advantages for polymer solution degassing and during solvent recuperation [16].

The activity of the catalyst appears to be affected by the method of its preparation, in particular the catalyst preparation temperature. Although this does not seem to influence microstructure, some reported results appear contradictory [54].

At commercial scale, polymerization of isoprene with Ziegler–Natta catalysts is typically carried out in an aliphatic solvent such as *n*-butane, *n*-pentane or isopentane. The process is continuous and takes place in a battery of reactors connected in series. The flow chart of the process is presented in Figure 3.7. By circulating brine at −15°C through the reactor mantle, the temperature is maintained at 25°C or higher. As the reaction medium passes from one reactor to another, conversion increases and implicitly the

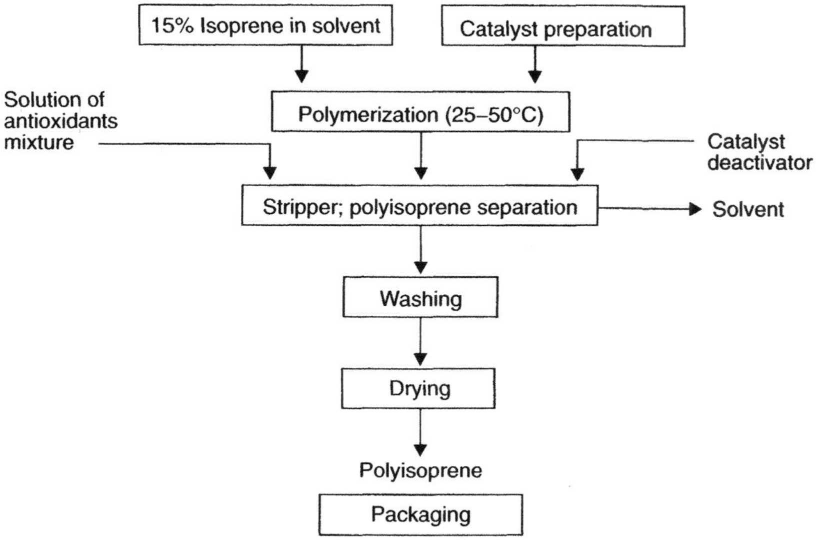

Figure 3.7 Isoprene polymerization.

medium viscosity. The process has to be controlled in order to obtain at the end a polyisoprene solution with 15–16% polymer. This is because at higher than 17% concentration the medium viscosity increases suddenly. Currently at a 93–94% conversion, the average polyisoprene concentration is 13%.

To avoid secondary reactions such as oxidation or isomerization, which may affect the polymer, the catalyst is deactivated with 20% methanol solution in toluene or isopentane. In the meantime, polyisoprene is treated with a mixture of antioxidants such as phenyl-β-naphthylamine and diphenyl-*p*-phenylene-diamine in a methanol–toluene mixture. To avoid *cis–trans* isomerization after catalyst decomposition, the polyisoprene solution is washed with water. By steaming polyisoprene solution, polymer separation takes place in the meantime with the release of the solvent and the unreacted monomer.

Drying the polymer through extrusion brings some advantages such as lower steam and electricity consumption, higher productivity and reduction of necessary space.

(b) Polymerization with catalysts based on lithium

As regards the stereospecificity of the propagation step, the most thoroughly investigated systems have been those involving lithium and organolithium catalysts. This is because the latter are the most versatile type, being soluble in a variety of solvents, and can be controlled to limit the number of side reactions (termination or transfer) [53].

Isoprene polymerization with lithium or alkyl-lithium catalyst takes place at temperatures in the range of 25–60°C in saturated hydrocarbons with a normal structure such as *n*-heptane. Polymerization starts after introducing the catalyst into the isoprene solution in *n*-heptane; excess heat is removed by circulating brine or cold water through the reactor walls as the reaction is exothermal. The produced polymer remains in solution, a fact which leads to the increase of the viscosity as conversion increases. Besides the catalyst, temperature plays an important role; at lower temperature, higher MWs are obtained, but the reaction rate is lower and the catalyst consumption is higher. Polymerization continues until a solid content of 25% is reached. Normally at 20% concentration the polyisoprene solution has a viscosity of 250 000 cP and even above this value. The other operations are identical with those of the procedure based on Ziegler–Natta catalysts.

(c) Polymerization with hydrides

In this technology, substituted aluminum hydrides are used in view of the reduction of $TiCl_4$ to $TiCl_3$. An efficient catalytic system contains aluminum aminate or alcoxylate halogenated hydride and $TiCl_4$. Maximum catalyst activity and the highest amount of 1,4-*cis* (minimum 96%) is obtained at a ratio of active H/Ti of *c.* 1.5 [55]. Polyisoprene produced in this way has 96% *cis* units. From many points of view such as abrasion resistance, heat development and service life, it is better than natural and other synthetic polyisoprenes.

3.5.4 Properties

The properties of synthetic *cis*-1,4-polyisoprene are very similar to those of the natural type and the two materials are essentially interchangeable. Some analytical data are presented in Table 3.8.

MW is readily controlled during isoprene polymerization and consequently synthetic polyisoprene is considerably more uniform in this regard than natural rubber. Polyisoprenes are polydisperse; the breadth of the distribution of the MWs and the nature of individual fractions have significant effects on processing and on the properties of the end-use articles. Like natural polyisoprene, the Ziegler–Natta type contains an appreciable quantity of gel, i.e. crosslinked polymer which is insoluble in common rubber solvents. The gel content of this PI is up to 20% and it breaks down with mastication, similarly to that of natural one.

Stereoregularity is the most important property of synthetic polyisoprene. By stretching, it crystallizes causing self-reinforcement; this change does not happen in the case of butadiene based rubbers. The self-reinforcement of natural and synthetic rubbers results in high tensile and tear strengths.

Table 3.8 Analytical data of different polyisoprenes [28]. (Reprinted with permission of the publishers, Butterworth-Heinemann Ltd from *Rubber Technology and Manufacture* edited by C.M. Blow and C. Hepburn, 1982)

	Synthetic polyisoprene		
Property	*Ziegler–Natta*	*Lithium*	*Natural rubber*
1,4 *cis* units (% wt)	96	92	98–100
Limiting viscosity number (dl/g)	2.5–4.5	8–11	6–7
Gel content (% wt)	10–20	0	High level depending on age
Geometric shape	Branched	Linear	Branched
Stabilizer content (pphr)	1	0.5	2–3
Ash content (% wt)	0.15–0.30	0.05	Approx. 0.5
Total metal content (ppm)	400–3000	70	Approx. 1000
Mooney viscosity [ML(1+4) 100°C]	60–90	–	Approx. 120
Color	White	White	Dark

The unvulcanized PI flows more readily; this feature is advantageous in injection molding.

Polyisoprenes have a low hysteresis, giving low heat build-up during flexing. Typical properties of polyisoprenes compared with some butadiene based rubbers are presented in Table 3.9.

Practical results show that unfilled compounds based on Ziegler–Natta polyisoprene and natural rubber reach higher tensile strength. From this point of view *cis*-1,4-polyisoprene obtained through Ziegler–Natta catalysis behaves like natural polyisoprene. But carbon black filled polyisoprene vulcanizates have better elongation, oxidation stability and cracking resistance than PB and styrene based rubbers.

Unlike natural polyisoprene, the synthetic one does not contain non-rubber ingredients which act as vulcanization activators; a slightly higher level of added activator is therefore required [6].

Table 3.9 Some vulcanizate properties [28]. (Reprinted with permission of the publishers, Butterworth-Heinemann Ltd from *Rubber Technology and Manufacture* edited by C.M. Blow and C. Hepburn, 1982)

	Type of polymer				
Property	*Ziegler–Natta PI*	*Lithium PI*	*Natural PI*	*SBR*	*PB*
Tensile strength (MPa)	30	25	31	25	17
Tear strength (N/mm)	70	60	110	45	35
Heat build-up (°C)	23	21	23	40	33

Due to the characteristics of the raw polymers, for synthetic polyisoprenes, calendering and extrusion are usually carried out at a temperature some 15°C lower when compared with natural rubber. The good flow properties result in extrusion at lower pressure and in calendering to a more uniform thickness; also higher speeds can be used [28].

The miscibility of PI with PB depends on the microstructure of these two elastomers. PI–PB block copolymers can be used as compatibilizers in natural rubber–PB blends [56,57].

3.5.5 Applications

Currently, the tire market accounts for 60% of the USA's polyisoprene consumption. The largest share is used in the carcass and tread of truck, off-the-road, and aircraft tires, where the low hysteresis properties of polyisoprene are an advantage. Polyisoprene is employed at a lesser percentage in passenger tire carcasses. Non-tire applications include extruded gaskets, sheeting, footwear and sponge rubber. An important application for the polyisoprenes with low gel content is their use as base materials for chemical modification to chlorinated and cyclized elastomers for the coatings. It is also suited for the medical field, nipples for babies' bottles, milk tubing, hospital sheeting and pharmaceutical sundries.

Oil extended PI is used in many of the same areas as the non-extended types where maximum physical properties are not required [58].

3.6 POLYCHLOROPRENE (NEOPRENE)

3.6.1 Introduction

The monomer, chloroprene (2-chlorobutadiene), can be obtained from acetylene and hydrochloric acid via monovinylacetylene or through butadiene chlorination and dehydrochlorination of 3,4-dichloro-1-butene. Neoprene is produced by the emulsion polymerization of chloroprene [4]. The sales of this rubber have decreased from 88 000 tons to 70 000 tons in the USA and Canada within 10 years [59].

3.6.2 Emulsion polymerization

Polychloroprene is manufactured commercially on a large scale by continuous emulsion polymerization. This procedure has made possible the production of newer types of polychloroprene as well as copolymers of chloroprene with other monomers. There are today about 15–20 different types of chloroprene homo- and copolymer latices available. Reactivity ratios of a number of comonomers are given in Table 3.10 [60].

A typical polymerization formulation is shown in Table 3.11.

Table 3.10 Reactivity ratios of a number of comonomers

Comonomer M_2	r_1	r_2
AN	6.93	0.03
MAA	2.68	0.15
VAc	50.00	0.10
MA	11.10	0.08

Table 3.11 A typical polymerization formulation for polychloroprene

	Parts by weight
Chloroprene	100
Water	150–180
Rosin	4
Sodium hydroxide (stabilizer)	0.8
Sodium salt of naphthalene sulfonic acid–formaldehyde condensation product (emulsifier)	0.7
Sulfur (modifier leading to polysulfide linkages)	0.6
Potassium persulfate (initiator)	0.2–1.0

Polymerization is carried out at 50°C and is allowed to continue until a 90% conversion is reached. The latex is cooled to 20°C, and tetraethylthiuram disulfide is added for controlling the plasticity and also acts as a stabilizer in the finished product.

Acetic acid is introduced into latex to a point just short of coagulation and then coagulated by freezing on the surface of a large rotating drum. The film is stripped from the freeze roll, washed, passed through squeeze rolls and dried. After drying, the film is gathered into rope form, cut in short lengths, and packed (Figure 3.8). This technique is used because polychloroprene is tacky and it is difficult to coagulate the latex directly to crumb [16].

The obtained elastomer contains about 85% *trans*-1,4 units, an amount which might be increased by lowering the polymerization temperature. To obtain a sulfurless rubber, an alkyl-mercaptan is used as an MW regulator. At conversions above about 70%, unmodified polychloroprene is quite highly crosslinked and is difficult to process. In order to obtain polychloroprene of limited MW which is readily processable, a sulfur-tetraethylthiuram disulfide modifier is used. A sulfur–chloroprene copolymer with the following microstructure is obtained through this method:

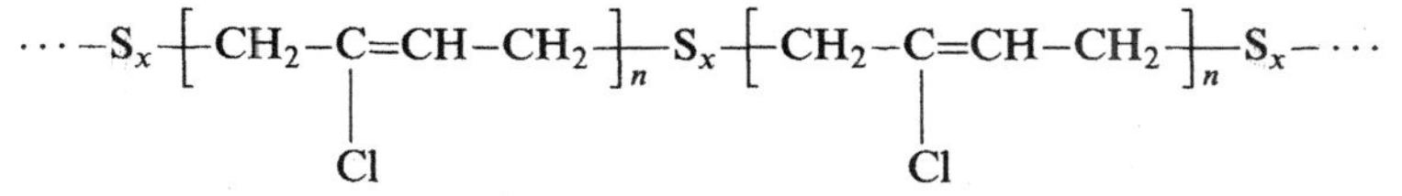

where x is 2–6 and n is, on average, 80–100x.

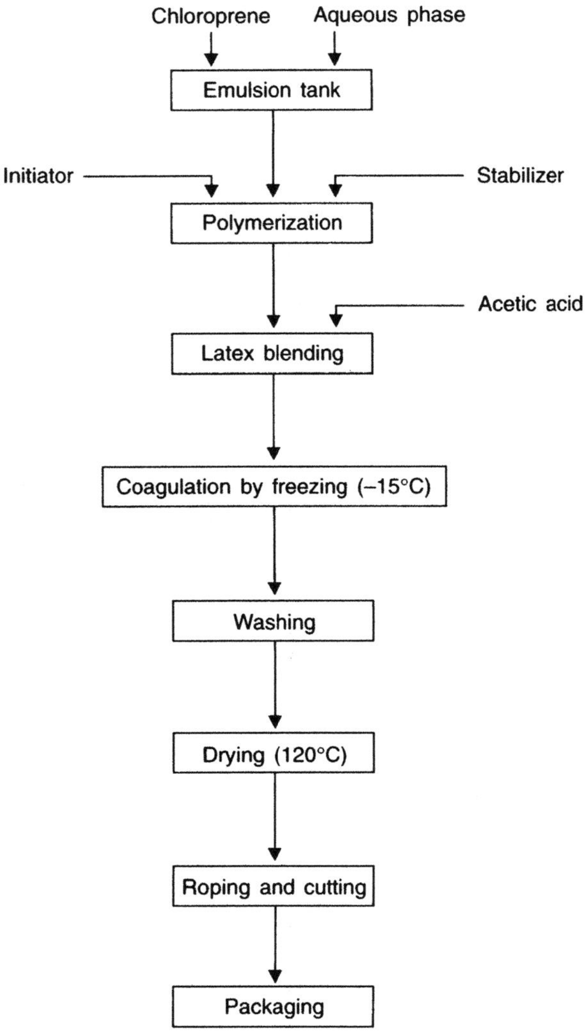

Figure 3.8 Continuous emulsion polymerization of chloroprene.

When the polymer is treated with the disulfide, interchange reactions occur at the polysulfide links to give stable thiuram terminated polymers of reduced MW [6]. Many of the comments concerning the manufacture of dipped goods using nitrile latex apply to polychloroprene, but according to K. Sellars and G.R. Brown [61], it is worth noting the following points.

1. The small particle size of the polychloroprene and the inherent polymer property of low permeability makes its latex more difficult to dry and vulcanize.

2. For satisfactory vulcanization, the latex film (for latex dipped goods) has first to be dried and then cured at a temperature of at least 140°C.
3. It is preferable to use an alcoholic coagulant to ensure satisfactory coagulant penetration.
4. Although such latices are non-carboxylated, it is sound to include 5–10 phr of zinc oxide in the formulation. This will maximize physical property development and rate of cure as it acts as a hydrochloric acid acceptor.
5. Storage of the latex will result in a progressive lowering of its pH, which has to be kept in the 10.5–11.0 region by adjusting with dilute KOH solution.
6. The finished product is generally of lower strength than natural rubber but has a similar modulus characteristic and excellent oil, fat, chemical, light and ozone resistance.

3.6.3 Properties

The microstructure of polychloroprenes obtained through radical polymerization consists of *trans*-1,4-2Cl-2-butylidene with a minor contribution of the units *cis*-1,4, 1,2 and 3,4 [20]. The amount of these different units depends on the polymerization conditions; i.e. the contribution of *cis*-1,4 and 3,4 units increases with the increase of polymerization temperature from 5% when the process takes place at −40°C to approximately 30% at *c*. 100°C.

A detailed NMR study has permitted the quantitative identification of isomers resulting from the monomer's head to tail, head to head, or tail to tail addition (Table 3.12).

V.A. Kargin *et al.* [62] studied the polychloroprene crystallization tendency. Polychloroprene obtained at −40°C has a crystallinity degree of 38% (and $T_m = +73$°C) and that synthesized at +40°C has a crystallinity degree of only 12% (and $T_m = +45$°C).

Most neoprene compositions are made by selecting the base elastomer, adding a variety of ingredients, and then processing and vulcanizing. The choice of additives affects the physical properties of the finished product. As the ingredients are in high number a wide variety of compounds of different composition and properties is available [63].

Chlorine presence increases in some cases the stability towards chemicals, especially towards oxygen and ozone. No vulcanization agent is necessary, although in commercial practice, compounding ingredients are used that seem to act in a similar manner. Metallic oxides such as magnesium oxide, zinc oxide and lead oxide (litharge) are generally used for this purpose. Sulfur acts as an accelerator. Organic accelerators have little, if any, effect. Carbon black has little effect on polychloroprene vulcanizates except to increase their stiffness. The resistance of reinforced vulcanizates to swelling in organic solvents and mineral oils is far superior to natural rubber and

Table 3.12 Polychloroprene microstructures [2]. (Reprinted by permission of C. Hanser Verlag, from *Introduction to Industrial Polymers* by H. Ulrich, 1982)

Type of polymer	*Structure*
1,2-Isotactic 1,2-Syndiotactic	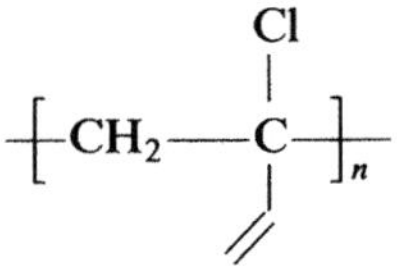
3,4-Isotactic 3,4-Syndiotactic	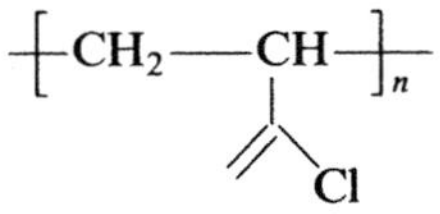
1,4-*Trans*, head to tail	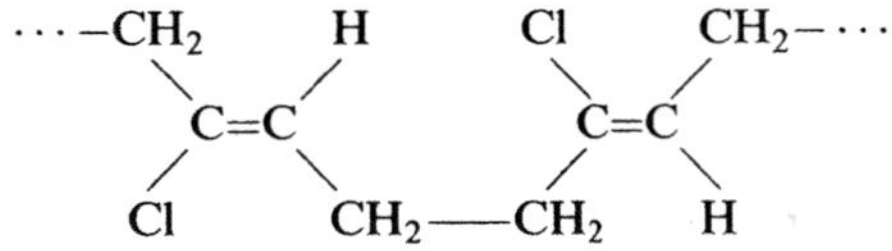
Head to head combination, *trans* form	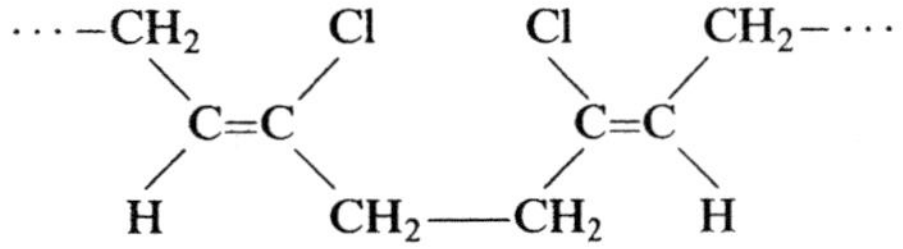
Tail to tail combination, *trans* form	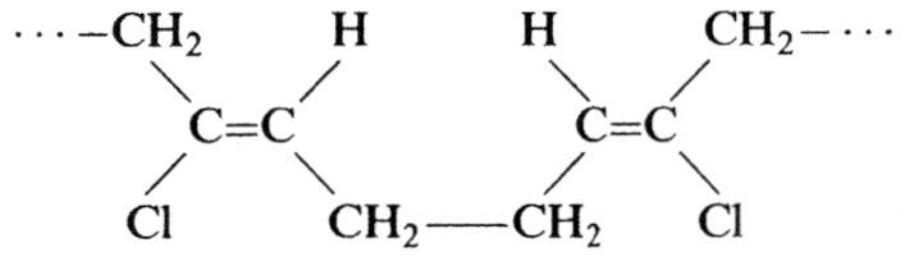
1,4-*Cis*, head to tail	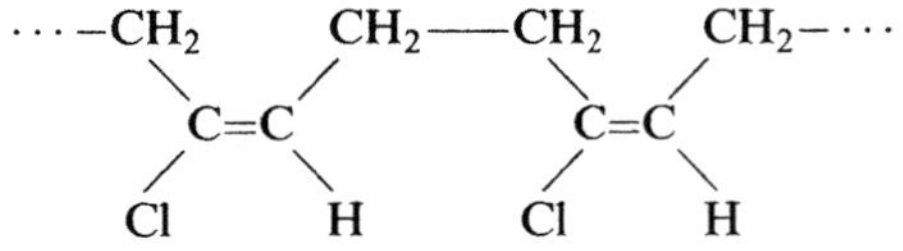
Head to head combination, *cis* form	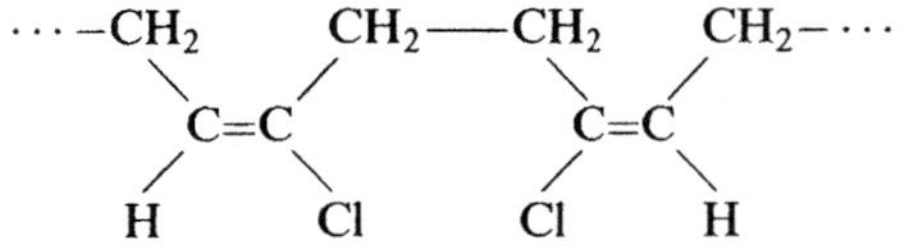
Tail to tail combination, *cis* form	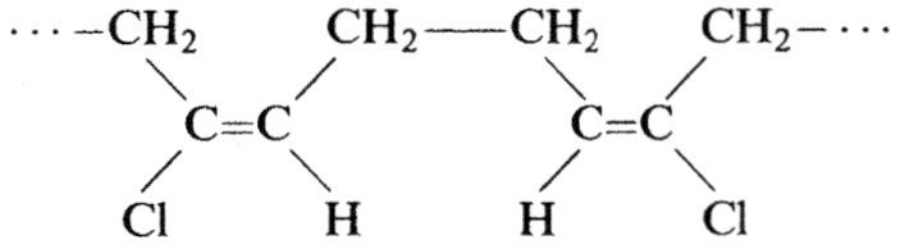

SBR, but somewhat less than that of nitrile and polysulfide rubbers. Polychloroprenes are also greatly superior to natural rubber in resistance to sunlight, abrasion, gas diffusion, heat and flame. However, long-term storage under the influence of heat and/or light leads to the release of hydrochloric acid.

The drawbacks which are retained include a relatively high density (1.25 g/cm^3); low stability at low temperature; crystallization tendency of uncured and vulcanizates at ambient temperatures; slightly lower tensile strength, lower tear resistance and slightly poorer electrical insulating properties than natural rubber; and a sudden decrease of mechanical properties during heating. Incorporation of certain resinous plasticizers and the use of sulfur in the vulcanizing system also serve to control crystallization [64].

Polychloroprene processing has to be done thoroughly to avoid pre-vulcanization.

3.6.4 Applications

The marked resistance of polychloroprene to severe conditions such as high temperatures, ozone, oils, organic solvents and chemicals is the main property that makes it valuable for many applications.

Several chemicals like alkalis (sodium hydroxide) have been found to accelerate the plasticization by milling. Material that has been softened by milling under water, or with chemical agents, can be kept for months at ambient temperature without greatly changing plasticity and other characteristics.

Flexible tubes and hoses for handling petroleum products comprise a large outlet for this elastomer. Large hose such as that used on tankers, hoses for hydraulic systems with oil, adhesives, gaskets, belts, conveyors, and foam rubber items used in contact with oils, waxes and greases, as well as most aliphatic hydrocarbons, are made of polychloroprene. Storage tanks for petroleum products and chemicals are lined with this polymer. It is also used because of its good chemical resistance to cover rolls in industries like paper, textile, etc. Polychloroprene vulcanizates can be used for lining equipment and vessels for the chemical industry, due to the fact that they resist the action of mineral acids, especially hydrochloric acid and other aggressive media for a long time.

Due to the high chlorine content (*c.* 40%), polychloroprene is non-combustible and recommended for the production of cables. Polychloroprene, as a consequence of its chlorine content (40%) and its tendency to form substantial amounts of crosslinked char, inherently burns less readily than hydrocarbon rubbers. This superiority in flame resistance has made this elastomer the material of choice in many applications demanding reduced flammability [65].

An important amount of this polymer is used as a sealant. There are two types of polychloroprene sealants: one-component and two-component. The

Table 3.13 Polychloroprene two-component sealants [66]. (Reprinted with permission of Van Nostrand Reinhold Publ. Co. from *Sealants* by A. Damusis, 1967)

Designation of formulation	*1* (parts by weight)	*2*
First component		
Polychloroprene–neoprene FB	100	100
Antioxidant – 'Neozone A'	2	2
Thermal black	100	100
Hard clay	30	30
Dioctyl sebacate – plasticizer	50	50
Zinc oxide	10	10
Second component		
'Accelerator 833', a butyraldehyde–butylamine condensation product	7–10	–
Tetraethylenepentamine	–	5–8
Pot life (h)	6–4	25–1
Physical properties		
Tensile strength	6.5	3.9
Elongation (%)	400	320
Shore A hardness	38	44

one-component sealants cure very slowly to a Shore A hardness of 35–40. Their shelf life can be prolonged by storage at low temperatures. The cured sealants become tough, have good water immersion properties, have a movement capability in the order of ±12.5% and have good chemical stability. Details of a typical two-component sealant are presented in Table 3.13. This is a solvent-free, non-shrinking, slump resistant sealant. It sets fast but cures slowly.

Table 3.14 Advantages and disadvantages of polychloroprene sealants [67]. (Reprinted by permission of J. Wiley & Sons Inc. from *Construction Sealants and Adhesives* by J.R. Panek and J.P. Cook, 1984, Copyright © J. Wiley & Sons)

Advantages	*Disadvantages*
1. Compatible with bitumens	1. High shrinkage, up to 40% by volume
2. Compatible with asphalts	2. Only dark color
3. Compatible with neoprene gaskets	3. Very slow cure
4. Low cost	4. Not recommended for dynamic joint movement
5. Good water resistance	5. No specifications
6. One component	6. Stains stone
7. May have ±12.5% movement capability	7. Stains wood
8. Good adhesion to bitumens	
9. Good adhesion to asphalts	
10. Good adhesion to metals	

Polychloroprene sealants are made with a solvent content of approximately 40%, which results in high shrinkage, and a limited adhesion to some surfaces. Their main desirable property is that they are one of the few sealants compatible with asphaltic concrete and bitumen [67].

As can be seen from Table 3.13, the formulation of polychloroprene sealant contains the macromolecular compound, as well as the plasticizer, the filler and metallic oxides which regulate the cure rate and serve as acceptors for the trace amounts of hydrogen chloride released during processing, curing or aging, and also as antioxidants and accelerators.

The advantages and disadvantages of polychloroprene sealants are listed in Table 3.14.

REFERENCES

1. Hanmer, R.S. and Railsback, H.E. (1974) in *Rubber Technology*, 2nd edn (ed. M. Morton), Van Nostrand Reinhold Publ., New York, p. 199.
2. Ulrich, H. (1982) *Introduction to Industrial Polymers*, C. Hanser Publ., Munich.
3. Reisch, M. (1991) *Chem. Eng. News*, **69**(14), 19.
4. Seymour, R.B. and Carraher Jr, C.E. (1992) *Polymer Chemistry*, Marcel Dekker, New York.
5. Greek, B.F. (1991) *Chem. Eng. News*, **69**(19).
6. Saunders, J.K. (1988) *Organic Polymer Chemistry*, 2nd edn, Chapman & Hall, London.
7. Hargis, J.G., Livigni, R.A. and Aggorwall, S.L. (1982) in *Elastomers and Rubber Elasticity* (eds J.E. Mark and J. Lal), ACS Symposium Series 193, American Chemical Society, p. 73.
8. Morton, M. (1983) *Anionic Polymerization: Principles and Practice*, Academic Press, New York.
9. Elias, J.G. and Vohwinkel, F. (1986) *New Commercial Polymers*, Gordon and Breach Science Publ., New York.
10. Motoyuki, K., Suzuki, M. and Akigawa, K. (1994) Jpn Kokai Tokkyo Koho JP 0625,312 (94 25,312), (Cl. C08 F2/18), 01 Feb. 1994, JP Appl. 92/165,284, 15 May 1992; *CA*, **120**, 324452b.
11. Hertz Jr, D.L. (1984) *Elastomerics*, **1**, Nov.
12. Reisch, M.S. (1994) *Chem. Eng. News*, **72**(35), 11.
13. Blackley, D.C. (1975) *Emulsion Polymerization: Theory and Practice*, J. Wiley & Sons, New York, p. 419.
14. Brydson, J.A. (1988) *Rubbery Materials and Their Compounds*, Elsevier Applied Science, London, p. 102.
15. Nappea, D.H. and Gilbert, R.G. (1990) in *Scientific Methods of the Study of Polymer Colloids and Their Applications* (eds F. Candau and R.H. Ottewill), Kluwer Academic Publ., Dordrecht, The Netherlands, p. 159.

16. Feldman, D. (1974) *Macromolecular Compounds Technology*, Ed. Technica, Bucharest.
17. Alger, M.S.M. (1987) *Polymer Science Dictionary*, Elsevier Applied Science, London, p. 258.
18. Odian, G. (1981) *Principles of Polymerization*, 2nd edn, J. Wiley & Sons, New York, p. 337.
19. Bovey, F.A., Kolthoff, J.M., Medalia, A.J. and Meehan, E.J. (1955) *Emulsion Polymerization*, Interscience Publ., New York, p. 55.
20. Kennedy, J.P. and Tornquist, E.G.M. (1968) *Polymer Chemistry of Synthetic Elastomers*, Part I, Interscience Publ., New York.
21. Mark, H., Bikales, N.M., Overberger, C.G. *et al.* (1985) *Encyclopaedia of Polymer Science and Engineering*, 2nd edn, Vol 2, J. Wiley & Sons, New York.
22. Henderson, J.N. (1987) in *Rubber Technology*, 3rd edn (ed M. Morton), Van Nostrand Reinhold, New York.
23. Thompson, M.W. (1985) in *Polymer Colloids* (eds R. Buscall, T. Corner and J.F. Stageman), Elsevier Applied Science, London, p. 1.
24. Buscall, R. and Ottewill, R.H. (1985) in *Polymer Colloids* (eds R. Buscall, T. Corner and J.F. Stageman), Elsevier Applied Science, London, p. 141.
25. Blackley, D.C. (1985) in *Polymer Colloids* (eds R. Buscall, T. Corner and J.F. Stageman), Elsevier Applied Science, London, p. 247.
26. Brydson, J.A. (1978) *Rubber Chemistry*, Applied Science Publ., London.
27. Brodhead, T.O., Hamielec, A.E. and MacGregor, J.F. (1985) in *Emulsion Copolymerization* (eds J. Guillot and C. Pichot), Hüthig and Wepf Verlag, Basel, p. 105.
28. Van der Bie, G.J., Rellage, J.M. and Vervolet, C. (1982) in *Rubber Technology and Manufacture*, 2nd edn (eds C.M. Blow and C. Hepburn), Butterworths, London, p. 93.
29. Saltman, W.M. (1974) in *Rubber Technology*, 2nd edn (ed. M. Morton), Van Nostrand Reinhold, New York.
30. Krol, L.H. (1968) *Rubber J.*, **150**(12), 20.
31. Feast, A.A.J. (1982) in *Polymer Latices and their Applications* (ed. K.O. Calvert), Applied Science Publ., London.
32. Edwards, D.C. (1990) *J. Material Sci.*, **25**, 4175.
33. Gharavi, F. and Katbab, A.A. (1990) *Prog. Rubber Plast. Technology*, **6**(2), 129.
34. Dyson, R.W. (1990) *Engineering Polymers*, Blackie, Glasgow.
35. Berhstein, V.A., Vileser, A.D., Vinogradova, L.V. *et al.* (1990) *Vysokomol. Soedin. Ser. A.*, **31**(12), 2324.
36. Morton, M., Lee, N.C. and Terrill, E.R. (1982) in *Elastomers and Rubber Elasticity* (eds J.E. Mark and J. Lal), ACS Symposium Series 193, American Chemical Society, Washington, p. 101.
37. Cohen, R.E. (1982) in *Elastomers and Rubber Elasticity* (eds J.E. Mark and J. Lal), ACS Symposium Series 193, American Chemical Society, Washington, p. 483.
38. Holden, G. and Legge, N.R. (1987) in *Thermoplastic Elastomers* (eds N.R. Legge, G. Holden and H.E. Schroeder), Hanser Publishers, Munich, p. 48.

39. Lee, N.S. and White, J.L. (1990) *Kautschuk Gummikunststoffe*, **43**(10), 876.
40. Hofman, W. (1968) in *Polymer Chemistry of Synthetic Elastomers*, Part I (eds J.P. Kennedy and G.M. Tornquist), Interscience Publ., New York, pp. 127–185.
41. Rodriguez, F. (1982) *Principles of Polymer Systems*, 2nd edn, McGraw-Hill Book Co., New York, p. 397.
42. Daniels, C.A. (1989) *Polymers: Structure and Properties*, Technomic Publ. Co., Lancaster, p. 67.
43. Nakajima, N., Harell, E.R., Kumler, P.R. *et al.* (1984) *Advances in Polymer Technology*, **5**(4), 267.
44. Ozawa, O. (28 June 1989) Ger. Offen. DE 4,020,797, Cl. C08L9/02, 10 Jan. 1991, JP Appl. 89/166,169; *CA*, **114**, 230476m.
45. Stagraczynski, R. (1993) *Kautschuk Gummikunststoffe*, **46**(10), 795.
46. Hall, C. (1989) *Polymer Materials*, 2nd edn, J. Wiley & Sons, New York.
47. Lauhus, W.P. (1993) *Kunststoffe German Plastics*, **83**(10), 63.
48. Semegen, S.T. (1973) in *Rubber Technology* (ed. M. Morton), Van Nostrand Reinhold Co., New York.
49. Blackley, D.C. (1985) in *Polymer Colloids* (eds R. Buscall, T. Corner and J.F. Stageman), Elsevier Applied Science, London, p. 247.
50. Subramanian, A. (1987) in *Rubber Technology*, 3rd edn (ed. M. Morton), Van Nostrand Reinhold Co., New York.
51. Gelling, J.R. (1986) in *Renewable-Resources Materials. The Polymer Sources* (eds C.E. Carracher Jr and L.H. Sperling), Plenum Press, New York, p. 261.
52. Kuzma, L.J. (1987) in *Rubber Technology*, 3rd edn (ed. M. Morton), Van Nostrand Reinhold, New York, p. 235.
53. Morton, M. (1983) *Anionic Polymerization Principles and Practice*, Academic Press, New York.
54. Brydson, J.A. (1978) *Rubber Chemistry*, Applied Science Publ., London.
55. Bruzzone, M. (1968) *Hydrocarbon Processing*, **44**(11), 179.
56. Massie, J.M., Halasa, A.F., Thudium, R.N. and Burkhart, C.W. (1992), *50th Annu. Tech. Conf. Soc. Plast. Eng.*, **1**, 1043.
57. Zanzig, D.J., Magnus, F.L., Hsu, W.L. *et al.* (1993) *Rubber Chem. Technol.*, **66**(4), 538.
58. Himes, G.R. (1973) in *Rubber Technology*, 2nd edn (ed. M. Morton), Van Nostrand Reinhold Co., New York, p. 274.
59. Anon. (1994), *Chem. Eng. News*, **72**(27), 30.
60. Feast, A.A.J. (1982) in *Polymer Latices and their Applications* (ed. K.O. Calvert), Applied Science Publ., London.
61. Sellars, K. and Brown, G.R. (1982) in *Polymer Latices and their Applications* (ed. K.O. Calvert), Applied Science Publ., London.
62. Kargin, V.A., Sogolova, T.I. and Saphchnikova, T.K. (1964) *Visokomol. Sold.*, **6**, 1022.
63. Graff, R.S. (1987) in *Rubber Technology* (ed. M. Morton), Van Nostrand Reinhold Co., New York, p. 339.

64. Fletcher, W.P. (1982) in *Rubber Technology and Manufacture*, 2nd edn (eds C.M. Blow and C. Hepburn), Butterworths, London, p. 123.
65. Lawson, D.F. (1989) in *Handbook of Polymer Science and Technology*, Vol. 2 (ed. N.P. Cheremisinoff), Marcel Dekker, New York, p. 203.
66. Damusis, A. (ed.) (1967) *Sealants*, Van Nostrand Reinhold Co., New York.
67. Panek, J.R. and Cook, J.P. (1984) *Construction Sealants and Adhesives*, 2nd edn, J. Wiley & Sons, New York.

PART TWO

Step Growth Polymers

Polyesters 4

The group of polyester polymers is one of the most versatile. They contain widely different materials with a large spectrum of characteristics and applications which are produced by a variety of manufacturing techniques. They have in common the ester group –CO–O– in the main polymer chain, which is a polar group and brings inter- and intramolecular chain interactions that are reflected in some properties such as the mobility of the chains and melting characteristics. From the chemical point of view the ester group imparts increased sensibility to hydrolysis that can cause chain breakage.

Due to their large diversity, polyesters will be classified throughout this chapter in two main categories, thermoplastic and thermosetting, and will be presented accordingly.

A. THERMOPLASTIC POLYESTERS

4.1 POLY(ETHYLENE TEREPHTHALATE) (PET)

4.1.1 Introduction

PET is a linear thermoplastic polyester which has widespread commercial use as the major synthetic fiber as well as a film and molding material. The overall volume of PET produced in the USA at the beginning of 1990 was 2.5 million tons of which 1.5 millions tons was fibers, staples and tows [1].

The repeating unit of PET is the ester of terephthalic acid and ethylene glycol:

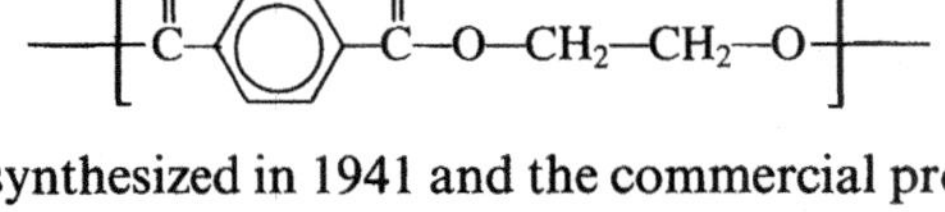

PET was first synthesized in 1941 and the commercial production of textile fibers started in 1953. At the beginning, the polymer was considered unsuitable for molding as it crystallizes from the melt which makes it brittle. It was not until the late 1960s that the polymer was produced in the amorphous state that provided toughness and transparency and enabled PET to be extensively used for blow molded bottles for carbonated soft drinks.

Today PET is produced as homopolymers or as copolymers. The copolymers have lower T_g and T_m and are amorphous and more flexible materials. They are obtained by using either a mixture of acids (terephthalic and isophthalic) or a mixture of glycols (ethylene glycol and 1,4-cyclohexanedimethanol).

4.1.2 Synthesis

At the beginning, PET was prepared exclusively by the transesterification of dimethyl terephthalate with ethylene glycol and not by the direct acid–glycol esterification. The reason for that was the insufficient purity of the acid available at that time. Today the terephthalic acid can be prepared with fiber grade purity and both processes, i.e. direct esterification and transesterification, are used commercially for the preparation of the polymer.

Terephthalic acid is a white solid that sublimes at 300°C and is quite insoluble in most common solvents. It is produced by the oxidation of *p*-xylene which is obtained largely from petroleum. Dimethyl terephthalate is also obtained from *p*-xylene by a stepwise process of oxidation and esterification. It melts at 142°C and is more easily purified than terephthalic acid.

From either of the starting materials, the preparation of the polymer is carried out in two steps. In the first step, the so called 'monomer', i.e. bis(2-hydroxyethyl) terephthalate is produced along with some oligomers:

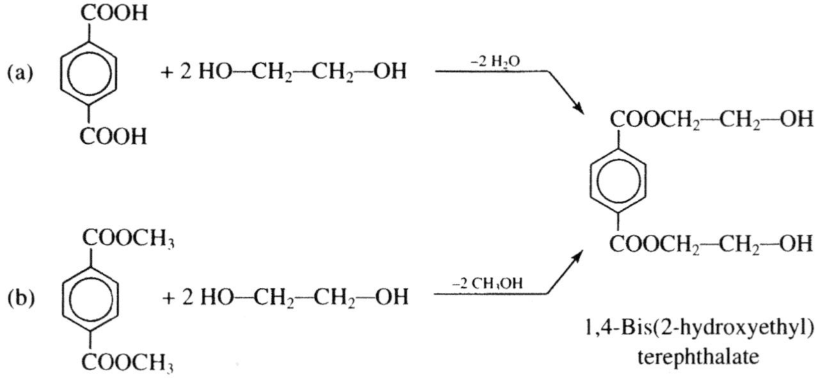

Conditioned terephthalic acid and ethylene glycol in the molar ratio 1 : 1.5 are introduced into the reactor and heated up to 235°C. Water from the reaction is eliminated continuously. When dimethyl terephthalate and ethylene glycol are used as starting materials (1 : 2.2 molar ratio), the transesterification is conducted in the presence of catalysts such as antimony trioxide, cobaltous or manganese acetate. Methanol formed as by-product is removed by the means of a distillation column [2–4].

The second stage of the process is the same regardless of the starting materials. The bis(2-hydroxyethyl) terephthalate formed in the first stage is condensed in the interchange reaction that occurs in the presence of antimony trioxide catalyst. The pressure is reduced down to 0.13 kPa (1 mm of Hg) in order to eliminate the ethylene glycol liberated in the reaction and to increase the molecular weight:

(c)

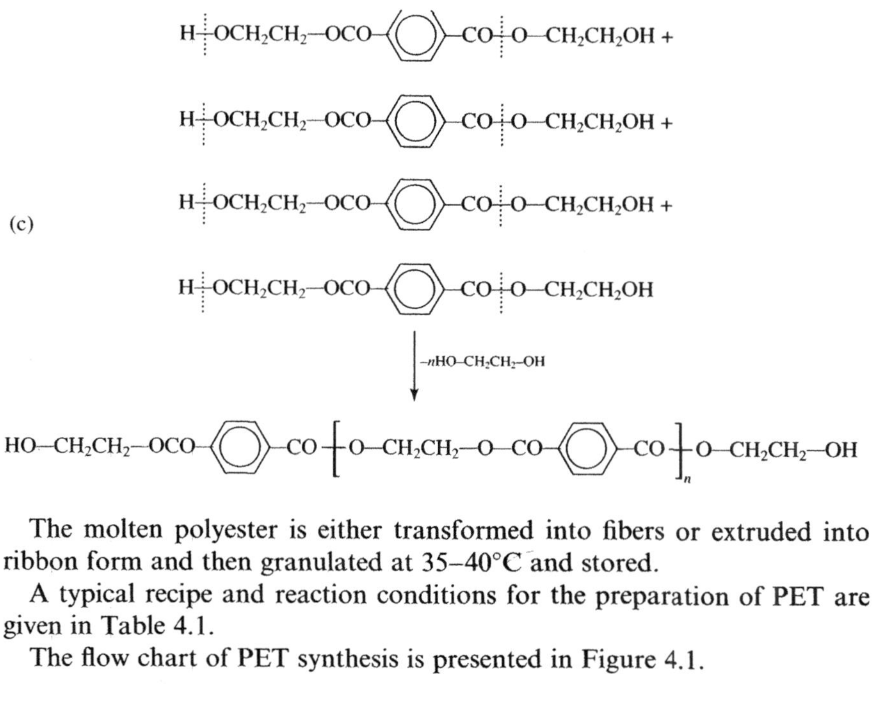

The molten polyester is either transformed into fibers or extruded into ribbon form and then granulated at 35–40°C and stored.

A typical recipe and reaction conditions for the preparation of PET are given in Table 4.1.

The flow chart of PET synthesis is presented in Figure 4.1.

Table 4.1 Typical recipe and reaction conditions for the preparation of PET

		Parts by weight	
Terephthalic acid	100		–
Dimethyl terephthalate	–		115
Ethylene glycol	40		40
First step conditions			
Temperature	250°C		140–220°C
Pressure	0.4 MPa		
Second step conditions			
Temperature		270–290°C	
Pressure		0.13 kPa	

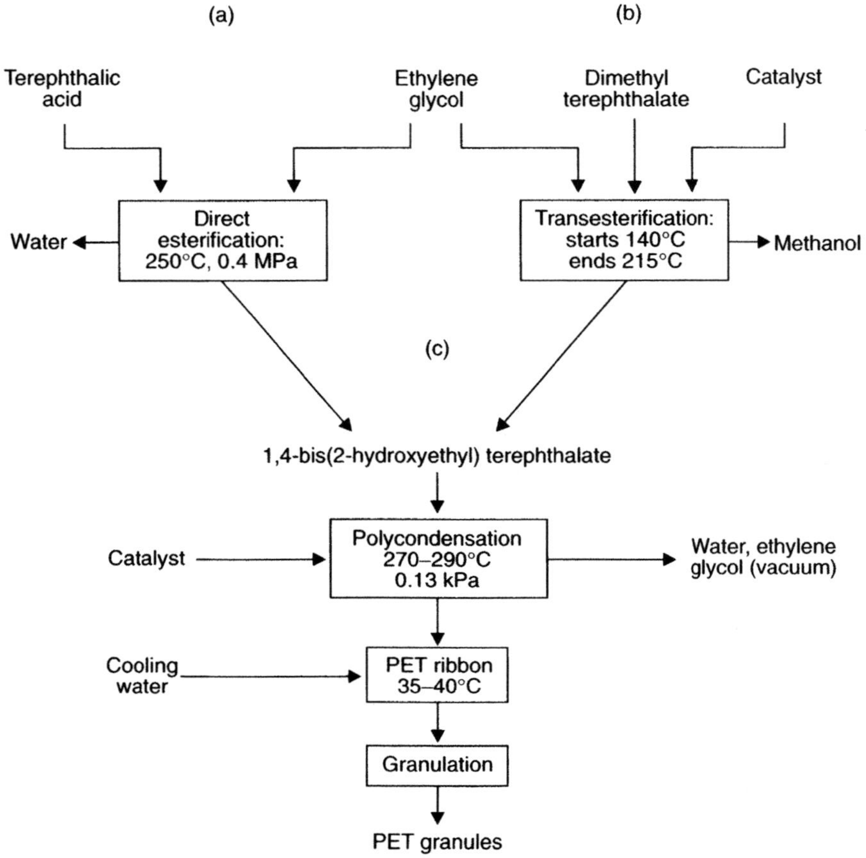

Figure 4.1 Preparation of poly(ethylene terephthalate).

Discontinuous or continuous polycondensation processes can be used for the preparation of PET.

A process based on ethylene oxide instead of ethylene glycol has also been commercialized [5]. Ethylene oxide reacts rapidly with terephthalic acid to form the same intermediate as above, namely bis(2-hydroxyethyl) terephthalate, which is subsequently condensed to PET:

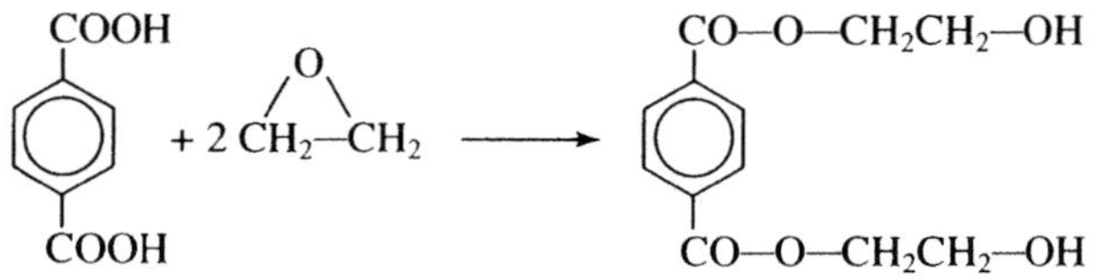

The desired MW depends on the end use of the material: it is 20 000 for textile fiber and film applications, and 30 000 for bottles and tire cord.

4.1.3 Properties

PET has the ability to exist in either an amorphous or a crystalline state. The degree of crystallinity can range from 0 to 60% and is determined by the history of the material [6]. The crystalline melting temperature T_m is 265°C and the glass transition temperature T_g is rather high, being reported about 80°C. When the amorphous polymer is maintained at a temperature between T_g and T_m it crystallizes spontaneously. The highest rate of crystallization occurs at 170–180°C.

These features are very important for the processing conditions of the material. The fiber and film are produced by extrusion and quenching at ambient temperature. Because of their thin sections, cooling occurs rapidly and the material remains amorphous and of low resistance. In order to impart good strength, the fiber or film is submitted to a process of hot drawing at a temperature above T_g (80–100°C) when orientation and crystallization are induced. The film is biaxially stretched approximately threefold in each direction when the crystallinity is raised to 25%. The tendency to shrink is then reduced by annealing the stretched fiber and the film by heating at 200°C under tension. This treatment increases the crystallinity to 40% and reduces the tendency to shrink on heating. The tensile strength of PET films is 117–173 MPa.

Molded parts can be obtained with amorphous or crystalline structure depending on the temperature of the mold: when the mold is maintained at temperatures below T_g (at around 50°C) and the parts are thin enough to permit a rapid cooling, the material remains amorphous and the moldings are transparent. These parts should not be used above T_g in order to avoid crystallization which leads to shrinkage and clouding. For crystalline and opaque moldings, the polymer is previously compounded with ionomer nucleating agents (substances which induce crystallization) and with plasticizers that decrease T_g. The temperature of the mold is maintained at 130°C to permit the material to rearrange in a crystalline form. Crystalline moldings may be used at temperatures above T_g as they are shape stable up to temperatures near T_m.

A remarkable success on the market was realized with the two liter bottle manufactured for soft drinks where the biaxial stretching from film technology was applied. In the first step, an amorphous parison is produced by injection into a cold mold. Then the parison is heated above the T_g and is biaxially stretch blown in the bottle mold. The bottles are thin walled and possess high tensile and flexural strength and toughness together with a low permeability to carbon dioxide and oxygen. They allow packaging of carbonated beverages under pressures up to approximately 0.4 MPa. Bottles for beer which need an even more reduced permeability to gases, have been produced using a multilayer parison from PET and copolymers of vinylidene chloride [7].

Although the polymer is polar, electrical insulating properties at room temperature even at high frequencies are good because the orientation of the dipoles is restricted below T_g. PET has good resistance to water and dilute mineral acids but is sensitive to bases and is degraded by concentrated nitric and sulfuric acids. The resistance to photochemical degradation is very good.

PET can be reinforced with glass fibers up to 55% for injection molded composite products with heat distortion temperatures as high as 220°C. Self-reinforcing PET has been produced by the addition of *p*-hydroxybenzoic acid which forms liquid crystals in the composite [8].

A number of copolymers has been prepared in order to lower the T_g value and the tendency to crystallization. They are made by partial replacement of terephthalic acid with isophthalic acid or of ethylene glycol with 1,4-cyclohexane dimethanol.

4.1.4 Applications

PET has found a wide range of applications because of its excellent balance of properties and to the possibility of controlling the degree of crystallinity and the level of orientation.

PET is used as fibers for clothing; as film for electrical insulation, packaging, recording tape and a number of drawing office applications; as stretch blown moldings for bottles and food containers and as engineering plastics for precision molded parts.

In the last years, PET has acquired an increasing use as refillable grade products and as recycled material [9]. Scraps from bottles and films are transesterified with glycol (e.g. diethylene glycol) at temperatures above 200°C. Patent literature describes the transesterification/glycolysis of PET [10] and that of the by-products from dimethyl terephthalate distillation bottoms [11], while the chemistry of the transesterification is described in Reference [12]. The lower MW polyesters obtained from these waste materials have been used as polyols in the preparation of polyurethane rigid foams bringing a combination of reduced flammability and advantageous cost [13].

4.1.5 Terephthalate copolymers with 1,4-cyclohexane dimethanol

1,4-Cyclohexane dimethanol is obtained from dimethyl terephthalate by a two step hydrogenation process in the presence of selective catalysts.

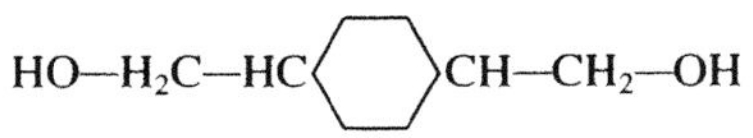

Several copolyesters have been prepared starting with this glycol [7].

1. The polyester with the terephthalic acid has rather stiff chains, $T_g = 130°C$ and $T_m = 290°C$, higher than those of PET. It has good weathering and water resistance and is used for film and fibers.
2. The copolyester with a mixture of terephthalic and isophthalic acids is amorphous, has a high clarity and good mechanical and processing characteristics.
3. A similar copolyester is prepared from terephthalic acid and a mixture of 1,4-cyclohexane dimethanol and ethylene glycol.

The copolyesters have less regular structures, are amorphous, have high clarity, good melt strength and are well processed by blow and injection molding into film and bottles for packaging. Glass reinforced grades are also available.

4.2 POLY(BUTYLENE TEREPHTHALATE) (PBT)

PBT has a similar structure with PET; the ethylene glycol moiety is replaced by butylene glycol:

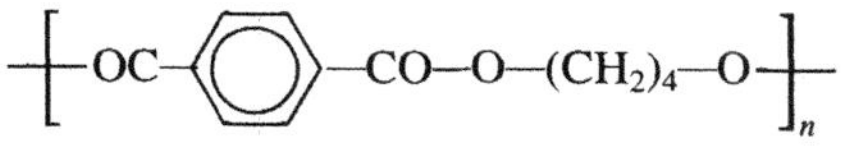

PBT is a partially crystalline engineering thermoplastic produced at an annual rate of 25 000 tons at the beginning of the 1990s [8]. The polymer is produced by the polycondensation of 1,4-butanediol with terephthalic acid or with dimethyl terephthalate in the presence of catalysts by a process that is similar to that described for the preparation of PET [14]. The longer hydrocarbon chain of the butylene glycol in the repeating unit renders the polymer molecules more flexible and less polar than those of PET. As a consequence, both T_m and T_g are lower when compared to PET [7]. The polymer may thus be processed at lower temperatures and the material can easily crystallize when cooling in the mold. About 95% of the polymer is processed by injection molding where the low melt viscosity allows easy filling of thin section and complicated cavities [15]. Prior to processing, PBT should be dried as it is sensitive to water like all polyester based materials.

The moisture absorption of the polymer is lower than in the case of PET owing to its less polar nature, resulting in excellent dimensional stability and good electrical insulating properties. PBT is resistant to detergents, weak acids and bases, aliphatic hydrocarbons, oils, motor oil and gasoline.

The material is available unmodified, blended with other polymers, reinforced and flame retarded. Reinforcing is done with 10–40% glass fibers or beads, talc and mica [14]. Some physical and mechanical properties of unmodified and reinforced PBT are given in Table 4.2.

Table 4.2 Some properties of PBT

	Unfilled PBT	*40% Glass filled*
T_m (°C)	224	–
T_g (°C)	22–43	–
Tensile strength (MPa)	55	147
Flexural modulus (MPa)	2280	10 340
Notched Izod impact strength (J/m)	53	187
Heat deflection temperature at 0.5 MPa (°C)	154	217
Water absorption (24 h at 23°C) (%)	0.08	0.08

As can be seen, mechanical and thermal properties are much improved for the reinforced grades.

Various composite PBT based materials with improved properties, reduced crystallization and increased impact resistance are manufactured by blending PBT with polybutadiene rubber, with silicone–polycarbonate block copolymers, with polyamides and, more recently, with acrylic ester–styrene–acrylonitrile copolymers (PBT–ASA blends) [15].

Major markets are electrical and electronic engineering as well as automotive applications (75% of the total consumption), industrial components and different consumers items [15].

4.3 THERMOPLASTIC POLYESTER ELASTOMERS (TPEs)

TPEs are block copolymers prepared from terephthalic acid and two glycols, a short chain 1,4-butanediol and a long chain poly(tetramethylene oxide)glycol (PTMG) (MW = 600–3000). They are composed of a hard thermoplastic phase or 'hard blocks', made up of butylene terephthalate repeating units and a soft elastomeric phase or 'soft blocks' from poly(tetramethylene ether) terephthalate units [16,17]. The hard blocks alternate with the soft blocks along the chain:

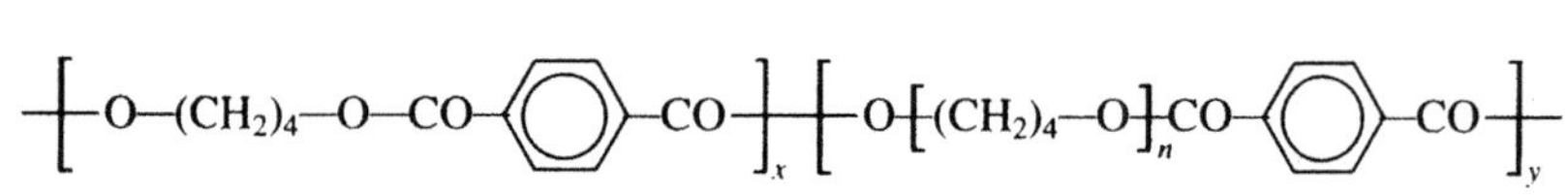

Crystalline hard block   Soft polyether-ester block

TPEs are prepared by a similar process to that used for the synthesis of PET, i.e. the two step transesterification of dimethyl terephthalate with a mixture of the two glycols in the presence of tetrabutyl titanate as catalyst [18]. The stoichiometry favors the formation of long sequences of tetramethylene terephthalate units (hard blocks). The values of M_n of the TPEs are in the range 25 000–30 000 [7].

From the one million tons that represent the total worldwide production volume of different thermoplastic elastomers in the mid 1990s, polyesters share about 6% of the market [19,20].

The polymers contain spherulitic structures composed of radial lamellae formed by the crystallized hard blocks and inter-radial amorphous regions that are mixtures of soft blocks and non-crystalline hard blocks [21]. A three dimensional network is formed resembling conventionally crosslinked elastomers at ambient temperatures. The spherulitic structures reinforce the more mobile soft blocks. At the T_m of the crystalline regions the polymer becomes viscous and can be processed and shaped by conventional thermoplastic methods. On cooling, the polymer recovers its elastomeric properties. The crystalline hard blocks contribute to the strength of the polymer and the amorphous soft segments account for the flexible elastomeric nature [22,23].

Physical properties can be varied by changing the ratio of hard and soft blocks. Commercial TPEs have a content of hard blocks in the range 33–76%. The T_g of the soft amorphous phase is about −50°C whereas the T_m of the crystalline hard phase is about 200°C (depending on the hard block length). This combination of a low glass transition temperature and high melting point makes the polymers suitable for applications in a wider temperature interval (−40 to 130°C) than many general purpose rubbers.

These polar rubbers have good oil and petrol resistance, superior tensile and tear strength, high resilience, good flex fatigue resistance and abrasion resistance. Increasing the polyether content results in more elastic polymers of lower modulus.

Some properties of TPE with various contents of hard blocks are presented in Table 4.3. The properties can be further modified by incorporation of various fillers such as carbon black, clays and fiber glass which increase the modulus.

The polymers can be processed by injection, blow, compression, transfer or rotational molding as well as by fluidized bed sintering and electrostatic

Table 4.3 Some properties of thermoplastic polyester elastomers

	Hard block content (%)		
Property	*33*	*58*	*76*
Specific gravity	1.15	1.20	1.22
Melting temperature (°C)	176	202	212
Glass transition temperature (°C)	−78	−50	−2
Shore hardness	92A	55D	63D
Tensile strength (MPa)	39.3	44.1	47.5
10% Modulus (MPa)	3.58	10.0	16.9
Flexural modulus (MPa)	44.8	206	496
Elongation at break (%)	810	760	510

spraying. They have a relatively low melt viscosity, good melt stability and rapidly harden from the molten state.

Despite their higher cost when compared to most other rubbers, TPEs were first commercialized in 1972 and have found uses as engineering rubbers in different applications such as hydraulic hoses, seals, belting, electrical cable connectors, rotationally molded low pressure tires and a variety of automotive, machinery and construction equipment parts. TPEs are not suitable for use in pneumatic tires for motor vehicles.

TPEs are toxicologically safer than thermoset rubbers in uses involving direct contact with living tissues. They have found applications in the medical field as synthetic materials for vascular grafts and reconstruction of damaged organs [24].

TPEs are completely recyclable at the end of their service lifetime without significant loss of properties [20]. This is simply done by melting and subsequently reshaping.

4.4 POLYCARBONATES (PCs)

4.4.1 Introduction

Polycarbonates (PCs) are high performance thermoplastic engineering polymers. They are polyesters in which the repeating unit corresponds to an ester of carbonic acid and a dihydroxy compound:

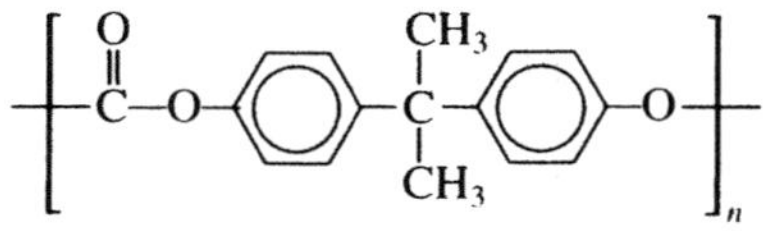

They are prepared by polycondensation reactions of dihydroxy compounds, more frequently 2,2-bis-(4-hydroxyphenyl)propane known as bis-phenol A and carbonic acid derivatives (carbonic acid itself, HO–CO–OH, does not exist in the free state). Polycarbonate polymers may be tailor made to suit specific requirements [25,26]. Besides the general purpose polycarbonate, other newer materials include copolymers and block copolymers of polyester–polycarbonates and silicone–polycarbonates as well as blends of polycarbonates with other thermoplastics [27–29].

In 1992 the worldwide production of polycarbonates was 660 000 tons [29] and they become the second major engineering thermoplastic after polyamides.

By laboratory experiments, the first polycarbonates were obtained as early as 1898; full commercial production started in 1959 in Germany at Farbenfabriken Bayer and in 1960 at General Electric in the USA, both companies holding today about 75% of the market.

Despite the rather high cost, polycarbonates imposed themselves on the market by their outstanding impact strength, and excellent optical and thermal properties. They have found various applications in automotive, electric, electronic, building and many other industries.

4.4.2 Preparation

The two components of a polycarbonate are a dihydroxy compound and a derivative of carbonic acid. Many hydroxy compounds have been tested for PC concerning transparency, toughness, melt flow, hydrolytic stability and production cost. The only one that has received commercial importance from the beginning is 2,2-bis-(4-hydroxyphenyl)propane or bis-phenol A (BPA). BPA is produced by the condensation of phenol with acetone under acidic conditions:

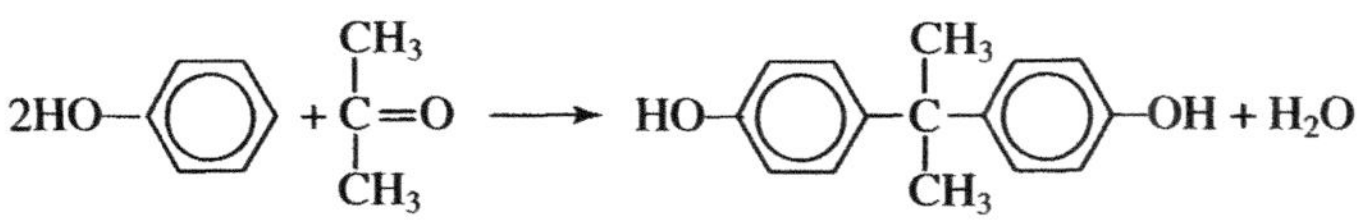

The reaction conditions, the ratio of reactants and the catalysts used are all of great importance for the degree of purity of BPA, as *o,p*- and *o,o*-isomers can be formed as by-products [7]. Polycarbonate grade BPA should have the melting point values in the range 154–157°C. Standard PC is based on BPA.

Phosgene ($COCl_2$) is used as a derivative of carbonic acid. An alternative source of carbonate group is diphenyl carbonate prepared by the condensation of phenol with phosgene in an aqueous NaOH solution. A new technology was introduced in 1993 for the manufacture of diphenyl carbonate avoiding the use of phosgene [29]. The process is realized in two steps: dimethyl carbonate is first prepared from methanol, carbon monoxide and oxygen; in the second step dimethyl carbonate is transesterified with phenol to diphenyl carbonate:

$$4CH_3OH + 2CO + O_2 \longrightarrow 2CH_3{-}O{-}CO{-}O{-}CH_3 + 2H_2O$$

An alternative synthesis route to PC by the copolymerization of epoxides with carbon dioxide is highly attractive due to low monomer cost and was studied intensively [30,31]:

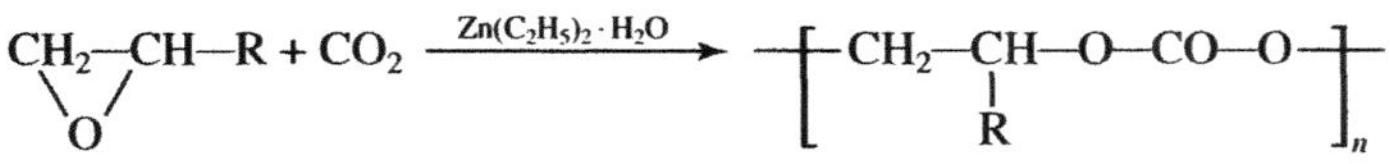

However, the only polycarbonate of commercial importance at the moment is based on BPA.

Two processes are available for the production of PC, (a) ester exchange and (b) phosgenation.

(a) Ester exchange process

This process is based on the catalyzed transesterification of diphenyl carbonate with BPA in the melt. Hydroxides or hydrides of alkaline metals and zinc or antimony oxides are used as catalysts. A molar excess of twice the amount or even more of diphenyl carbonate is used in the first stage of the reaction in order to use up the BPA that is not very stable in the reaction conditions. The reaction mixture is heated at 180–200°C under 2.6–3.9 kPa vacuum until 90% of the liberated phenol has been removed. In the second stage the temperature is raised up to 290–300°C and the pressure is dropped to 0.13 kPa or less in order to promote the polycondensation to PC and the elimination of low volatile diphenyl carbonate:

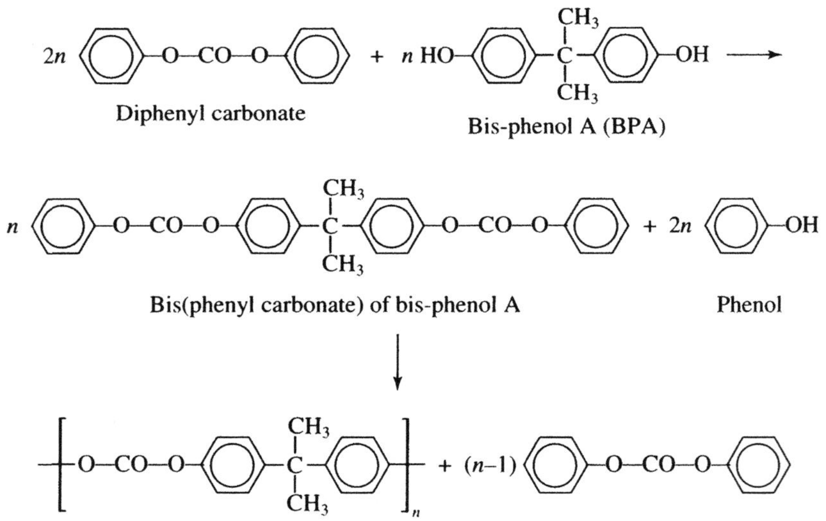

The flow chart of the ester exchange process for preparing PC is presented in Figure 4.2.

The molten resin has a very high viscosity that presents difficulties for mixing and heat transfer, thus limiting the possibility of obtaining MW higher than 50 000. The high temperatures and vacuum required in the process together with the very high melt viscosity impose difficult technical conditions and the process is of a long duration.

(b) Phosgenation process

The reaction of BPA and phosgene proceeds with the elimination of hydrochloric acid so that an acid acceptor has to be used:

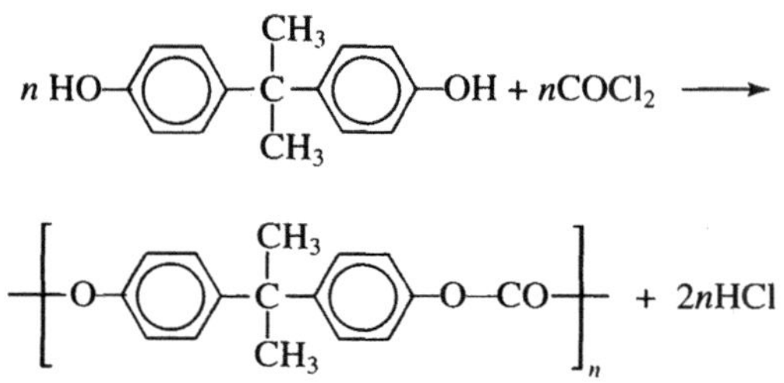

The reaction is carried out in organic solvents that are also HCl acceptors such as pyridine or a mixture of pyridine and a cheaper solvent (chloroform, methylene dichloride or tetrachloroethane) that can dissolve both the complex pyridine–HCl and the polymer. The BPA is dissolved in the solvent and phosgene is bubbled under vigorous stirring at 25–35°C. The reaction proceeds very fast, the pyridine–HCl complex precipitates whereas the polymer dissolves in the solvent. The reaction is stopped when the viscosity of the solution becomes too high for stirring. Methyl alcohol is then added to precipitate the polymer and to dissolve the pyridine complex.

The flow chart of the solution phosgenation process is illustrated in Figure 4.3.

At present the phosgenation is almost exclusively carried out as a phase boundary interfacial polycondensation [29]. Two non-miscible solvents are used and they form two phases. An aqueous caustic soda is used as solvent for BPA whereas an organic solvent is used to dissolve the phosgene and

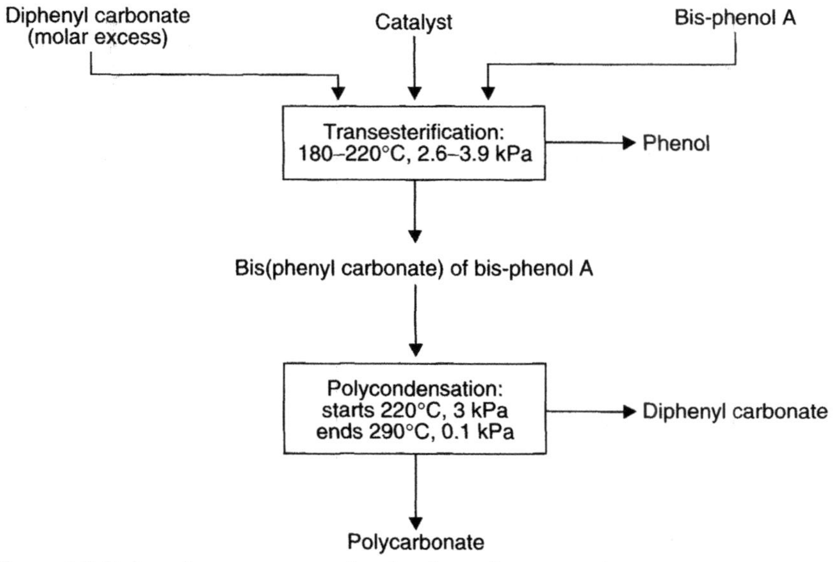

Figure 4.2 Polycarbonate preparation by the melt ester exchange process.

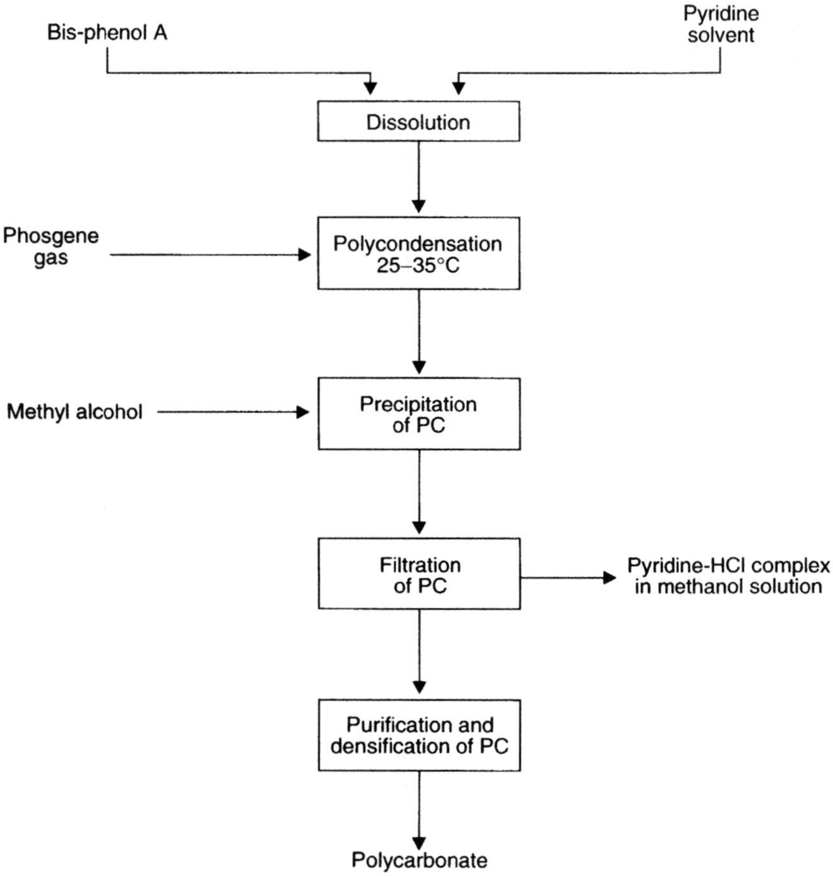

Figure 4.3 Polycarbonate preparation by the solution phosgenation process.

the polymer [32]. Phosgene is introduced into the reactor and the reaction occurs at the boundary interface of the two phases under vigorous stirring. Catalysts such as quaternary ammonium compounds are used to promote the reaction. Anion exchange resins containing trimethylammonium groups are useful catalysts that can be easily separated from the PC [33]. The polymer is purified by washing the organic phase with water, neutralizing the caustic soda and then separating the polymer by precipitation or by evaporation of the solvent. The flow chart of the interfacial phosgenation process is illustrated in Figure 4.4.

The MW of PC prepared by phosgenation is too high and it can be controlled by adding monofunctional compounds or, better, by using non-equivalent ratios of reactants [7]. With this method PC can be made very economically and its MW and structural uniformity can be tailored to the needs of the application and the processing method [29]. The disadvantages

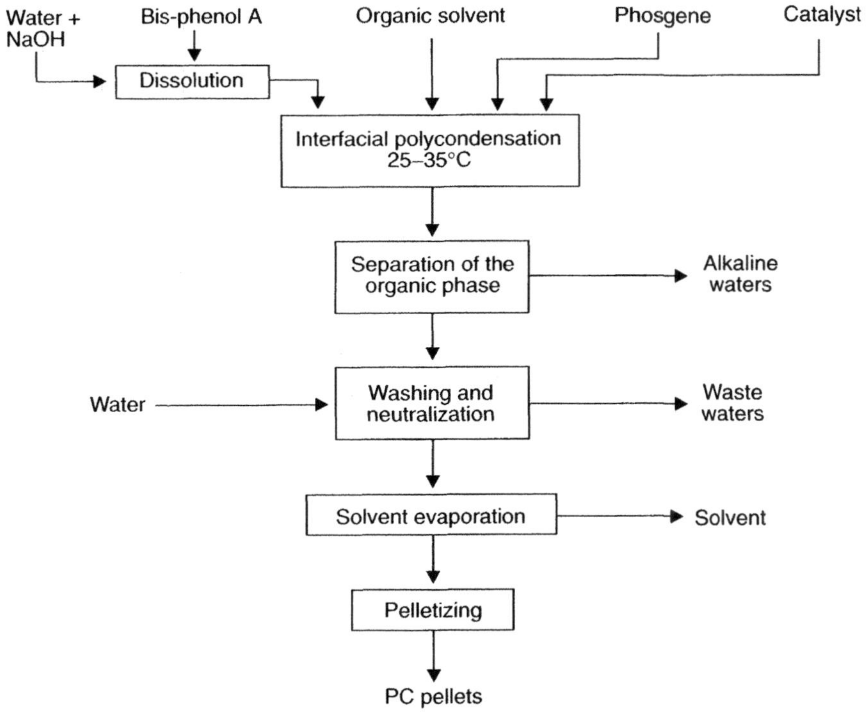

Figure 4.4 Polycarbonate preparation by the interfacial phosgenation process.

of this method consist of the hazards associated with the utilization of phosgene and organic solvents, and their recovery.

4.4.3 Properties

The values of M_n of commercial PCs are in the range 20 000–50 000 and they have a pronounced effect on the melt index [7]. For example, for M_n values of 23 000 and 39 100, the melt indices at 300°C are 23.9 and 1.5 g/10 min, respectively.

There will be processing difficulties when the viscosity is too high. So, polymers with M_n values of 70 000 may be processed into film only by solution casting.

Commercial polymers have a limited degree of crystallinity even if crystalline spherulitic structures may be induced into PC by thermal treatments. The flexibility of the chain is restricted by the presence of carbonate polar groups and of aromatic benzene rings leading to high T_m and T_g. Melt processing temperatures are high (250–300°C) and degradation can easily occur.

The outstanding characteristic of PC is the high impact strength. A 3 mm thick sheet stops a 0.22 caliber bullet. PC exhibits a specific feature called 'critical thickness' which is 4–5 mm for molded parts [7]. Below this thickness high values of notched Izod impact strength and high energy absorbing ductile failures are obtained but above the critical thickness the impact strength drops to much lower values and the failure becomes brittle. The temperature has the same effect on impact strength by producing a sharp drop at about −10°C to −15°C.

Similar to most of the engineering materials, the shape of the stress–strain curve of PC simulates those of aluminum and steel, in accordance with Hooke's law. Therefore, the performance of a part can be predicted by standard equations [34].

The creep limit or deformation under continuous load for PC is 14 MPa at 23°C in air, one of the highest values for any unreinforced thermoplastic [34]. At lower stress levels creep becomes negligible. Elevated temperatures, many chemical environments, dynamic conditions and aging of the specimen, all contribute to the diminishing of the creep limit when cracking and crazing may occur.

Optical properties of PCs are excellent: they exhibit light transmittance of 86–89% and only 1–2% of haze [34]. The refractive index is 1.586. This makes possible the extensive use of polycarbonate for lens and glazing applications. In order to improve the abrasion resistance of the surface, so that it approaches inorganic glass, a silicone scratch resistant coating is applied.

PC resins have good electrical insulation properties due to their low power factor, high volume resistivity, low water absorption and good heat resistance.

The hydrolytic resistance of polycarbonates is better than generally expected for polyesters because the ester linkage is protected by the two hydrophobic benzene rings. Resistance to oxidizing agents and to ozone is good. Chloroform, 1,2-dichloroethylene and 1,1,2-trichloroethane are good solvents. From all the environmental factors, UV light and water produce the most significant deteriorations. UV stabilizers are used, especially for outdoor applications.

PC may be processed by extrusion, injection molding, blow molding, rotational molding, structural foam molding and film casting [35]. The low degree of crystallization renders molding shrinkage to be low, of the order of 0.006–0.008 cm/cm for unreinforced material and 0.003–0.005 cm/cm for glass filled resins.

Additives such as pigments, UV stabilizers, heat stabilizers, fire retardants and blowing agents have to be carefully selected because of the high processing temperatures of PC resins. Stabilizers such as metaphosphoric acid and boron phosphate are used in amounts as small as 0.005%.

Composite materials based on reinforced PC are prepared using different types of fibers. Glass fibers, used up to 40%, more than double the creep resistance and increase the tensile modulus up to five times; the flexural, compressive and tensile strength are improved as well. The fiber surface is

Table 4.4 A comparison of some properties of unfilled and glass filled PC

Property	*Unfilled*	*35% glass filled*
Specific gravity	1.2	1.44
Melting temperature (T_m) (°C)	230–250	–
Glass transition temperature (T_g) (°C)	149	–
Tensile strength (MPa)	65	127
Elasticity modulus (MPa)	2400	10 000
Elongation at break (%)	110	2.7

previously treated with a silane coupling agent in order to provide a good adhesion between the glass and the PC. Carbon fiber reinforced PC has the highest notched Izod impact strength of any fiber reinforced thermoplastic, and the flexural modulus is seven times higher than that of the standard resin. Some properties of PC resins – unreinforced and glass filled – are presented in Table 4.4.

Copolymers known as poly(co-carbonates) have been prepared with a variety of bis-phenols in mixture with BPA. Tetrabromo-bis-phenol A is used for flame retardance. Tetramethyl-bis-phenol A enhances the resistance to hydrolysis and contributes to higher heat distortion temperatures (196°C compared to 100–130°C for standard PC). The best combination of properties is obtained in PC based on trimethyl-cyclohexanone–bis-phenol (TMC–BP) and its copolymers and blends with BPA–PC [29]:

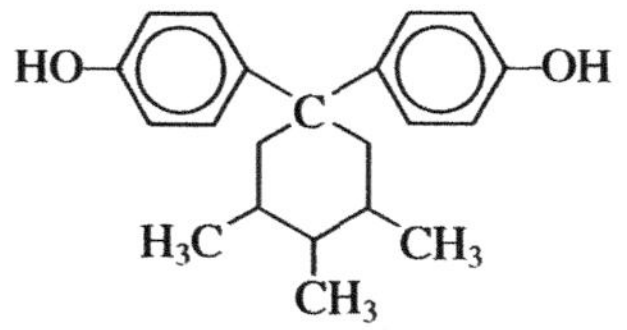

The restriction to chain flexibility introduced by the cycloaliphatic substituent group is reflected in the values of T_g compared to that of the BPA based PC, as can be seen below:

- PC from BPA: $T_g = 149$°C;
- PC from TMC–BP: $T_g = 239$°C;
- Copolymer from 45% BPA and 55% TMC–BP, $T_g = 205$°C.

The transparency of these co-PCs is 90% light transmission at 1 mm thickness. They are easily processable, and have good light stability and impact strength. Depending on the content of TMC–BP, the heat deflection temperatures of these co-PCs range between 160°C and 205°C. They represent a heat resistant alternative to standard PC.

Other types of copolymers are the polyester-carbonates that contain carbonate units and ester units from BPA and terephthalic or isophthalic acids. They have intermediate properties between polycarbonates and polyarylates. Block copolymers of polycarbonates and silicone polymers have improved low temperature properties. High flowability is achieved by the incorporation of aliphatic 'soft block' segments that allow the use of higher M_n resins with excellent impact strength [29].

Alloys of PC with 2–9% styrene based terpolymers such as ABS, have reduced notch sensitivity and enhanced resistance to environmental stress cracking [27]. Blends of PC with PBT have improved low temperature properties and are used for car bumpers [28].

4.4.4 Applications

The major field of application is in electronic and electrical engineering; together they share 55% of the market. Moldings from PC are used for housings and internal parts of printers, copiers, printed circuit boards, terminals and other computers and business equipment, for telephone connectors, all kinds of compact disks and a host of other applications.

In the construction industry, polycarbonate sheeting is being used more and more to replace glass and glazing materials in factories, schools and commercial buildings because of its greater resistance to breakage (the impact strength is 250 times that of safety glass of the same thickness) and high transparency [36,37]. PC sheets are used for door and window components, skylights, sunscreens, wallcoverings, solar collectors and greenhouse applications, bus shelters and telephone kiosks. Other uses are lamp housings for street, traffic and automobile lights, safety goggles, helmets and even for the visors worn by lunar astronauts. Other high security fields are in applications requiring bullet resistance. Flame retardant sheet is used for aircraft interiors.

In the domestic and medical field they are used for microwave cookware, food processor bowls, tableware, baby bottles and transparent filter bowls for transferring blood. They have also replaced some sterilizable stainless steel surgical equipment. One of the major areas of application is for very light, unbreakable and returnable water and milk bottles [29].

PC structural foams with densities as low as $0.6\,g/cm^3$ have a flexural strength/weight ratio which is twice that of metals, and may be nailed and screwed [7].

4.5 POLYARYLATES

Polyarylates are linear aromatic polyesters from bis-phenols and aromatic dicarboxylic acids [38,39]. They were first marketed in 1974. Commercially produced polyarylate is based on 2,2-bis(4-hydroxyphenyl)propane or BPA and a 1:3–3:1 mixture of isophthalic and terephthalic acids. The repeating unit has the structure:

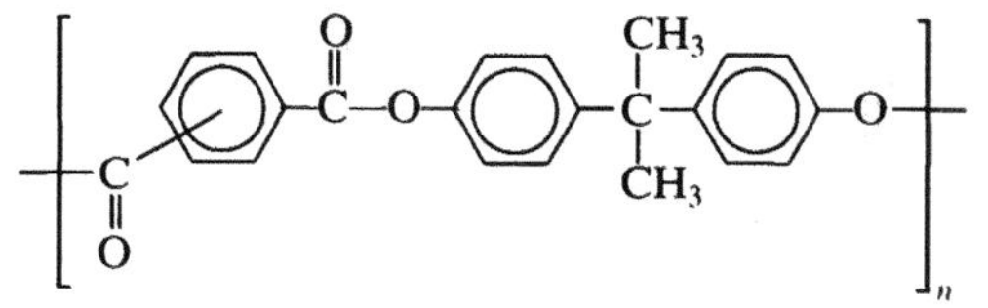

Polyarylates are difficult to prepare by direct polycondensation of diacids with diphenols because the high temperature necessary to achieve practical MW frequently leads to decarboxylation and to ether formation. Attempts have been made to conduct direct polycondensation at moderate temperatures (120°C) in the presence of activating agents such as diphenyl chlorophosphate or *p*-toluenesulfonyl chloride which convert the acid into a highly reactive ester intermediate [40,41].

Two types of manufacturing processes are used for polyarylate production.

1. Low temperature solution methods based on interfacial polycondensation of acid chlorides, namely isophthaloyl and terephthaloyl chlorides, with a dialkali metal salt of bis-phenol A [42,43]:

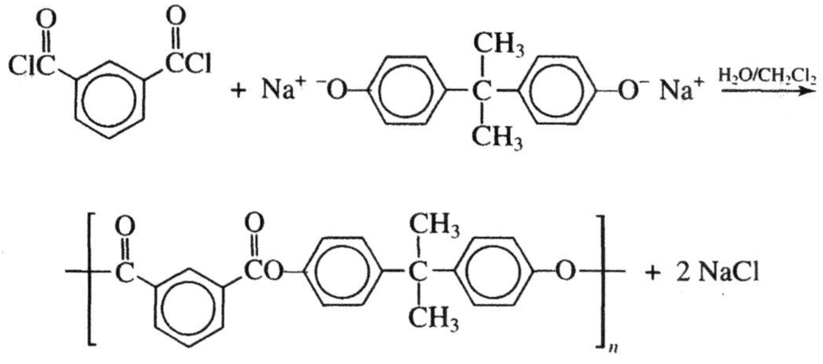

   The phenol is dissolved in aqueous alkali whereas the acid chlorides are dissolved in chlorinated solvents, such as methylene chloride. The two non-miscible solvents form phases which are stirred vigorously and the condensation takes place at the interface. As the reaction progresses, the forming polymer is dissolved in the organic solvent and the metal halide by-product is dissolved in the aqueous phase. The reaction is conducted at room temperature in the presence of phase transfer catalysts such as quaternary ammonium or phosphonium salts which accelerate the reaction. At the end, the two phases are separated and the polymer is isolated either by the evaporation of the solvent or by coagulation in a non-solvent. The flow chart of the interfacial polycondensation is similar to that presented in Figure 4.4 for the preparation of PC with the difference that phosgene is replaced by the acid chlorides of iso- and terephthalic acids.
2. High temperature polycondensation processes run in melt or in slurry are based on two kinds of transesterification reactions: (a) the acidolysis of bis-phenol A diacetate with the mixture of iso/terephthalic acids with

the elimination of acetic acid as by-product [44,45]; (b) an alternative route is the phenolysis of phenyl esters of phthalate acids with bisphenol A with the removal of phenol [46].

(a) $HOCO{-}R{-}COOH + CH_3COO{-}R_1{-}OCOCH_3 \longrightarrow$

$$-\!\!\left(CO{-}R{-}COO{-}R_1{-}O\right)\!\!_n\!- + CH_3COOH$$

(b) $PhOCO{-}R{-}COOPh + HO{-}R_1{-}OH \longrightarrow$

$$-\!\!\left(CO{-}R{-}COO{-}R_1{-}O\right)\!\!_n\!- + PhOH$$

where R = 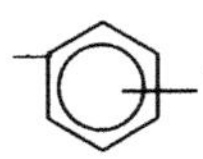, R_1 = 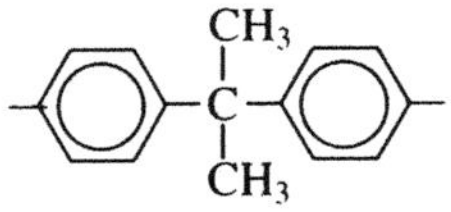 and Ph = phenyl.

These processes are carried out at 250°C and in the final stage, vacuum is applied in order to facilitate the removal of by-products. The process may alternatively be conducted in a vented extruder. Transition metal catalysts have been used in both processes [47].

Polyarylates can be processed by injection molding and extrusion and can be formed as film, sheet, rod, tubing and other profiles. They can be reinforced with glass fibers with loadings of up to 40%.

Polyarylates are classified as engineering polymers. The use of the mixture of two phthalic acids results in a less regular structure of the chains which precludes crystallization. This allows much lower processing temperatures and decreases shrinkage on molding to as low as 0.009 mm/mm.

The high concentration of aromatic rings assures high T_g (185–195°C) yielding materials with high heat distortion temperatures (170–180°C), low creep, high rigidity and high recovery after deformation. The impact strength is as high as 220 J/m, which is between that of a polycarbonate and a polysulfone. The abrasion resistance of polyarylates is better than that of PC and they exhibit good electrical insulation properties. Polyarylates have high resistance to ignition and flame spreading and are self-extinguishing. Their mechanical properties are retained up to about 140°C [48–50].

Exposed to UV light the aromatic ester group rearranges to *o*-hydroxybenzophenone structures in the main chain [51]. These structures are known as UV stabilizers. For this reason the light stability of polyarylates is excellent and the resins can be used as low cost UV stabilizers for other polymers, particularly polyesters and polycarbonates [52].

Polyarylates have found applications in aircraft interiors, fire protection helmets, safety equipment housings, and electrical connectors and appliances. Outdoor applications based on their excellent light stability include solar energy collectors and greenhouse glazing, automotive lenses, safety reflectors, traffic lights and many others [53].

The environmental stress cracking resistance of polyarylates in organic solvents such as aromatic hydrocarbons, esters, ketones and chlorinated hydrocarbons is limited, and exposure to such media should be avoided. The hydrolytic stability is another limitation of polyarylates.

4.6 LIQUID CRYSTAL POLYESTERS (LCPs)

The name of this group of polymers is based on the tendency of the polymeric chains to maintain a high degree of crystalline order in the melt or in solution [54–56]. Generally, the crystallinity in liquid phase occurs when molecules become aligned in ordered regions called mesomorphic phases and the liquid exhibits anisotropic behavior. The liquid crystal polymers are classified as lyotropic when the mesomorphic phase is formed in appropriate solvents (poly-*p*-benzamide) and as thermotropic when the mesomorphism is brought about in the melt (liquid crystal polyesters).

LCPs are thermoplastic polymers based on aromatic monomers that form long, rigid, rod-like macromolecules. The ordered arrangement of polymer molecules in the melt is retained upon cooling and is manifested in a number of exceptional properties, unsurpassed by any other thermoplastics.

The first LCPs were produced commercially in 1972 but most of their major modifications have occurred after 1985 especially with the development of adequate fillers and blends. It is estimated that in 1992 the worldwide production volume was 10 000 tons per year [57]. Electrical and electronic applications share 70% of the market, audio/video 24%, aeronautical, aerospace and automotive 5% and fiber optics 1%.

Liquid crystal polyesters are prepared from monomers with long molecules that are rigid along their major axis. The linkages of the polymer chains are parallel or coaxial [58]. *p*-Linked benzene rings are typical for such rigid structures.

LCPs have been divided into three categories, as seen in Table 4.5, on the basis of their thermal performance (heat deflection temperature) which, in turn, depends on their composition [57,59].

Table 4.5 Classification of LCP [57]

Type of LCP	*Heat deflection temperatures (°C)*	*Monomers*
Type I	60–216	*p*-Hydroxybenzoic acid Terephthalic acid Ethylene glycol
Type II	210–260	*p*-Hydroxybenzoic acid 2-Hydroxy-6-naphthoic acid Terephthalic acid
Type III	260–354	*p*-Hydroxybenzoic acid Terephthalic acid *p,p′*-Dihydroxybiphenyl

The various types of LCP are based on the monomers presented below:

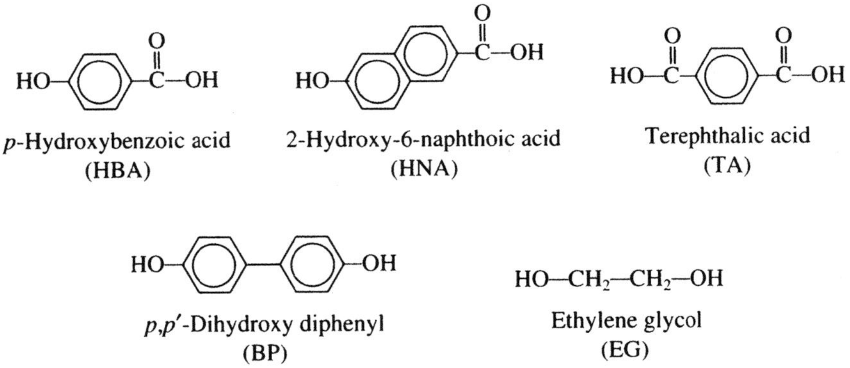

Little detail has been disclosed on the manufacture of these materials. Polymerization methods similar to those used for polyarylates have been applied for the preparation of LCPs. Either of the two general methods is used. (1) The polycondensation of diacyl chlorides with diphenols and the elimination of HCl as by-product. The reaction may be carried out in solution, by interfacial polycondensation or in melt. (2) The second commercial method is based on the transesterification and polycondensation in the melt and under vacuum of either (a) diacetoxy compounds of hydroxy acids and the release of acetic acid or (b) phenyl esters of the acids with the release of phenol:

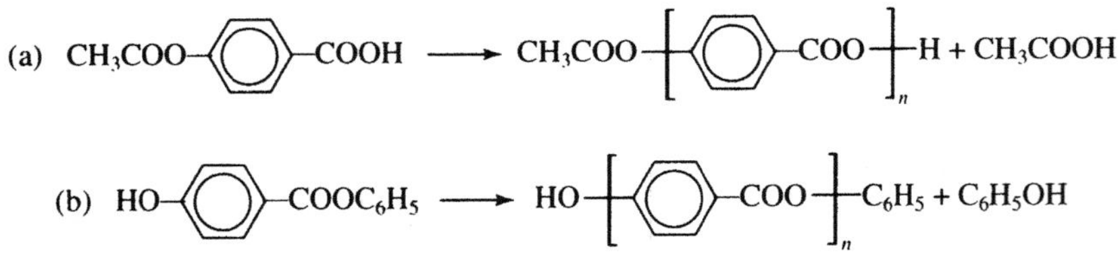

Commercially, LCPs are manufactured in a stepwise polycondensation either by batch or continuous process [57]. In the first step of polycondensation, which is run in solution or as melt, rather medium MW products are obtained. In the second step the MW is increased by heat treatment in solid form [59]. The raw polymer is mixed with additives and pelletized. Recycled material can be added to virgin resin at levels up to 50% without adversely affecting the properties of the final product [57].

LCP can be processed by extrusion or injection molding, with very short molding cycles and rapid solidification of the melt. The material manifests very low shrinkage. The low viscosity of the melt permits the easy filling of very thin walled and intricate parts of precision machinery such as watch gears and platines having diameters of 3 mm and only 0.1 mm thickness [57]. Some rather unusual methods of processing have also been used with

LCP. Compressive sintering and forging consist of molding the ground polymer at 370–455°C and 70 MPa [39], and coatings are produced by plasma spraying in which powdered polymer is fed into a plasma arc and sprayed onto metal substrates [53].

A property of major interest in LCP is the viscosity of the melt. This presents significant variations with the content of *p*-hydroxybenzoic rigid units [58]. At levels of about 30 mol% *p*-hydroxybenzoate units, the melt viscosity begins to decrease, reaching a minimum value at about 60–70 mol% when the material can be easily processed on conventional equipment. At the same time the melt appearance changes from clear to opaque. The decrease in viscosity and the opaqueness arise from the onset of a thermotropic mesomorphism. These low viscosities of the melt are explained by the absence of entanglements and by the alignment of rigid molecules in the direction of flow, thus mimimizing frictional drag. Low viscosities present obvious advantages from the standpoint of processing.

The ordered arrangement of polymer molecules in the melt is retained upon cooling (thermotropic LCP) giving a self-reinforcing character to the material. This is manifested in the exceptionally high level of mechanical properties. The material is a strong, stiff and brittle thermoplastic. The properties are determined by the morphology of the mesomorphous phase which develops during the processing and which consists of oriented bundles of fibrils formed by the polymeric chains [57]. The difference in properties between LCP and conventional crystalline thermoplastics is based on the essential difference of their morphologies and is schematically illustrated in Figure 4.5.

The material in finished parts presents different morphologies in the surface layer and at the center. At the surface, molecules align in the direction of the flow; in the center the molecular orientation is perpendicular to the direction of the flow with much higher crystallinity [59]. Between them is a transitional region where the orientation gradually shifts by 90°. The surface skin layer is highly flow oriented and the interior core is cross flow oriented. The strength and stiffness of LCP are provided by the skin layer [59,60]. LCPs are inherent anisotropic materials, properties being higher along the direction of the flow and only 50–70% along the perpendicular direction. Some properties of LCP are given in Table 4.6; the higher values correspond to Type III resins.

Type III LCP provides superior performance at elevated temperatures with tensile strength as high as 34 MPa at 302°C and is the only thermoplastic material that can be used at this temperature level. Poly(*p*-hydroxybenzoic acid) retains its highly crystalline structure up to 330°C and it does not melt below 550°C, when it decomposes rapidly [39].

LCPs have very low flammability, low smoke emission and UL 94 V–O ratings at thicknesses as low as 0.4 mm. Type III resin resists for 15 minutes when exposed to a 1095°C flame by building a char on the surface that protects the layers beneath [59].

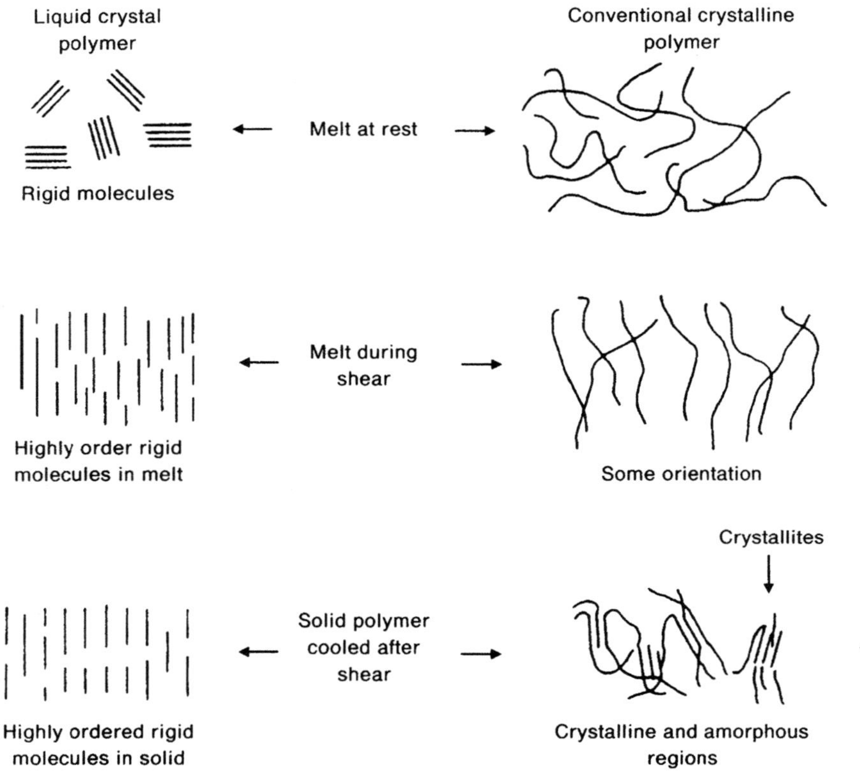

Figure 4.5 Comparison of crystalline structures in LCP and in conventional thermoplastics [7]. (Reprinted by permission of J.A. Brydson from *Plastic Materials*, by J.A. Brydson, 6th edn, Butterworth-Heinemann, Oxford, 1995.)

Table 4.6 Some properties of LCP

Property	*Value*
Density	1.70 g/cm^3
Melting temperatures	370–455°C
Heat deflection temperatures	260–355°C
Tensile strength, unfilled	165–230 MPa
filled	149–207 MPa
Flexural strength, unfilled	169–256 MPa
filled	207–317 MPa
Flexural modulus, unfilled	9–15.2 GPa
filled	13.8–27.6 GPa
Elongation	1–3%
Izod impact strength	53–160 J/m
Coefficient of thermal expansion	
in flow direction	(2–5) × 10^{-5} cm/cm · °C
in perpendicular direction approximately twice these values	

LCPs are good electrical insulating materials and have high transparency to microwaves, especially at elevated temperatures. This renders the material very suitable for microwave cookware and for military tracking devices [59]. Weatherability and chemical resistance are also very good. The material is attacked only by very strong acids and bases, such as concentrated H_2SO_4 and 30% NaOH.

About 70% of the total amount of LCP consists of filled materials. The degree of anisotropy can be reduced by the addition of 30–40% of various fillers. The polymer has a high affinity for a variety of fillers such as milled glass, glass and carbon fibers, graphite flakes, mineral fillers and powdered metals. This allows the tailoring of the mechanical, thermal, electrical and processing characteristics of the material. LCP is considered as a matrix for the reinforced composite parts. LCPs are very expensive polymers; this is a typical feature of very high performance materials.

Blends of LCP with other semicrystalline or amorphous thermoplastics such as PTFE, PEI, PC, PET and polyethylene sulfides have been developed in which the mesomorphous domains of LCP act as reinforcing agents for the other plastics [61].

LCPs have replaced many metal parts in industrial applications because of lower costs. They have been used for precision machinery parts, for fiber optic sheathing, data storage rigid disks, molded circuit boards, connectors, chip carriers and for the replacement of ceramic brittle packing materials in industrial distillation columns.

Plasma sprayed coatings are self-lubricating, corrosion resistant, thermally stable and are applied for jet aircraft engine parts [53]. Blends with up to 25% PTFE are self-lubricating and wear resistant and are used for bearings, seals and rotors.

LCPs are of special interest for non-linear optical materials due to the possibility of fabrication of thin films and to the strong response to electric and magnetic fields. Because of these benefits, LCPs have potential applications in many areas of optoelectronics including telecommunications, optical switching, sophisticated laser elements and integrated optics [62].

4.7 POLYESTERS AS PRECURSORS AND ADDITIVES FOR OTHER POLYMERS

As described in Section 7.2.2 saturated polyesters of low MW, in the range 2000–4000 are used as precursors for polyurethanes. They have hydroxyl functional groups at each end of the molecule and constitute the macroglycol which is one of the three principal components entering the polyurethane composition. They are based on dibasic acids, mostly adipic, sebacic and sometimes phthalic and are linear or branched depending on the glycol component.

Linear, difunctional polyesters are prepared from glycols such as ethylene glycol, diethylene glycol or 1,4-butane diol whereas branched polyester polyols are obtained from glycerol or trimethylolpropane. The condensation reaction is conducted in glycol excess at 200–250°C in an inert atmosphere with the removal of water. Reduced pressure of about 0.1 kPa is applied at the end and the reaction is stopped when the desired MW is reached. These polyesters are low melting waxy solids or viscous liquids.

Other types of polyesters used as macroglycols for polyurethanes are obtained from ε-caprolactone and methyl-ε-caprolactone by ring opening polymerization with dihydroxy or trihydroxy initiators in the presence of titanium catalysts [63,64]:

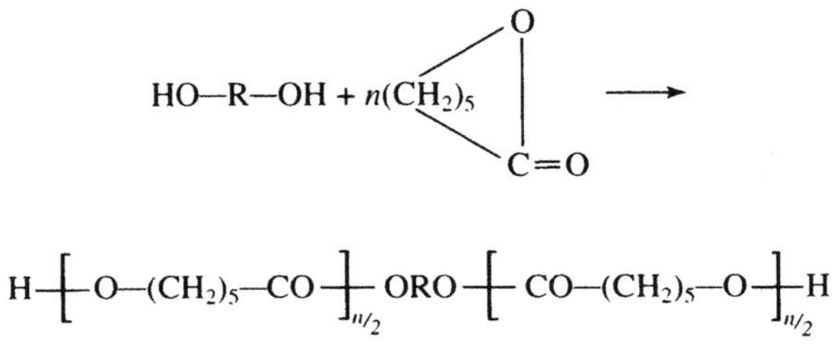

Polycaprolactone polyesters may also be used as additives for other polymers up to 10%. They improve some characteristics such as gloss or dyeability of PE, PP or even polyester laminating resins. They are also used as polymeric plasticizers for PVC as they have the advantage of being less volatile than monomeric plasticizers.

Polycaprolactone has achieved some commercial significance in the medical field as biodegradable surgical sutures and as a delivery medium for the controlled release of drugs [65].

Polymeric plasticizers may also be prepared by the condensation of diols with dicarboxylic acids in the presence of monofunctional alcohols or acids that serve to control the MW by blocking the functional end groups in the macromolecules. Average MWs of polyester plasticizers are in the range 500–8000. Common diols are ethylene glycol, propylene glycol, butanediol, and diethylene and triethylene glycols. The acid component is adipic, azelaic, sebacic or phthalic. These polymeric plasticizers are viscous liquids and are mainly used with PVC.

B. THERMOSETTING POLYESTER RESINS

Polyester resins are thermosetting crosslinked polyesters. They are of several types.

- Unsaturated polyester resins that contain reactive double bonds in the polyester molecules and which are subsequently polymerized to crosslinked structures. They are based on maleic acid or anhydride.

- Saturated crosslinked polyesters in which polyesterification and crosslinking occur in the same step.
- Poly(allyl ester)s.

The first two types are commonly known as polyester resins or alkyds.

4.8 UNSATURATED POLYESTER RESINS BASED ON MALEIC ANHYDRIDE

4.8.1 Introduction

Unsaturated polyester resins are among the four most important thermosetting resins, besides phenolic, amino and epoxy resins and they represent about 20% of the total volume of thermosets [66]. Their world production volume in 1990 was 1.6 million tons [67].

They are prepared in two steps. A linear polyester of low MW of about 700–4000, containing carbon–carbon double bonds in the polymer chain, is synthesized in the first step and is dissolved in a vinyl monomer and stabilized with an inhibitor. This composition may be stored for months. When used, the resin composition is mixed with a catalyst system (peroxide), compounded with fillers, reinforced with glass fibers and crosslinked by the copolymerization of the unsaturated double bonds in the polyester with the vinyl monomer used as solvent. The polymerization takes place during the molding or laminating operation.

Unsaturated resins were first prepared in the USA in 1946 from maleic anhydride and various glycols, and cured by reaction with styrene. The polyester compositions served for the preparation of glass fiber reinforced laminates used for large structures such as boat hulls and sport car bodies. Later, bulk molding compounds (BMC) and sheet molding compounds (SMC) were developed [68].

4.8.2 Raw materials

Unsaturated polyester resins are based on several components that may be combined in different ratios [69,70]. Each component may be replaced by any of several alternative components resulting in as much as fifty or sixty different resin formulations. The choice of components is dictated by the properties needed for a particular application [71].

The categories of components are:

(a) glycols;
(b) unsaturated acids and anhydrides;
(c) saturated modifying acids and anhydrides;
(d) crosslinking monomers.

(a) Glycols

1,2-Propylene glycol, $HO{-}CH(CH_3){-}CH_2{-}OH$, is the most widely used diol in the manufacture of unsaturated polyesters for several reasons. It gives resins which have no tendency to crystallize and which are compatible with styrene (the most frequently used crosslinking monomer) and it is available at low cost. It is obtained by the oxidation of propylene to propylene oxide that is subsequently hydrated to the diol. Polyesters from ethylene glycol have the tendency to crystallize.

Other diols are used for special requirements. Diethylene (**1**) and triethylene glycols impart greater flexibility but bring about an increased water sensibility. High MW diols improve the flexibility but slow down the rates of curing. 2,2-Dimethyl-1,3-propandiol (neopentyl glycol) (**2**) improves the resistance to thermal and UV degradation. The resistance to alkali is improved by glycols with larger molecules (when the concentration of ester groups is reduced), e.g. the diether of bis-phenol A with propylene oxide called bis-glycol (**3**) as well as the hydrogenated bis-phenol A (**4**):

$$HO{-}CH_2{-}CH_2{-}O{-}CH_2{-}CH_2{-}OH$$

(**1**)

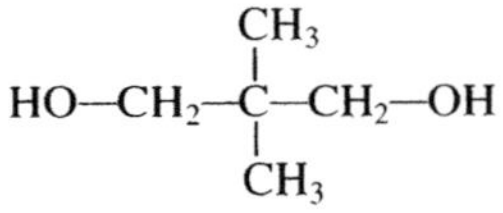

(**2**)

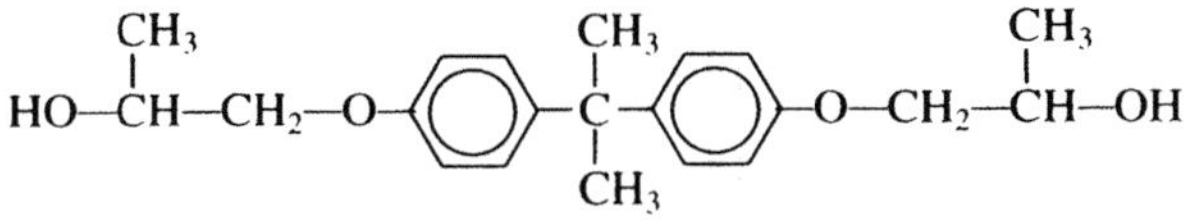

(**3**)

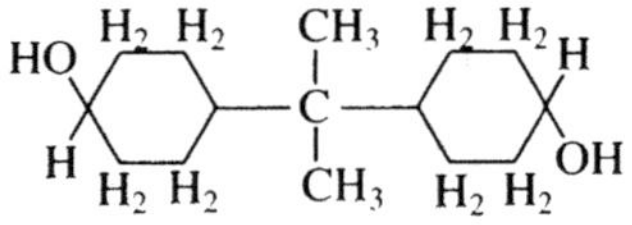

(**4**)

One technology increasing in importance is the use of terephthalate diols obtained from waste poly(ethylene terephthalate) that is hydrolyzed with ethylene or diethylene glycol at 200–240°C to various MWs [72]. They bring about not only cost improvements but also some good features from the aromatic content.

(b) Unsaturated acids

The unsaturation can be introduced into the polyester chain by the use of either an unsaturated glycol or an unsaturated acid. In practice, the unsaturated acid variant was chosen for economic reasons. Maleic acid (**5**) or more usually its anhydride (**6**) or fumaric acid (**7**) are mostly used [73]. Maleic anhydride, which is more reactive than the acid, is obtained by the oxidation of either benzene or *n*-butane. Its melting point is 52–53°C; maleic acid melts at 130°C. The *cis* isomer copolymerizes poorly with styrene whereas the *trans* form participates in alternating copolymerization leading to more uniform products. For this reason fumaric acid, even though it is more expensive and more difficult to handle, is sometimes preferred to maleic anhydride as it gives lighter colored products with improved heat, chemical and moisture resistance. Fumaric acid is prepared by heating maleic acid. Its melting temperature is 284°C.

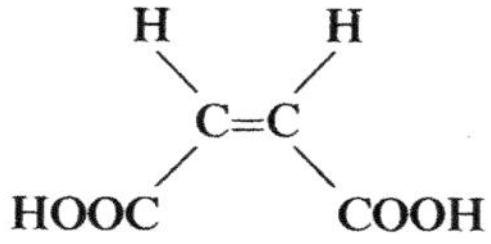

(**5**)

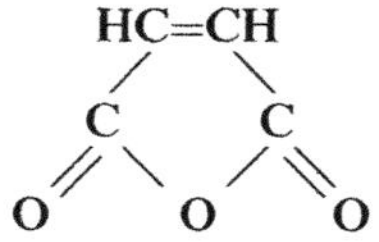

(**6**)

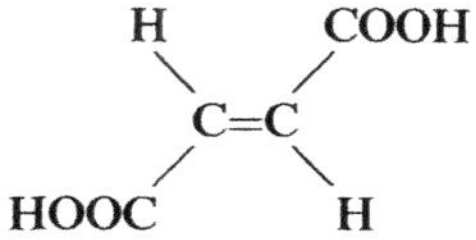

(**7**)

(c) Saturated modifying acids

Saturated modifying acids are used in order to increase the distance between the double bonds along the polymer chain. Unsaturated and saturated acid units would be randomly distributed. Thus, the resins have reduced cross-linking density and consequently are less brittle. Optimum physical properties are obtained when the ratio of saturated acid/unsaturated acid is in the range 40–60/60–40 [69].

The most widely used modifying compound is phthalic anhydride, m.p. 131°C. The resulting polyesters are compatible with styrene monomer and the final resins are hard and rigid. The anhydride is obtained by the oxidation of *o*-xylene and has a reasonably low price.

Isophthalic acid is used where higher heat distortion temperatures and better alkali resistance are required. Succinic anhydride, adipic and sebacic acids are used for more flexible cured resins for coats.

Some other acids are used for special purposes. A resin with improved heat and chemical resistance is obtained with nadic anhydride (the Diels–Alder reaction product of cyclopentadiene and maleic anhydride) (**8**). Flame resistant resins are prepared with chlorinated acids such as tetrachlorophthalic anhydride (**9**) or chlorendic acid (the Diels–Alder reaction product of hexachlorocyclopentadiene with maleic anhydride) (**10**).

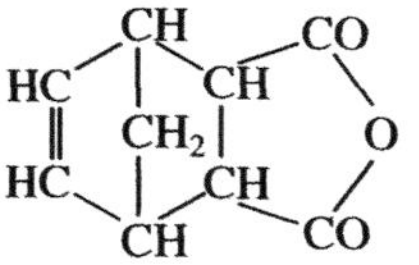

(**8**)

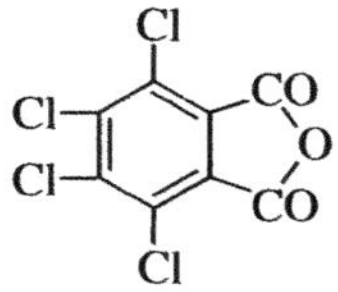

(**9**)

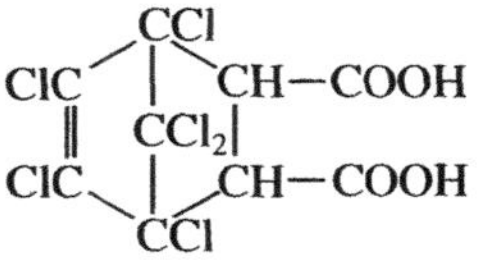

(**10**)

(d) The crosslinking monomer

The most widely used crosslinking monomer is styrene. It serves as solvent for the unsaturated linear polyester in order to reduce the viscosity and to facilitate the impregnation of glass fiber. By subsequent copolymerization of polyester double bonds with styrene a crosslinked structure is obtained in which polyester chains are linked by polystyrene bridges.

The best properties are obtained using styrene/fumarate ratios 1.9–2.4. The drawback of styrene is its high vapor pressure, namely 0.58 kPa at 20°C that leads to 5–15% losses during spray-up and lay-up operations and to pollution problems. To meet strict regulations that impose a limit of 20 ppm of styrene in air, environmentally friendly resins have been prepared with reduced styrene emissions by the addition of barrier forming agents into the resin [74]. Alternative higher boiling monomers such as vinyl toluene and divinyl benzene can replace styrene at a higher resin cost [75].

If styrene is the less expensive monomer for general purpose resins, other monomers are employed for special property resins. Methyl methacrylate in 50% mixtures with styrene gives better weather resistant and translucent products. Improvements in heat resistance are also obtained by replacing styrene with diallyl phthalate [**11**) and especially with triallyl cyanurate (**12**).

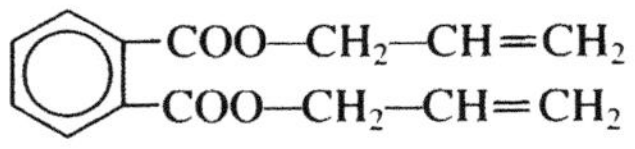

(**11**)

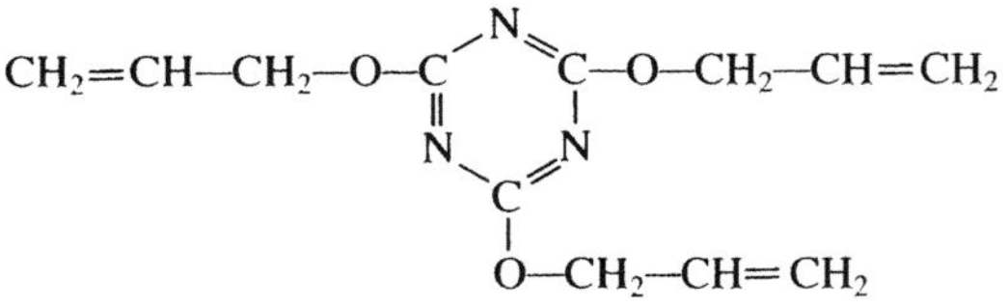

(**12**)

4.8.3 Synthesis and curing

The preparation of unsaturated polyester resins is made by batch processes which allow an easy changeover from one formulation to another, as the variety of resins to be prepared is rather large. Continuous processes are used only for the preparation of large volume, general purpose resins [75].

Stainless steel, glass lined reactors are used, jacketed for heating and cooling, and fitted for distillation. The components (glycol, maleic anhydride and phthalic anhydride) are introduced and the air is replaced by carbon dioxide or nitrogen. A typical formulation in parts by weight and the molar ratio of components for a general purpose resin might be as given in Table 4.7.

The slight excess of glycol serves to compensate for losses due to evaporation and side reactions as well as to restrict the MW of the polyester.

The standard method called the fusion process proceeds in two stages. The first stage is the formation of the half ester (**13**); it takes place almost spontaneously at ambient temperature by the reaction of primary hydroxyl groups of the glycol and the maleic anhydride. As maleic anhydride has a

Table 4.7 A typical formulation for a general purpose polyester resin

	Parts by weight	*Molar ratio*
Propylene glycol	100	1.1
Maleic anhydride	59	0.33
Phthalic anhydride	78	0.67

greater reactivity toward glycol than phthalic anhydride or isophthalic acid, in order to obtain a more uniform resin with improved properties, the fusion is started by reacting first the glycol with the phthalic component and then adding the maleic anhydride [71]. The reactor is then heated up to 190–220°C for the second stage which consists of the esterification of the free acid groups in the half ester (**13**) with the hydroxyl groups of the other half ester molecules. Catalysts such as *p*-toluenesulfonic acid or dibutyltin oxide are sometimes used to accelerate the esterification.

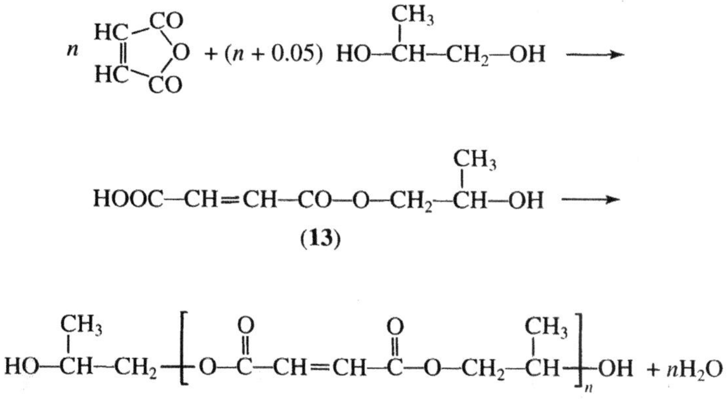

(**13**)

Water from the reaction is continuously removed by the flow of inert gas. Sometimes xylene or toluene are added, followed by azeotropic distillation that removes water.

The progress of the esterification is tested periodically by measuring the acid number and the viscosity. The reaction is stopped when the MW reaches values of 1000–2000 (acid number 25–50). The reaction mass is cooled to temperatures below the boiling point of the vinyl monomer to be used (about 90°C) and then it is mixed with a polymerization inhibitor (hydroquinone) and pumped into the blending tank where it is dissolved in the vinyl monomer that acts as solvent for the resin. The weight of the styrene is about half that of the polyester. The complete production cycle takes 10–20 hours. The dissolved resin may be stored for months or used immediately for curing.

The heat applied during the esterification is also responsible for a *cis–trans* isomerization of maleic units to fumaric structures. At the end, the polyester

Table 4.8 A typical resin–styrene blend prepared for storage

	Parts by weight
Polyester resin	200
Styrene	100
Benzyltrimethylammonium chloride	0.26
Hydroquinone	0.034
Quinone	0.0034

resin contains 70–100% fumaric groups and 0–30% maleic groups as was evidenced by NMR [71]. This constitutes an advantage for the next step of curing as fumarate structures are much more reactive toward vinylic monomers.

The main resin types produced by the fusion process are gel coats and resins for casting, glass reinforced applications and compression molding.

A typical resin–styrene blend prepared for storage is given in Table 4.8.

A flow chart of the preparation of polyester resin by the fusion process is presented in Figure 4.6. Additives such as fire retardants, UV stabilizers and fillers are usually used with polyester resins.

Curing the polyester is a delicate balance between the effects of the catalyst, inhibitor and accelerator. Curing the unsaturated polyester resins is a free radical copolymerization reaction initiated by peroxides which decompose to free radicals. Curing may be effected in two ways: at elevated

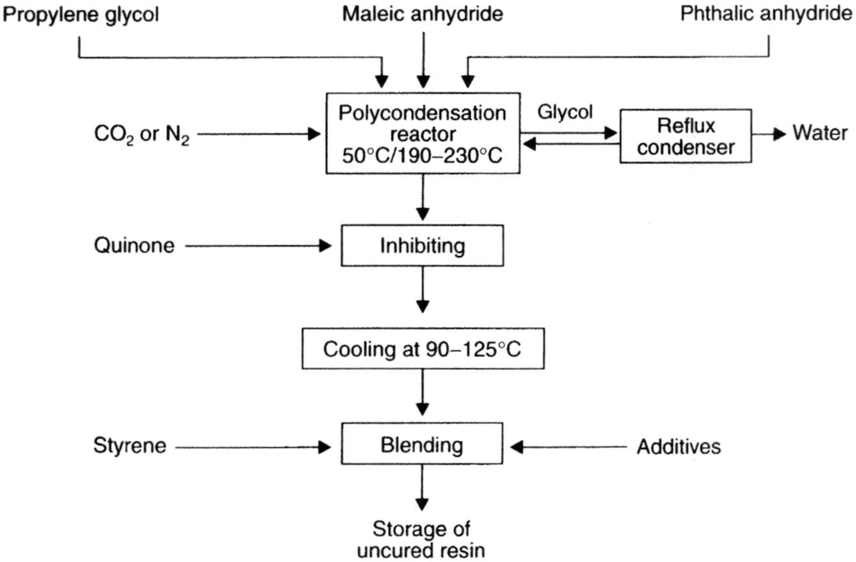

Figure 4.6 Preparation of unsaturated polyester resins.

temperatures (140–160°C) for molding products or at ambient temperature for large hand lay-up structures. Initiating systems are chosen to be effective for specific polymerization conditions (temperature and gel time) [76].

Peroxides are the initiators used for elevated temperature curing (70–150°C). They are decomposed to free radicals by heat and curing is effected in several minutes. Besides the most commonly used benzoyl peroxide (**14**), other peroxides are employed such as 2,4-dichlorobenzoyl peroxide (**15**), di-*tert*-butyl peroxide (**16**) and lauroyl peroxide (**17**).

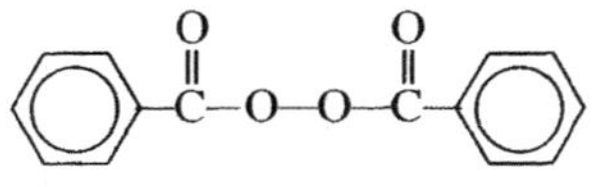

(**14**)

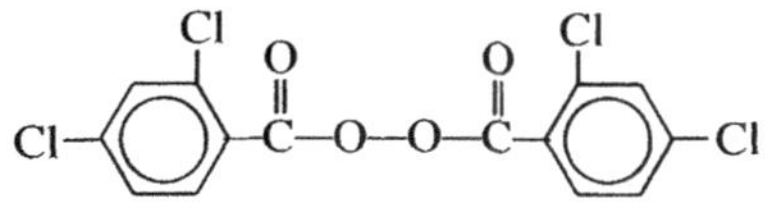

(**15**)

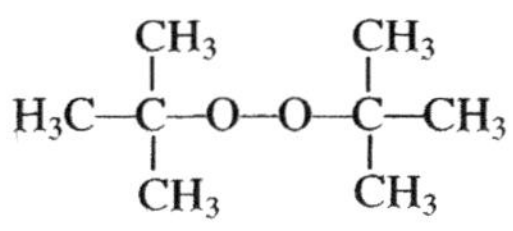

(**16**)

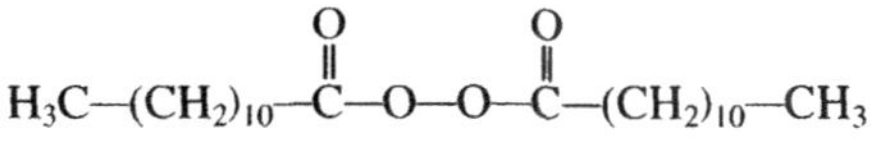

(**17**)

Peroxides used as initiators at elevated temperatures may also be used at room temperature in the presence of tertiary amine accelerators. Such amines are dimethylaniline, diethylaniline or dimethyl-*p*-toluidine. On the other hand, the initiating systems at room temperature are composed of hydroperoxides and an accelerator (activator) that promotes the decomposition of the hydroperoxide into free radicals without any heat. The most frequently used are methyl ethyl ketone peroxide and cyclohexanone peroxide. Both of them are actually mixtures with variable composition of several peroxides and hydroperoxides.

Hydroperoxides are used in conjunction with metal soaps such as cobalt or vanadium napththenates or octoates. However, they should not be mixed together as they can react violently with explosion and they are added to the resin separately, making sure that one is completely dissolved before the other is added. The decomposition of a hydroperoxide by a metal salt proceeds by the redox reaction sequences below:

$$ROOH + Co^{2+} \rightarrow RO^{\cdot} + OH^{-} + Co^{3+}$$

$$ROOH + Co^{3+} \rightarrow ROO^{\cdot} + H^{+} + Co^{2+}$$

It should be mentioned that metal accelerators used for the decomposition of hydroperoxides do not decompose the peroxides.

Various monomers and a wide range of catalyst–accelerator–inhibitor systems are available for the crosslinking. The structure of unsaturated polyester resin cured with styrene is represented in Figure 4.7. The crosslinks in general purpose resin contain about one to three styrene units.

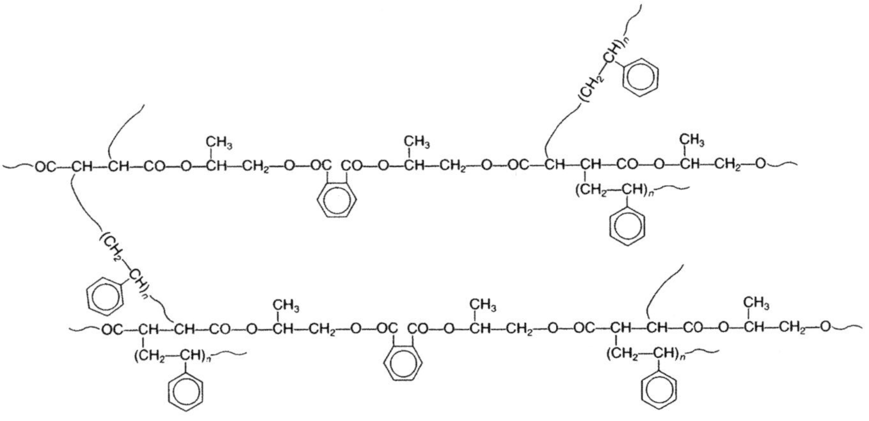

Figure 4.7 Styrene cured polyester resin.

Curing and gelation are exothermic and the reaction may be followed by the temperature curve. At the beginning the temperature rises slowly as during this time the inhibitor is consumed by the free radicals released by the initiator. Then the polymer starts to form, the system quickly gels as the polymerization accelerates and the temperature rises rapidly reaching a peak and then falls when the cured plastic cools back [75].

Depending on the amount of unsaturation, a substantial shrinkage (4–12%) occurs on hardening of unfilled resins. High performance molding composite materials based on unsaturated polyester matrix and different reinforcing fillers have very reduced, near-zero shrinkage.

Modified polyester resins have been prepared by the use of dicylopentadiene as modifying agent [71]. Besides the economic reasons, improvements in resin properties are obtained, such as reduced volumetric shrinkage, higher temperature performance, better moisture and corrosion resistance and improved surface characteristics.

Dicyclopentadiene (**18**) is a by-product in the manufacture of ethylene and it functions as an olefin in addition reactions. The modified resin is prepared *in situ* in the following manner. Water and dicylopentadiene are added to molten maleic anhydride at 70°C. The ratio of maleic anhydride/dicyclopentadiene is 1/0.6. Water hydrolyzes the anhydride to maleic acid which adds to the double bond of dicylopentadiene at the strained ring (**19**):

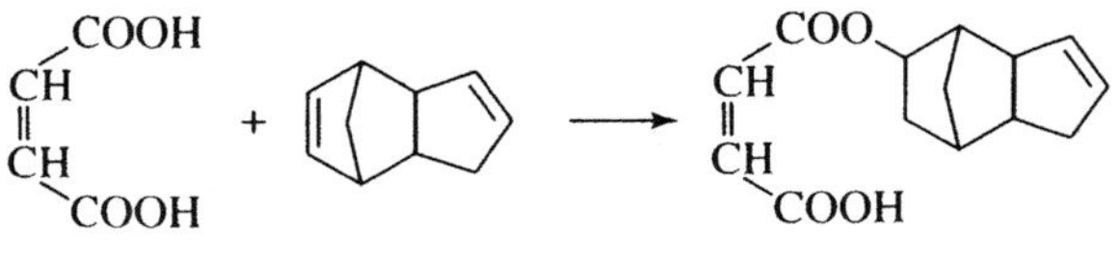

(**18**) (**19**)

The resultant dicylopentadiene maleate (**19**) is subsequently esterified with glycol and isomerized to fumarate structures. Dicyclopentadiene molecules are the predominant end groups in the modified resin and the residual unsaturation in the cyclopentyl group participates at the curing process.

In 1990 136 000 tons of modified resin were prepared in the USA [71].

Other modified resins have been prepared using isocyanates or polyurethane (PU). They can be either in the form of alloys (or interpenetrating networks) [77] or of reaction products of unsaturated polyesters and isocyanates [71]. The resultant resins have a combination of properties that are superior to either of the polymers alone.

4.8.4 Properties and applications

Cured polyester resins are thermosetting, infusible and insoluble materials. Many varieties of unsaturated polyester resins are now commercially available as they can be formulated to produce cured plastics ranging from hard and brittle to soft and flexible. Almost every property can cover a broad range of values.

General purpose unsaturated polyester resins provide a rigid thermostable matrix that makes them excellent binders for resin–fiber glass composites. This can bring about large variations of properties. Impact strength, for example, may be increased by as much as 50 times for reinforced resins. Compared to most metals, reinforced polyester resins exhibit high strength to weight ratios.

Many mechanical properties are dependent on the density of crosslinks and on the flexibility of the molecules between crosslinks. The crosslinking density may be controlled by the ratio of maleic anhydride to saturated acid whereas the rigidity/flexibility depends on the structure of the employed glycol and saturated acid. Mechanical properties are maintained up to 100°C and the maximum service temperature is about 150°C for reinforced general purpose resins and about 200°C for heat resistant grades. Table 4.9 presents a comparison of some properties of fiber glass, and unfilled and glass reinforced general purpose resins.

The ester groups are the weakest links and they provide sites for hydrolytic degradation especially in alkaline environments. Reducing the concentration of ester groups by the utilization of larger glycols and maximizing the steric hindrance at the ester group result in improved alkali resistance. The material is resistant to most inorganic and organic acids and solvents with the exception of strong oxidizing acids, chlorinated hydrocarbons, esters and ketones.

The use of resins as high frequency electrical insulators are limited because of the relatively high power factor and dielectric constant due to polar ester groups.

Table 4.9 A comparison of some properties of unfilled and glass reinforced resin and glass fiber

Property	*Unfilled casting*	*Glass chopped strand mat laminate (hand lay-up)*	*Glass woven cloth laminate (hand lay-up)*	*Glass fiber*
Glass content (% weight)	0	30	55	100
Specific gravity	1.2	1.37	1.7	2.54
Tensile strength (MPa)	62	140	340	3455
Flexural strength (MPa)	120	210	410	–
Compressive strength (MPa)	140	140	240	–
Impact strength (unnotched) (J/m)	110	1100	1300	–

Specific fillers, called thixotropic, such as fine silicas and silicates are employed in order to increase the viscosity of the resin on standing and thus to avoid drainage from vertical surfaces when hand lay-up techniques are used. Mineral fillers such as calcium carbonate reduce not only the cost but also the shrinkage on curing.

Glass fibers are the most employed reinforcement for polyester resins and various types of glass are available. The glass should be previously treated with a finish based on silane coupling agents in order to achieve better adhesion between the resin and the glass.

Water extended polyesters consist of almost equal amounts of resin and water which is dispersed in very tiny droplets (2–5 μm diameter) and forms a creamy water-in-oil emulsion. After curing the system, cellular white materials are obtained and are used as plaster of Paris and as wood substitutes [7].

Hand lay-up and spray-up processing techniques are used for large composite objects. The resin is formulated for room temperature curing and is applied to a mold with layers of fiber glass cloth using a brush or a roller. In the spray-up techniques commonly used for the manufacture of boats and household sanitary ware, the resin is catalyzed as it is being applied. The resin spray carries the chopped glass fiber and they are simultaneously applied by gun to the mold surface. In molding techniques, the fiber glass mat reinforcement previously shaped is placed in the mold, the liquid resin is poured over it and the pressure is applied for curing.

The major markets for unsaturated polyesters include transportation, marine, construction, electrical and corrosion [75].

During the past few years the use of polyester molding compositions has grown significantly, particularly in the transportation field. Land, sea and air transport applications account for almost half the polyester resin produced [78]. Common applications are for boat hulls, sport car bodies, truck cabs and public transport vehicles.

Crosslinked unsaturated polyesters constitute the basic matrix for polymer–concrete composites as well as for sheet molding compounds used for roofing and building insulation and they account for about one third of the resin applications [36]. Also included in the construction category are such corrosion resistant and underground corrosion resistant products as pipes, water tanks and swimming pools.

Other applications include such diverse items as decorative furniture castings, simulated ornamental stone, buttons, trays, bowling balls, skis, surfboards, safety helmets, chemical processing equipment, missile shells, rocket motor cases and many others.

Most unsaturated polyesters are not suitable as coatings because atmospheric oxygen inhibits the crosslinking reaction. To circumvent this difficulty, a number of photochemical methods have been developed to cure unsaturated polyester coatings. Among them is the incorporation into the polymer of photosensitive groups such as chalconyl or cinnamyl or of dyes [79,80]. Photochemical curing represents a promising alternative to heat curing.

The process of reconditioning and reusing wastes and scrap parts has been introduced since 1992 and manufactured components with up to 10–15% recyclate content present the same good mechanical properties [74].

4.9 SATURATED POLYESTER RESINS

4.9.1 Introduction

The saturated polyester resins are based on polyfunctional components, namely acids and polyols; natural oils containing triglycerides of unsaturated acids (known as drying oils) constitute one of the principal components of the resins. The formation of the final crosslinked network or the 'drying of the resin' is based on the polymerization of the double bonds of the unsaturated acids contained in the oil and takes place simultaneously with the polycondensation process [81,82].

Commonly known as 'alkyds' (derived from alcohol and acid), the saturated polyester resins have been estimated to represent more than a half of all the resins used as surface coatings, under the name of alkyd paints. The first prepared resin called glyptal was based on phthalic anhydride and glycerol but was too brittle to find any practical application. The commercial development of alkyds after 1933 was due to their chemical modification with drying oils. This has brought a favorable change in properties.

4.9.2 Raw materials

Three major components are employed to prepare an alkyd resin, besides a large variety of other reactants and additives that will impart complementary characteristics to the resin. The main components are:

(a) polyols;
(b) dibasic acids or anhydrides;
(c) modifying oils.

(a) Polyols

Glycerol (**20**) is by far the most used polyol for the preparation of alkyd resins due to its low cost and accessibility, high boiling temperature and good characteristics imparted to resins. Glycerol is obtained either as a by-product by the saponification of oils and fats (triglycerides) during the manufacture of soaps or by synthetic routes via propylene or propylene oxide in the petrochemical industry. Glycerol is a high boiling (290°C) and viscous liquid.

Pentaerythritol (**21**), sorbitol (**22**) and trimethylolpropane (**23**) are other polyols employed occasionally.

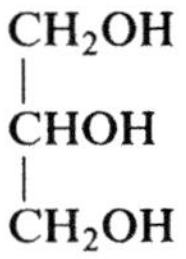

(**20**)

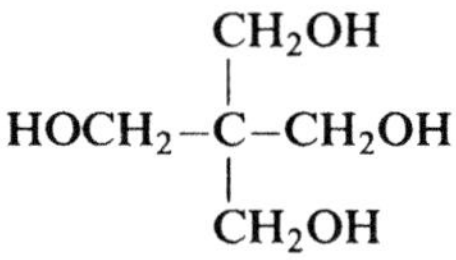

(**21**)

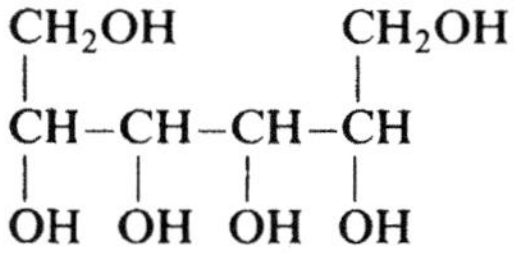

(**22**)

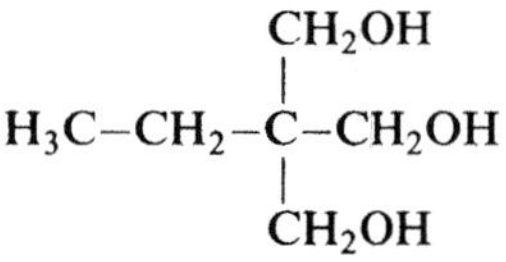

(**23**)

(b) Acids

The most common dibasic acid component is phthalic anhydride (**24**). For special purposes pyromellitic dianhydride (**25**), trimellitic acid (**26**), maleic anhydride (**27**), adipic and sebacic acids are also employed.

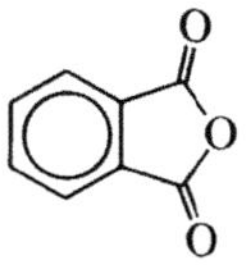

(**24**)

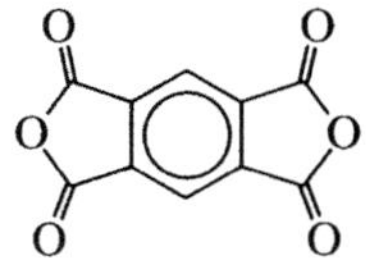

(**25**)

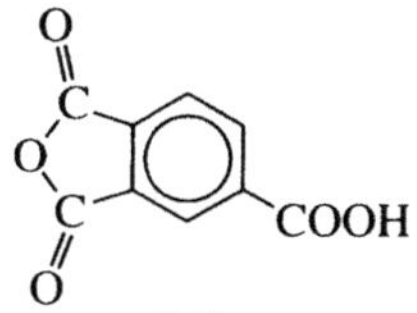

(**26**)

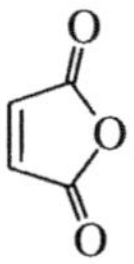

(**27**)

(c) Modifying oils

The introduction of long chain oil structures into the polyester provides the resin with better solubility and the resultant coatings are more flexible.

Mainly drying and semi-drying oils are used as modifying oils depending on the application of the resin.

The type of oil is based on the degree of unsaturation contained in the oil. The higher the degree of unsaturation the faster a film of oil will dry and will become insoluble when exposed to air (in contact with oxygen). Drying oils are mainly composed of triglycerides (**28**) of unsaturated acids such as linoleic (**29**) and linolenic acids (**30**) in linseed oil or eleostearic acid (**31**) in tung oil. Dehydrated castor oil also contains linoleic acid as a result of the dehydration of ricinoleic acid (**32**) in the original oil.

$$\begin{array}{l} CH_2{-}OCOR' \\ | \\ CH{-}OCOR'' \\ | \\ CH_2{-}OCOR''' \end{array}$$

(**28**)

$$CH_3(CH_2)_4CH{=}CHCH_2CH{=}CH(CH_2)_7COOH \quad (29)$$

$$CH_3(CH_2CH{=}CH)_3(CH_2)_7COOH \quad (30)$$

$$CH_3(CH_2)_3CH{=}CH{-}CH{=}CH{-}CH{=}CH(CH_2)_7COOH \quad (31)$$

$$CH_3(CH_2)_5\underset{\displaystyle OH}{\underset{|}{CH}}{-}CH_2{-}CH{=}CH{-}(CH_2)_7{-}CH_3 \quad (32)$$

Resins based on non-drying oils such as castor or olive oils, which mainly contain saturated acids, will always remain tacky and have little use other than as plasticizers for nitrocellulose coatings.

The amount of oil incorporated into the resin (oil length) also influences the global amount of unsaturation and the time of drying respectively, as well as the flexibility of the film. Short oil alkyds (30–50% oil) and medium oil alkyds (50–70% oil) are cured at elevated temperatures (baked) in order to shorten the time of curing whereas the long oil resins (70–80% oil) dry rapidly in air. Short oil resins give very hard finishes; the flexibility of the coating increases by incorporating larger amounts of oil.

4.9.3 Synthesis and curing

The preparation of the resin is carried out in two different ways, both of them starting with the transformation of the oil. If the oil itself is used directly in a mixture with the polyol and the acid, a heterogeneous two phase product results.

1. In the 'fatty acid process' the oil is first hydrolyzed to free fatty acids called 'split oils'. In the next step the split oil is mixed with glycerol and phthalic anhydride. Polycondensation takes place by heating the mixture at 200–240°C under an inert gas flow that removes water from the reaction. As a variant of the process, a small amount of solvent may be used that contributes to the removal of water by azeotropic distillation.
2. In the 'alcoholysis process' the oil is transesterified by treatment with glycerol at 240°C. This leads essentially to a monoglyceride which is then heated with phthalic anhydride either in the presence or absence of

a solvent, such as in the above process:

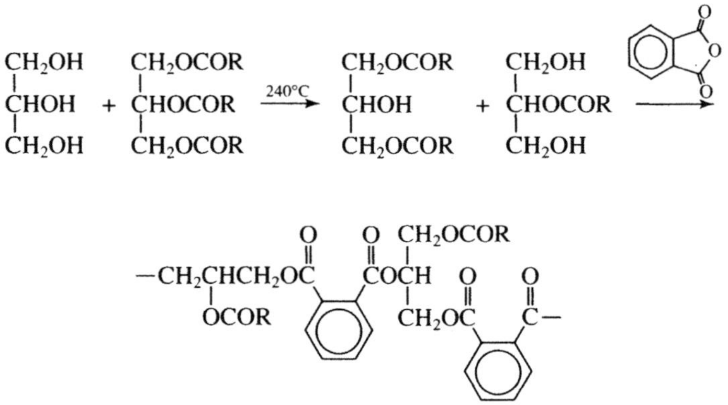

The initial esterification is followed by the polycondensation to a viscous liquid and the process is monitored by the control of the viscosity.

A flow chart of the two processes is presented in Figure 4.8.

Oil modified alkyd resins are viscous liquids and they are used for coatings in formulations that contain solvents and additives. The choice of the solvent depends on the oil content in the resin. Only aromatic solvents such as toluene or xylene can be used to dissolve short oil resins while long and medium oil alkyds are soluble in aliphatic solvents or aromatic–aliphatic mixtures.

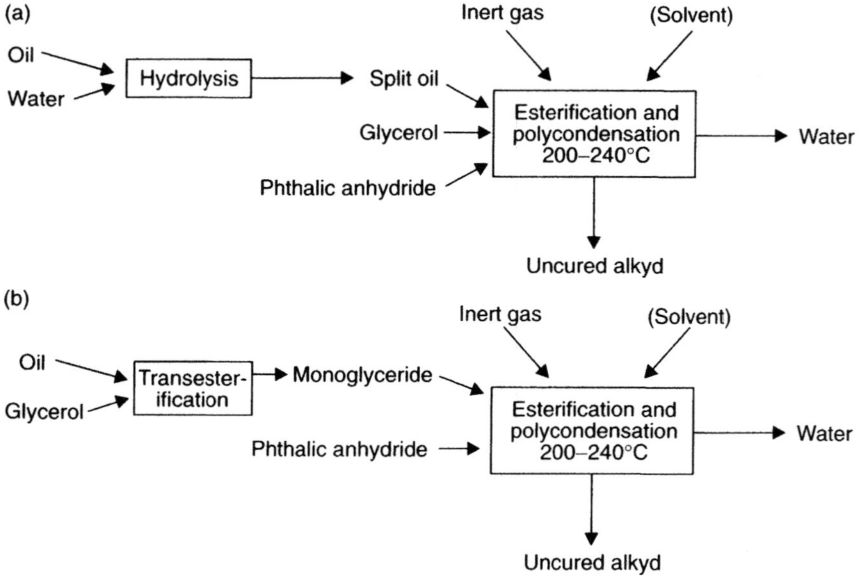

Figure 4.8 Preparation processes for alkyd resins: (a) fatty acid process; (b) alcoholysis process.

A paint formulation contains a catalyst, the so called dryer, which is a hydrocarbon-soluble organic salt of cobalt, lead or manganese such as naphthenates or octoates. The dryer has the function to speed up the formation of a solid film of coating when the liquid alkyd paint is spread on a surface. The drying process consists of a succession of free radical aerial oxidation reactions that take place in the unsaturated oil part of the resin in the presence of atmospheric oxygen. The reaction chain starts at the methylene groups adjacent to the double bonds and continues through peroxidic radicals leading to crosslinked networks.

The dryer may initiate the reaction chain as follows:

$$\sim CH{=}CH{-}CH_2\sim + Co^{3+} \longrightarrow \sim CH{=}CH{-}\dot{C}H\sim + Co^{2+} + H^+$$

As has already been mentioned above, the drying time depends on the total amount of unsaturation, which means the type and the quantity of oil used to prepare the alkyd resin.

It is general practice to use different additives for specific purposes. Styrenated alkyd is a chemically modified resin obtained by the copolymerization of the resin through the unsaturation contained in the oil portion with styrene monomer in the presence of a peroxide:

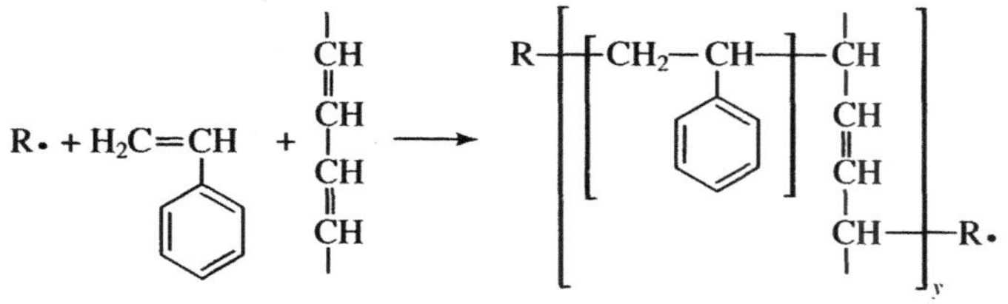

Styrenated alkyds give tougher and more durable film with better alkali resistance and color retention.

Silicone resins added to alkyds improve their heat and water resistance. Fire resistance is enhanced by using chlorinated rubber as an additive while cellulose nitrate contributes to speeding up the drying process.

Aklkyd resins are used as coating materials and as plasticizers for coating formulations.

4.10 POLY(ALLYL ESTER)S

Poly(allyl ester)s are thermosetting resins prepared by the polymerization of esters derived from allyl alcohol and dibasic acids. One of the first polyester materials introduced in 1941 was prepared from diethylene glycol bis-(allyl carbonate) [83]:

$$CH_2{=}CH{-}CH_2{-}O{-}CO{-}O{-}CH_2{-}CH_2{-}O{-}CH_2{-}CH_2{-}$$
$$-O{-}CO{-}O{-}CH_2{-}CH{=}CH_2$$

It is widely used today for spectacle lenses due to its dimensional stability, good abrasion resistance and light weight.

At present the most important allyl ester resins are prepared from diallyl phthalate (DAP) and diallyl isophthalate (DAIP):

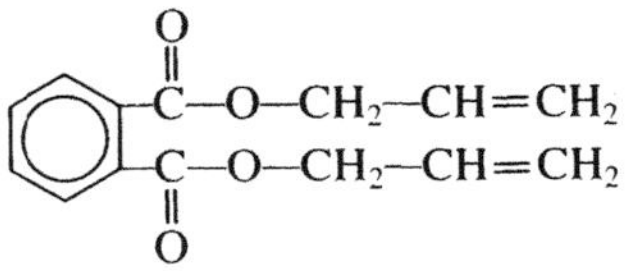

Diallyl phthalate (DAP)

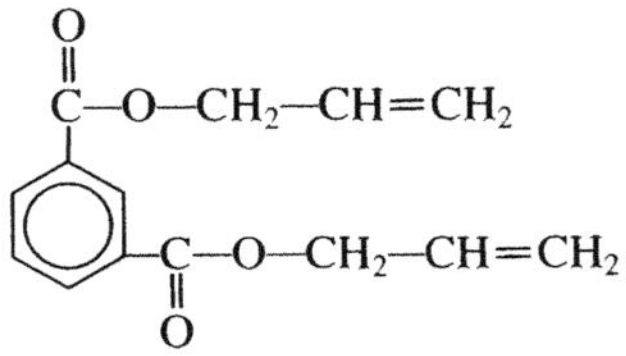

Diallyl isophthalate (DAIP)

Allyl esters are available as monomers and as B-staged prepolymers. In 1981 the USA's production of prepolymers and resins was 1350 and 3200 tons, respectively [84].

Diallyl phthalate and isophthalate esters are prepared by the esterification of phthalic anhydride or isophthalic acid, respectively, with allyl alcohol. Diallyl phthalate and isophthalate are colorless liquids boiling at 160°C/0.5 kPa and 181°C/0.5 kPa, respectively.

The polymerization of these monomers is effected with free radical initiators of the peroxide type in two stages. In the first stage a prepolymer (B-stage) is obtained by the polymerization of only one of the two allyl double bonds. The prepolymers have MWs in the range 10 000–25 000. They are compounded with mineral or organic fillers or glass fibers and molded by all conventional molding methods. During the molding, the second allyl group polymerizes leading to highly crosslinked thermosetting resins.

Allyl resins exhibit an exceptional dimensional stability and abrasion resistance. They are strong, heat and chemical resistant materials. They retain their electrical insulating properties under high heat and high humidity. Diallyl phthalate can withstand temperatures as high as 180°C for long periods, whereas diallyl isophthalate exhibits even more enhanced thermal stability with long-term operation temperatures at 260°C.

Moldings based on diallyl phthalate and isophthalate resins are expensive and are used in the very demanding electrical and electronic fields where the retention of electrical and mechanical properties under extreme conditions are very important. Applications include connectors, switches, relays, circuit breakers, etc.

Allyl resins have also been used as laminates for the manufacture of radomes and aircraft structural parts. Laminate constructions that are made in combination with plywood, particle board and other materials are used as decorative parts with superior surface.

Diallyl phthalate monomers are often used as crosslinking agents and flow promoters for other polyesters.

REFERENCES

1. Bennett, R.A. (1991) *Polym. Preprints*, **32**(2), 121 (Paper presented at the ACS Philadelphia, Pennsylvania meeting).
2. Feldman, D. (1974) *Macromolecular Compounds Technology*, Ed. Technica, Bucharest.
3. Korshak, V.V. and Vinogradova, S.V. (1965) *Polyesters*, Pergamon Press, New York.
4. Goodman, I. and Rhys, J.A. (1965) *Polyesters, Vol. 1, Saturated Polymers*, Iliffe Books, London.
5. Anon. (1970) *Chem. Eng. News*, **48**(49), 42.
6. McFarlane, F.E. (1990) in *Handbook of Plastic Materials and Technology* (ed. I.I. Rubin), Wiley-Interscience, New York, p. 639.
7. Brydson, J.A. (1995) *Plastics Materials*, 6th edn, Butterworth, London.
8. Seymour, R.B. and Carraher Jr, C.E. (1992) *Polymer Chemistry*, 3rd edn, Marcel Dekker, New York.
9. Macdonald, W.J. (1993) *Kunststoffe* (Engl.), **10**, p. 29.
10. Altenberg, M.J. (1986) US Pat. 4,604,410, Chardonol Corp.
11. Liebfried, R.T. (1975) US Pat. 3,892,796, Hercules.
12. Vaidya, J.R. and Nadkarni, V.M. (1988) *J. Appl. Polym. Sci.*, **35**(3), 775.
13. Friedly, H.R. (1992) in *Reaction Polymers* (eds W.F. Gum, W. Riese and H. Ulrich), Hanser Publishers, Munich, p. 80.
14. Westdyk, R. and McNally, D. (1990) in *Handbook of Plastic Materials and Technology* (ed. I.I. Rubin), Wiley-Interscience, New York, p. 629.
15. Berres, K., Homburg, B. and Hettrich, G. (1993) *Kunststoffe*, **10**, 30.
16. Brown, D. (1975) *Rubb. Ind.*, **9**(3), 102.
17. Hoeschele, G.K. (1974) *Polym. Eng. Sci.*, **14**(12), 848.
18. Wolfe Jr, J.R. (1977) *Rubber Chem. Technol.*, **50**(49), 688.
19. Reed, D. (1986) *Eur. Rubb. J.*, January, 16.
20. Rader, C.P. (1993) *Kunststoffe*, **10**, 36.
21. Cella, R.J. (1973) *J. Polym. Sci., Part C*, **42**, 727.
22. Knox, J.R. (1973) *Polymer Age*, October, 357.
23. Lilaonitkul, A. and Cooper, S.L. (1977) *Rubber Chem. Technol.*, **50**(1), 1.
24. Anand, J.N. and Li Wu, M. (1992) in *Reaction Polymers* (eds W.F. Gum, W. Riese and H. Ulrich), Hanser Publishers, Munich, p. 735.
25. Christopher, W.F. and Fox, D.W. (1962) *Polycarbonates*, Reinhold, New York.
26. Schnell, H. (1964) *Chemistry and Physics of Polycarbonates*, Wiley-Interscience, New York.
27. Sikdar, S.K. (1987) *Chemtech*, **17**, 112.
28. Hanrahan, B.D., Angeli, S.R. and Runt, J. (1985) *Polym. Bull.*, **14**, 399.
29. Uerdingen, W. (1993) *Kunststoffe*, **10**, 25.
30. Rokicki, R. and Kuran, W. (1981) *J. Macromol. Sci., Rev. Macromol. Chem.* **C21**, 135.
31. Inoue, S. (1976) *Chemtech*, **6**, 588.
32. Petrus, A., Mandasescu, L. and Matei, I. (1969) *Rev. Roumaine de Chimie*, **14**, 773.
33. Pakull, R., Fennhoff, G., Struever, W. and Kirsch, J. (1994) Ger. Offen. DE 4,233,345, Bayer A.-G.; *CA* (1994) **121**(10), 231,609.

34. Anon. (1990) in *Handbook of Plastic Materials and Technology* (ed. I.I. Rubin), Wiley-Interscience, New York, p. 253.
35. Kirchner, K. (1987) *Kunststoffe*, **77**, 993.
36. Feldman, D. (1989) *Polymeric Buildings Materials*, Elsevier Applied Science, London.
37. Montella, R. (1985) *Plastics in Architecture*, Marcel Dekker, New York.
38. Korshak, V.V. (1972) *Heat Resistant Polymers*, Halsted Press, New York.
39. Critchley, J.P., Knight, G.J. and Wright, W.W. (1983) *Heat Resistant Polymers*, Plenum Press, New York.
40. Higashi, F., Hoshio, A., Yamada, Y. and Ozawa, M. (1985) *J. Polym. Chem. Polym. Lett. Ed.*, **23**, 69.
41. Higashi, F., Akiyama, N., Takahashi, I. and Koyama, T. (1984) *J. Polym. Sci., Polym. Chem. Ed.*, **22**, 1653.
42. Temin, S.C. (1977) *Interfacial Synthesis*, Vol. 2, Marcel Dekker, New York, Chapter 12.
43. Elkin, L.M. (1980) *High Temperature Polymers*, Supp. A., SRI International, Menlo Park, California, p. 167.
44. Hamb, F.L. (1977) *J. Polym. Sci.*, **10**, 3217.
45. Riecke, E.E. and Hamb, F.L. (1977) *J. Polym. Sci., Polym. Chem. Ed.*, **15**, 593.
46. Blaschke, F. and Ludwig, W. (1968) US Pat. 3,395,119, Chemische Werke.
47. Robeson, L.M. (1990) in *Handbook of Polymer Materials and Technology* (ed. I.I. Rubin), Wiley-Interscience, New York, p. 237.
48. Prevorsek, D.C., Kesten, Y. and DeBona, B. (1979) *Polym. Preprints, Div. Polym. Chem.*, **20**(1), 187.
49. Domine, J.D. (1979) *Proceedings of 37th SPE ANTEC*, 7–10 May, New Orleans, USA, p. 655.
50. Freitag, D. and Reinking, K. (1981) *Kunststoffe*, **71**(1), 46.
51. Cohen, S.M., Young, R.H. and Markhart, A.H. (1971) *J. Polym. Sci.*, Part A-1, **9**, 3263.
52. Robeson, L.M. (1981) US Pat. 4,259,458, Union Carbide Corp.
53. Dickinson, B.L. (1984–1985) *Polyarylates, Modern Plastic Encyclopedia*, McGraw-Hill, New York, p. 45.
54. Cifferi, A., Krigbaum, W.R. and Meyer, R.B. (eds) (1982) *Polymer Liquid Crystals*, Academic Press, New York.
55. Varshney, S.K. (1986) *J. Macromol. Sci., Rev. Macromol. Chem. Phys.*, **C26**, 551.
56. Bailey, M. (1986) *Modern Plastic Encyclopedia*, McGraw-Hill, New York, **64**, 10 A.
57. Römer, M. (1993) *Kunststoffe*, **10**, 40.
58. Jackson Jr, W.J., and Kuhfuss, H.F. (1976) *J. Polym. Sci., Polym. Chem. Ed.*, **14**, 2043.
59. Anon. (1990) in *Handbook of Plastic Materials and Technology* (ed. I.I. Rubin), Wiley-Interscience, New York, p. 147.
60. Duska, J. (1986) Liquid crystal polymers, *Plast. Eng.*, December.
61. He, J. and Bu, W. (1994) *CA*, **121**(10), 232,044.
62. Lin, H.C. (1994) *CA*, **121**(10), 231,697.
63. Brode, G.L. and Koleske, J.V. (1973) in *Polymerization of Heterocyclics* (eds O. Vogl and J. Furukawa), Marcel Dekker, New York.
64. Lundberg, R.D. and Cox, E.F. (1969) in *Ring-Opening Polymerization* (eds K.C. Frish and S.L. Reegen), M. Dekker, New York.

65. Saunders, H.J. (1985) *Chem. Eng. News*, **63**(13), 30.
66. Anon. (1991) *Chem. Eng. News*, April, 36.
67. Manno, P.J. (1992) in *Reaction Polymers* (eds W.F. Gum, W. Riese and H. Ulrich), Hanser Publishers, Munich, p. 37.
68. Nelson, D.L. (1992) in *Reaction Polymers* (eds W.F. Gum, W. Riese and H. Ulrich), Hanser Publishers, Munich, p. 452.
69. Boenig, H.V. (1964) *Unsaturated Polyesters: Structure and Properties*, Elsevier, Amsterdam.
70. Bruins, P.F. (ed.) (1976) *Unsaturated Polyester Technology*, Gordon and Breach Science Publishers, New York.
71. Nelson, D.L. (1992) in *Reaction Polymers* (eds W.F. Gum, W. Riese and H. Ulrich), Hanser Publishers, Munich, p. 153.
72. Altenberg, M. and de Jong, J. (1987) US Pat. 4,701,477, Freeman Chemical.
73. Trivedi, B.C. and Culbertson, B.M. (1982) *Maleic Anhydride*, Plenum Press, New York, Chapter 12.
74. Dühl-Mühlbach, I. (1993) *Kunststoffe*, **10**, 55.
75. Updegraff, I.H. (1990) in *Handbook of Plastic Materials and Technology* (ed. I.I. Rubin), Wiley-Interscience, New York, p. 817.
76. *Encyclopedia of Chemical Technology* (1982) Vol. 13, Wiley-Interscience, New York, p. 358.
77. Kim, J. and Kim, S. (1987) *Polym. Eng. Sci.*, pp. 1243, 1252.
78. McQuarrie, T.S. (1984–1985) *Modern Plastic Encyclopedia*, McGraw-Hill, New York.
79. Sadafule, D.S. and Panda, S.P. (1979) *J. Appl. Polym. Sci.*, **24**, 511.
80. Alexander, I.J. and Scott, R.J. (1983) *Br. Polym. J.*, **15**, 30.
81. Martens, C.R. (1961) *Alkyd Resins*, Reinhold Publishing Co., New York.
82. Patton, T.C. (1962) *Alkyd Resin Technology*, Wiley-Interscience, New York.
83. Raech, H. (1965) *Allylic Resins and Monomers*, Reinhold Publishing Co., New York.
84. Wright, R.E. and Gardner, M.F. (1964) in *Handbook of Plastic Materials and Technology* (ed. I.I. Rubin), Wiley-Interscience, New York, p. 43.

5 Polyamides and polyimides

5.1 NYLONS

5.1.1 Introduction

The term nylon, originally a trade name, later became the generic name given to linear aliphatic polyamides. Nylons can be obtained from a single monomer such as amino acids or their cyclic lactams by self-polycondensation or by ring opening polymerization, respectively. Other nylons are based on two monomers, namely aliphatic diamines and dicarboxylic acids from which they are prepared by polycondensation. The polymeric chains of nylons contain only methylene and amide groups.

Nylons are designated by a numbering system. In the case of single monomer nylons the digits that follow the nylon name represent the number of carbon atoms between two successive amide groups, such as nylon 6 and nylon 12. For nylons based on diamines and dicarboxylic acids, the numeral is composed successively of the number of carbons in the diamine and diacid molecules, respectively. Thus, nylon 6,12 (or 612) is made of 1,6-hexamethylenediamine and 1,12-dodecanoic acid. Copolyamides are described by the numerals of their components such as nylon 6/6,6.

The USA's production of nylons was estimated at approximately 1.15 million tons in 1993 [1]. Nylons are used as fibers (80–90% of the total volume), as engineering plastics, adhesives and rubbers [2,3].

The most important nylons produced commercially are:

1. nylon 4,6: $-\!\!\!+\!NH(CH_2)_4NH-CO(CH_2)_4CO\!+\!\!\!-_n$, poly(tetramethylene adipamide);
2. nylon 6,6: $-\!\!\!+\!NH(CH_2)_6NH-CO(CH_2)_4CO\!+\!\!\!-_n$, poly(hexamethylene adipamide);
3. nylon 6,9: $-\!\!\!+\!NH(CH_2)_6NH-CO(CH_2)_7CO\!+\!\!\!-_n$, poly(hexamethylene azelamide);

4. nylon 6,10: $-[NH(CH_2)_6NH-CO(CH_2)_8CO]_n-$, poly(hexamethylene sebacamide);
5. nylon 6,12: $-[NH(CH_2)_6NH-CO(CH_2)_{10}CO]_n-$, poly(hexamethylene dodecanedioamide);
6. nylon 6: $-[NH(CH_2)_5CO]_n-$, polycaprolactam;
7. nylon 11: $-[NH(CH_2)_{10}CO]_n-$, poly(11-aminoundecanoic acid);
8. nylon 12: $-[NH(CH_2)_{11}CO]_n-$, poly(12-aminododecanoic acid).

Various copolyamides are also commercially available such as nylon 6,6/6 or 6,6/6,10.

The first synthesis of a polyamide, namely nylon 6,6 is due to Carothers and his coworkers at Du Pont in 1935 as a result of his intensive research work on condensation polymers. The polymer was produced commercially in 1939 and originally used for fibers and later, in 1941, it was used for moldings [4(a)]. Almost at the same time, P. Schlack at I.G. Farbenindustrie in Germany investigated fiber forming polymers not covered by the Du Pont patents and as a result nylon 6 was produced on a commercial scale in 1940 [4(b)]. The production of nylon 11 and 12 was reported in France in 1955 and 1960, respectively [4(c)]. Nylons 7 and 9 were synthesized and investigated as fiber forming materials by Russian chemists [5].

Of the total volume of the above-mentioned polymers, about 80% is represented by only two nylons, 6 and 6,6; these have achieved large scale worldwide production and uses. All other nylons are designed for specific applications and are produced in restricted geographic areas.

Based on raw materials and on polymerization processes, two types of nylons are produced even if their physical properties are very similar. Those based on a unique monomer (polyamides 6–8 above) and designated by a single number are obtained either by the self-condensation of ω-amino acids or by ring opening polymerization of cyclic lactams:

$$nH_2N{-}R{-}COOH \longrightarrow -(NH{-}R{-}CO)_n- + nH_2O$$

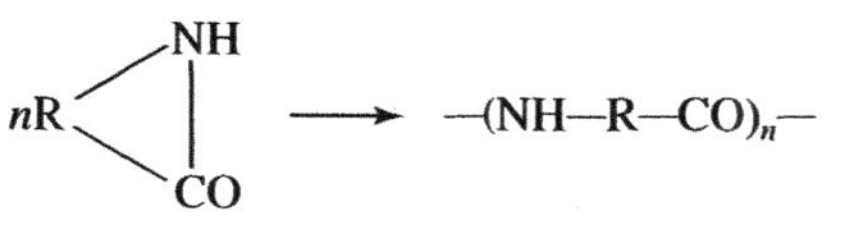

The second type of polyamides (1–5 above) designated by two numbers are synthesized by the polycondensation of aliphatic diamines with aliphatic dicarboxylic acids:

$$nH_2N{-}R{-}NH_2 + nHOOC{-}R_1{-}COOH \rightarrow$$

$$-(HN{-}R{-}NH{-}OC{-}R_1{-}CO)_n- + nH_2O$$

5.1.2 Synthesis of aliphatic polyamides

(a) Nylon 6

Caprolactam is the raw material for nylon 6. It is preferred to ω-aminocaproic acid because it is easier to produce and to purify. Caprolactam may be prepared by various procedures based on the reactions presented in Figure 5.1.

The bulk of manufacture starts from cyclohexane (obtained by the fractionation of petroleum and natural gas) which is air oxidized to cyclohexanone. Cyclohexanone is also produced by the catalytic hydrogenation of phenol. In the next step, cyclohexanone is treated with hydroxylamine sulfate to form cyclohexanone oxime. The oxime is stirred with oleum (sulfur trioxide in sulfuric acid) and undergoes Beckmann rearrangement to yield the caprolactam. In this process, a cumbersome amount of 4–5 tons of ammonium sulfate per ton of caprolactam is obtained as a by-product. To avoid the use of hydroxylamine sulfate, the oxime can be alternatively prepared by catalytic ammoxidation of cyclohexanone with ammonia and hydrogen peroxide [6]. Another method, which avoids such high quantities of ammonium sulfate, involves a photochemical reaction of nitrosyl chloride with cyclohexane and gives cyclohexanone oxime directly [7].

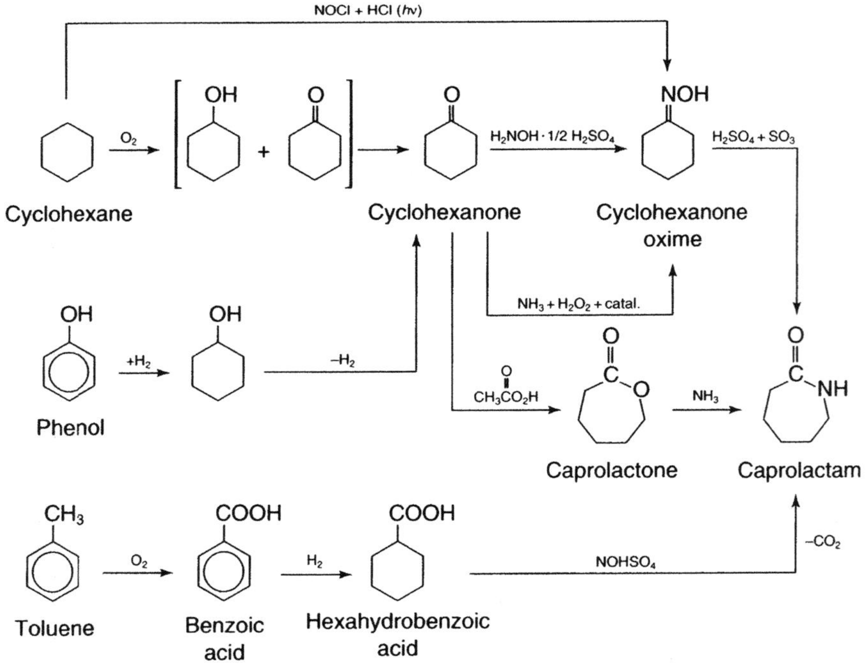

Figure 5.1 Various methods for caprolactam synthesis.

Caprolactam can be prepared from cyclohexanone via caprolactone. Cyclohexanone is oxidized with peracetic acid (made by air oxidation of acetaldehyde) to caprolactone which is then treated with ammonia to form caprolactam [7]. A different process involves oxidizing toluene to benzoic acid in air, followed by reduction to hexahydrobenzoic acid and treatment with nitrosylsulfuric acid to give caprolactam.

Pure caprolactam is a white, hygroscopic, crystalline solid melting at 69.3°C.

Nylon 6 is prepared by hydrolytic or anionic ring opening polymerization of caprolactam [8,9,10]. The water initiated polymerization is believed to begin by the hydrolysis of caprolactam to form the amino acid which subsequently reacts with lactam by ring opening:

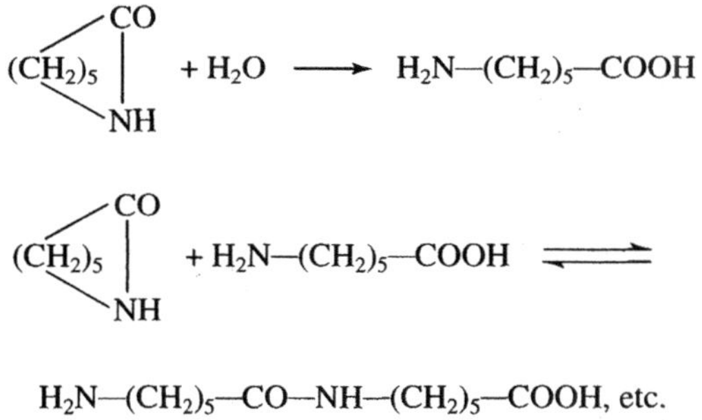

Water initiated polymerization can be brought about by both batch and continuous processes. In a typical batch procedure, the molten caprolactam is mixed with 5–10% water by weight, which constitutes the catalyst, and with 0.1% acetic acid as the MW regulator. All the operations are effected under a nitrogen blanket in order to prevent discoloration. The mixture is maintained at 250°C and 1.5 MPa for 12 hours. The resultant polyamide is extruded as ribbon and cut into chips. The product contains up to 10% low MW compounds (unreacted caprolactam, higher lactams and amino acids). A good quality product is obtained by the demonomerization of the raw polymer either by leaching with water at 85°C or by heating at 180°C and 5 kPa.

Similar reaction conditions are used in the continuous polymerization process. The reaction mixture is continuously fed and resides in the reactor for about 18 hours at 250°C.

The flow chart of the hydrolytic polymerization process of caprolactam is presented in Figure 5.2.

Rapid polymerization of caprolactam, achieved by the use of anionic initiators, was developed for the casting polymerization procedure *in situ* and for the reaction injection molding technique (RIM).

Anionic polymerization of caprolactam is carried out under anhydrous conditions using strong bases (B^-M^+) such as metal hydrides, metal

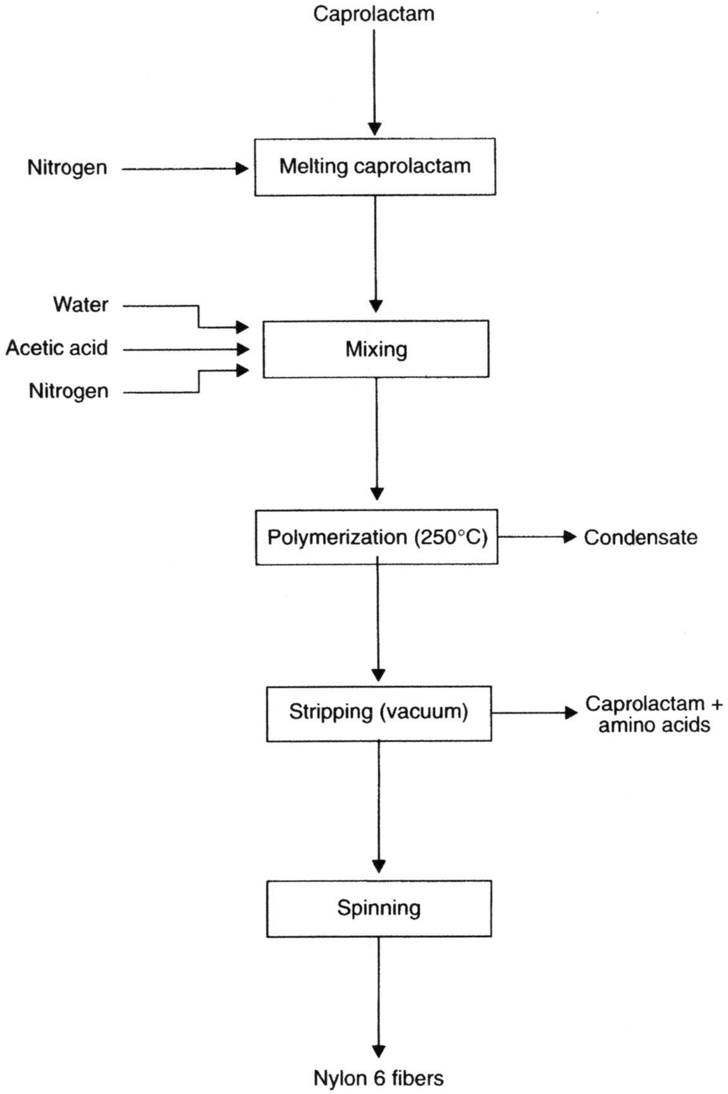

Figure 5.2 Hydrolytical polymerization of caprolactam.

amides and alkali metals as catalysts. The mechanism using base $B:^-$ involves initial formation of a resonance stabilized anion (**1**) which reacts slowly with the monomer and gives the amine anion (**2**) which is not stabilized by resonance.

Because the second step is slow, this type of ring opening polymerization by strong bases alone is characterized by an induction period [11]. Once the active amine anion (**2**) is formed it undergoes rapid proton transfer;

the acyllactam (**3**) which is formed participates in the propagation.

Initiation:

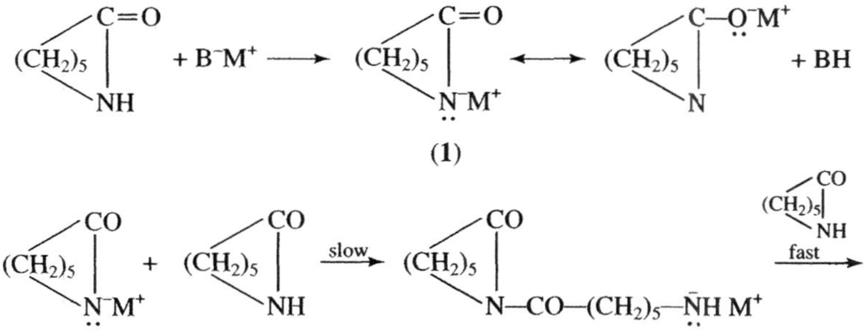

(**1**) (**2**)

Propagation:

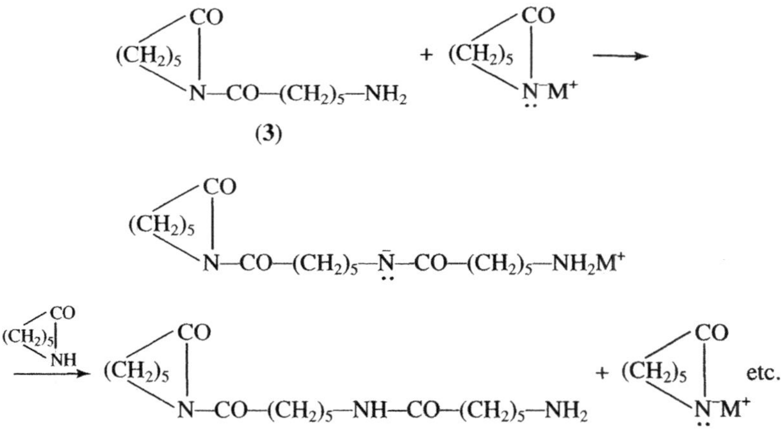

(**3**)

The induction period may be eliminated by the addition of co-initiators (called activators) such as *N*-acyllactams that ensure rapid polymerization. A typical initiating system consists of a mixture composed of 0.15–0.50 mol% catalyst and 0.1–1 mol% of initiator. Common catalysts are caprolactam magnesium bromide and sodium caprolactam. The initiators used most often are bis-acyllactams and reaction products of diisocyanates or of isocyanate terminated prepolymers with caprolactam [7,12]. The reaction temperature is initially about 150°C but during the polymerization it rises to about 200°C. The liquid mixture is poured into a mold and polymerized *in situ* in four minutes. Very large moldings, up to one ton in weight, have been produced by casting techniques [13]. A recent study on the kinetics of anionic polymerization of caprolactam demonstrated that the polymerization and crystallization time at 150°C is less than 90 s [14].

RIM technique, a variation of the casting polymerization technique initially developed for polyurethanes, has also been adapted for nylon 6 and used for making mechanical parts such as gears, bumper covers and housings for business machines [15]. Nylon RIM (discussed in Section 5.3). are block copolymers of caprolactam and polyols.

(b) Nylon 11

The polymer is prepared by the self-condensation of ω-aminoundecanoic acid, $H_2N-(CH_2)_{10}-COOH$, a crystalline solid melting at 189°C. The raw material for the amino acid is castor oil, a vegetable product composed of the triglyceride of ricinoleic acid. By a sequence of reactions ω-aminoundecanoic acid is obtained. This amino acid may also be produced by telomerization reactions [5].

The self-condensation of ω-aminoundecanoic acid is conducted in a continuous process at about 220°C withdrawing water continuously [16]. The latter stages of the process are conducted under reduced pressure. Small quantities of the order of 0.4–0.6% of 12-membered ring lactam are produced by intramolecular condensation but it is not normally removed since its presence has little effect on the properties of the polymer.

(c) Nylon 12

Nylon 12 is obtained by the hydrolytic polymerization of dodecanelactam $HN-(CH_2)_{11}-CO$ prepared from butadiene by a multistaged reaction. The process is conducted at about 300°C in the presence of aqueous phosphoric acid. It is not an equilibrium reaction such as the polymerization of caprolactam and hence the residual monomer concentration is low and the extraction step is eliminated [17].

(d) Preparation of nylon 6,6

Nylon 6,6 is prepared by the polycondensation reaction of adipic acid and 1,6-hexamethylenediamine. The raw material for adipic acid is cyclohexane obtained from petroleum or natural gas by fractionation. Another source of cyclohexane is benzene which is hydrogenated using Raney nickel as catalyst. Cyclohexane is oxidized with air using cobalt naphthenate as catalyst. The mixture of cyclohexanol and cyclohexanone obtained in the first step is subsequently oxidized in the second step with 50% nitric acid and an ammonium vanadate–copper catalyst:

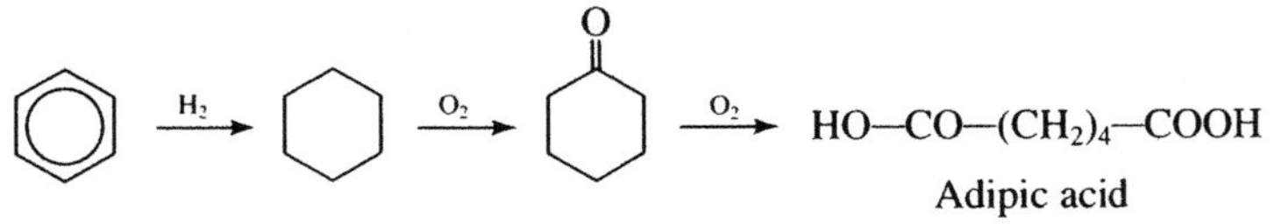

The crystals of adipic acid, $HOOC{-}(CH_2)_4{-}COOH$, are separated from the reaction mass by centrifuging and drying. Their melting point is 152°C.

1,6-Hexamethylenediamine, $H_2N{-}(CH_2)_6{-}NH_2$, is prepared by the hydrogenation of adiponitrile which may be obtained by several commercial procedures. The standard route to adiponitrile is by treatment of adipic acid with ammonia. By other more recent methods, adiponitrile is obtained either by the reaction of butadiene with hydrogen cyanide or by the electrolytic dimerization of acrylonitrile:

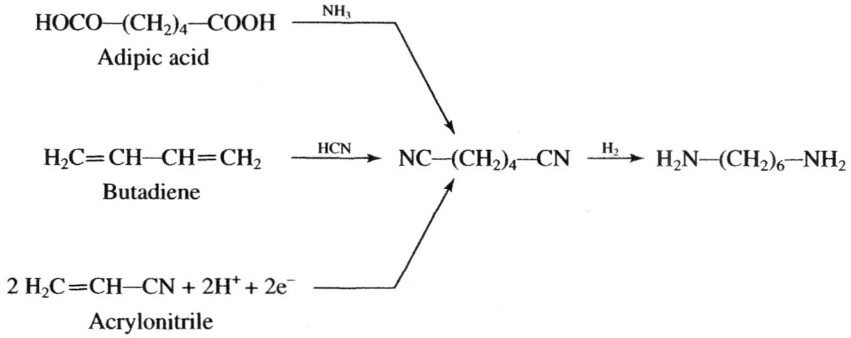

In order to obtain high MW nylon 6,6 (10 000–15 000), it is necessary to ensure the equivalence of the number of amino and acid groups. Equivalent amounts of adipic acid and hexamethylene diamine form the hexamethylene-diammonium adipate known as 'nylon salt':

$$HOOC{-}(CH_2)_4{-}COOH + H_2N{-}(CH_2)_6{-}NH_2 \longrightarrow$$

$$[^{-}OOC{-}(CH_2)_4{-}COO^{-}][^{+}H_3N{-}(CH_2)_6{-}NH_3^{+}]$$

Nylon salt

$$\longrightarrow -\!\!\left[CO{-}(CH_2)_4{-}CO{-}NH{-}(CH_2)_6{-}NH\right]_n + 2H_2O$$

The first step, prior to polycondensation, is the preparation of nylon salt. Two commercial procedures are employed. One of them consists of the exact neutralization, determined electrometrically, of an aqueous solution of adipic acid with hexamethylenediamine followed by its concentration to 75–80%. In the other method the salt is prepared by mixing the acid and the amine in boiling methanol from which the insoluble salt precipitates out.

The subsequent polycondensation of nylon salt may be carried out as a discontinuous or continuous process. In the batch process, the 75–80% solution of the salt together with acetic acid used as MW regulator are fed into the reactor and heated to 220°C at an autogeneous pressure of about 2 MPa for 1–2 hours. The temperature is then raised to 270–280°C and steam is bled off to maintain the same pressure. The polycondensation is completed by heating for two more hours while reducing the pressure to atmospheric. The molten polymer is extruded by nitrogen on a water cooled drum to

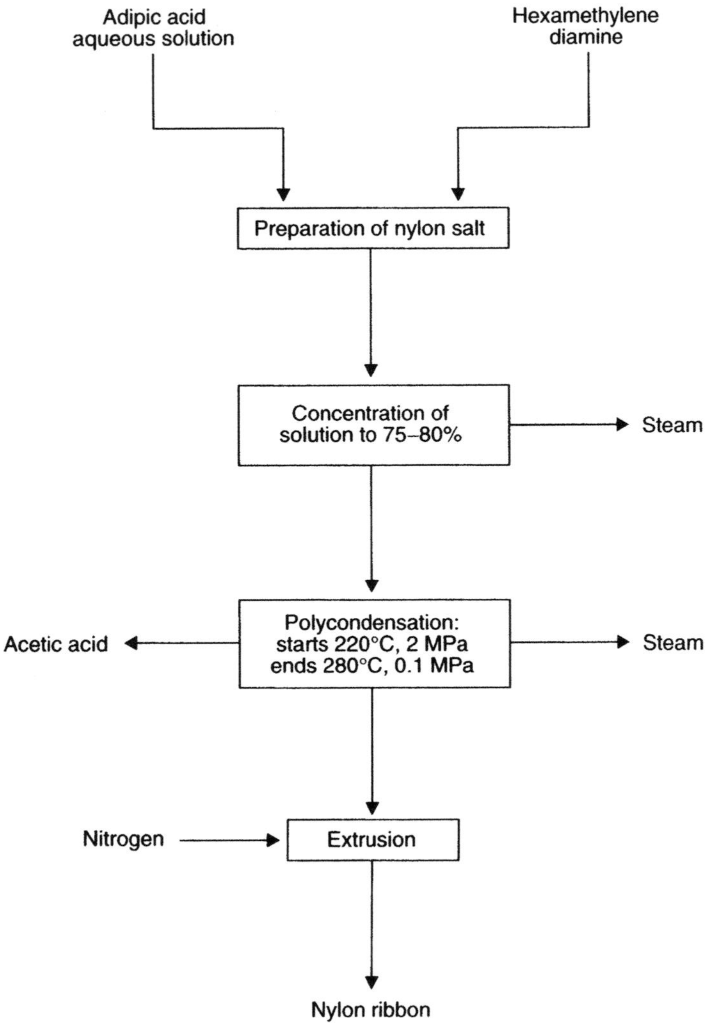

Figure 5.3 Preparation of nylon 6,6.

form a ribbon which is subsequently disintegrated. In the continuous process the reaction mixture traverses slowly different reaction zones in a coiled tube reactor and the polymer is generally directly melt spun into fiber.

The flow chart of the preparation of nylon 6,6 is presented in Figure 5.3.

(e) Nylons 4,6; 6,9; 6,10 and 6,12

Compared to nylon 6 and 6,6, these polymers are prepared in much smaller quantities and are designed for specific applications. They are prepared in an essentially analogous manner, mostly by discontinuous processes. The nylon salts are synthesized in the first step from hexamethylenediamine and the corresponding dicarboxylic acids [18]:

- azelaic acid, 9 C atoms (obtained from oleic acid) for nylon 6,9;
- sebacic acid, 10 C atoms (obtained from castor oil) for nylon 6,10;
- dodecandioic acid, 12 C atoms (obtained from butadiene) for nylon 6,12.

Nylon 4,6 introduced in the late 1970s is prepared from 1,4-diaminobutane and adipic acid [19].

5.1.3 Properties of nylons

Nylons are linear thermoplastic polymers containing highly polar amide groups –NH–CO– separated by flexible aliphatic hydrocarbon chain segments. Amide groups lead to strong intermolecular hydrogen bonding so that polymers are composed of crystalline regions with high cohesive energy density and amorphous zones that impart some flexibility to the polymer.

These structural characteristics determine the general properties of nylons such as:

- sharp melting points, at high temperatures;
- high tensile strength, rigidity, hardness and resistance to creep;
- high heat deflection temperatures;
- high abrasion resistance;
- low thermal coefficient of linear expansion;
- reduced solubility;
- tendency to absorb water from the environment.

The general properties of commercial nylons are presented in Table 5.1.

Nylons with higher concentrations of amide groups have higher T_m, tensile strength and water absorption but lower impact strength. Longer aliphatic segments between amide groups impart more flexibility, higher impact strength and are easier to process because of their lower T_m. The glass transition temperatures of nylons are below room temperature so that materials have a certain flexibility.

The length of aliphatic segments and the even or odd number of methylene groups determine the packing ability of the polymer chains and the frequency of hydrogen bonding, and consequently the percentage of crystallinity and the morphological structure of crystalline regions.

Processing conditions have also a very important influence on the crystallinity of nylons. Slowly cooled and subsequently annealed nylon 6 may reach 50–60% crystallinity compared to only 10% for a rapidly cooled material [13]. Small and uniform spherulites are formed upon slow cooling. Nucleating agents such as fine silica (in amounts of 0.1%) favor the formation of

Table 5.1 General properties of nylons

Property	*4,6*	*6,6*	*6,9*	*6,10*	*6,12*	*6*	*11*	*12*	*6,6/6,10 (35:65)*	*6,6/6,10/6 (40:30:30)*
Density (g/cm^3)	1.18	1.14	1.09	1.09	1.07	1.13	1.04	1.02	1.08	1.09
Melting temperature (°C)	280	264	205	215	210	220	185	175	195	160
Heat distortion temperature (°C)										
– at 0.5 MPa	–	199	–	160	160	160–180	150	140	–	–
– at 1.8 MPa	150–170	100–104	60	66–85	65	54–91	55	50	–	30
Tensile strength (MPa)	100	80	59	59	59	76	38	46	38	52
Tensile modulus (GPa)	3	3	–	2.1	–	2.8	1.4	1.4	1.4	1.4
Elongation at break (%)	30	80–100	–	100–150	150–300	50–100	280–300	200	>200	300
Izod impact strength (J/m)	–	27–53	59	85–107	53–101	53	96	101	107	–
Rockwell hardness	R123	R118	–	R111	R114	R112	R108	R107	–	R83
Water absorption at saturation (%)	14	8.0	–	2.4	3	9.0	2.0	1.8	6.5	10.7
Coefficient of linear expansion (10^{-5} cm/cm · °C)	9.2	10	–	15	9	9.5	15	12	–	–

such spherulites that give polymers with more uniform structures, enhanced mechanical properties and high abrasion resistance.

The melt viscosity of nylons is rather low due to their high melting temperatures and to relatively low MW.

The solubility of nylons is very reduced because of their crystalline structure. Only those solvents with similar high solubility parameter are capable of interaction with the polymer. These are strong proton donor solvents such as formic and acetic acids, concentrated mineral acids and fenols. Nylons are resistant to aliphatic and aromatic hydrocarbons, lubricants, engine fuels, hydraulic fluids, refrigerants and paint solvents. They are also resistant to aqueous solutions of alkalis and inorganic salts at room temperature.

On exposure to outdoor conditions, particularly in film and filament applications, discoloration and reduction in mechanical properties occur because of relatively poor UV resistance and stabilizers such as carbon black are recommended.

All nylons are hygroscopic to some extent, the higher the content of amide groups, the higher the water absorption. Thus, nylon 4,6 absorbs about 14% of water; it has the highest level of absorption among nylons. The absorbed water acts as plasticizer reducing the tensile strength, tensile modulus, hardness and resistance to creep but increasing the impact strength.

Another effect of water absorption is the dimensional instability: for each 1% of absorbed water there would be a 0.9% increase in volume [18]. In humid conditions of work, nylon 11 and 12 should be chosen as their water absorption is lower. Glass reinforced nylons (30%) retain their dimensions particularly well in a humid environment. At room temperature and under conditions of low humidity, nylons are good electrical insulators at low frequencies.

Polyamides exhibit good oxygen barrier properties but have a high permeability for carbon dioxide [20]. The combination of these two characteristics is put to profit in food packing applications.

Nylons exhibit molding shrinkage specific to crystalline polymers. For nylon 6,6 the molding shrinkage is 0.018 cm/cm [18]. An after-shrinkage effect, especially for nylon 6,6, may occur for periods up to two years due to a very slow relaxation of molding stresses and absorption of water which produces a plasticizing effect. In order to provide more dimensional stability, the after-shrinkage may be accelerated by annealing the parts at temperatures that correspond to maximum crystallization rates.

Various nylon random copolymers based on four, five or more aliphatic and cyclic monomers are developing rapidly, to afford ever-new copolyamide variations. They are prepared commercially either by copolymerizing more than two monomers or by mixing two or more nylons before processing by amide interchange reactions. By copolymerization, products with tailor-made property combinations and improved processability can be manufactured. As expected, the resulting structures are less regular, which considerably reduces or even cancels the crystallinity. Consequently, nylon copolymers have lower T_m, are easier to process, have lower tensile strength, are more flexible, tougher and more transparent.

Additives significantly change one or more of the properties of nylon resins. They are incorporated into nylons by the manufacturers and only general information is made available. The following additives are used in nylons.

1. Nucleating agents, such as fine silica or phosphorus compounds, are used to control the percentage of crystallinity and the morphological structure of the crystalline regions. They contribute to higher nucleating rates which lead to smaller spherulites. Consequently, polymers have more uniform structures and higher impact strength and abrasion resistance. High melting point polymers such as nylon 6,6 or poly(ethylene terephthalate) may be used as nucleating agents, especially for nylon 6 [13].
2. Heat stabilizers are used in order to obtain good resistance to high temperature aging. They are mercaptobenzimidazole, mercaptobenzothiazole, phenylnaphthylamine, phosphoric acid esters [21] and especially copper salts which are very effective.
3. Light stabilizers such as carbon black improve the performance of nylons in outdoor applications where their resistance to ultraviolet radiation is relatively poor.
4. Impact modifiers are incorporated into impact grade nylons. They are rubber-type materials such as polyurethanes, ethylene–propylene rubbers, ionomers, ABS polymers, acrylates and methacrylates. Rubber droplets are dispersed in the matrix of nylon without any interaction between the two phases.
5. The most common additives for nylon resins are the reinforcing fillers. They may be glass fibers, glass beads, carbon fibers or certain mineral fillers such as talc. They impart a significant increase in stiffness, impact tensile strength, hardness as well as in heat deflection temperature and maximum service temperature. Self-extinguishing properties of glass filled nylons are also improved.

A comparison of the properties for unfilled and filled nylon 6 and nylon 6,6 are presented in Table 5.2.

Glass fillers are added in the amount of 15–50% to the resin. They are treated with silane coupling agents before blending in order to improve the resin–glass bonding.

When glass beads are used as fillers, the composite nylon resin has an increased compressive strength up to eightfold as compared to unfilled material.

Nylon resins may also come with other additives such as plasticizers, lubricants, flame retarders, pigments, fungicides, etc. depending on their specific

Table 5.2 Effect of glass fiber reinforcement on properties of nylon 6 and 6,6

	Nylon 6		*Nylon 6,6*	
Property	*Unfilled*	*Filled*	*Unfilled*	*Filled*
Tensile strength (MPa)	76	115–220	80	160
Elongation (%)	50–100	2–3	80–100	3–5
Flexural modulus (MPa)	2410	5000–11 000	3000	8000
Izod impact strength (J/m)	53	60–190	53	–
Heat deflection temperature under load (°C)				
– 0.5 MPa	160–180	220	200	250
– 1.8 MPa	54–90	190–215	100–104	250
Coefficient of linear expansion (10^{-5} cm/cm · °C)	9.5	–	9.9	2.8
Water absorption at saturation (%)	9	–	8–9	5–6
Maximum service temperature (°C)	180	199	199	241

applications. The most frequently used halogen-free flame retarder is based on melamine derivatives [20].

Waste polyamides can be split into their monomers by hydrolysis or recycled and reused, maintaining a good level of properties [20].

Modified nylons with reduced hydrogen bonding, in which about 33% of amide hydrogen atoms have been substituted by methylmethoxy groups, are also available on the market. The methoxymethylation is accomplished by the treatment of the resin dissolved in 90% formic acid with formaldehyde and methanol, in the presence of phosphoric acid as catalyst:

$$\left[NH(CH_2)_6NH{-}CO(CH_2)_4CO\right]_n + CH_2O + CH_3OH \longrightarrow \left[NH(CH_2)_6N(CH_2O{-}CH_3){-}CO(CH_2)_4CO\right]_n + H_2O$$

This methylmethoxy nylon is then crosslinked by heating at 120°C with 2% citric acid:

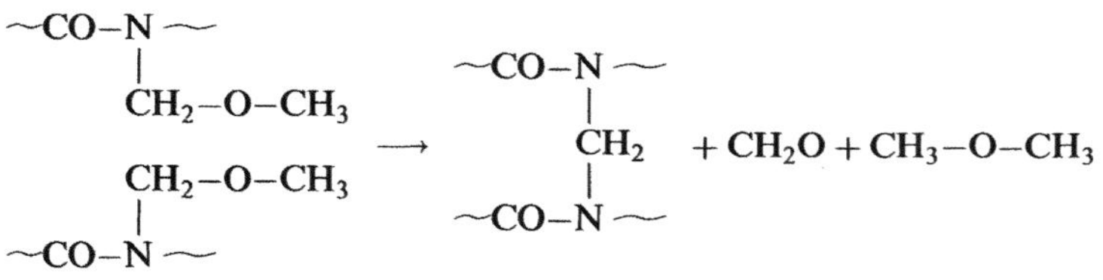

Methylmethoxy nylons have been used as flexible and abrasion resistant coatings.

5.1.4 Applications of nylons

The bulk of about 80% of the total production of nylons goes into fibers where high tensile strength polymers are required [1]. However, due to their outstanding properties, nylons have become indispensable in almost all branches of engineering. Since the price of nylons is almost three times that of general purpose materials (polyethylene and polystyrene), they are used as plastics only in those applications where high mechanical characteristics are required. In 1992, the worldwide market volume of nylons for non-fiber applications reached 1 million metric tons [20].

Several factors such as mechanical properties, dimensional stability and ease of processing have to be considered for the selection of a nylon resin for a specific application. Nylon 6 and 6,6 have the highest tensile strength and are mainly used for fibers. Nylon 6 is the best material for shock-proof parts since it is tough. High mechanical loads are well withstood by nylon 6,6, which has the greatest hardness and stiffness. Nylons 11 and 12 are preferred for cable sheathing, surgical sutures and for precision engineering parts as they are easier to process, have smaller hygroscopicity and are more dimensionally stable.

More than 80% of nylons go into synthetic fiber applications, for clothing, tire cords, ropes and mostly for carpets.

The major engineering field is in the automotive industry as parts for chassis, steering wheels, steering column bearings, bumpers, housings for rear view mirrors, parts for electrical equipment, for transmission systems, for fuel supply system, parts in the engine block and in the lubricating system and many others. For example, fuel tanks are composed of six coextruded plastic layers, with a central barrier forming polyamide layer.

In mechanical engineering they are used in a large variety of parts such as gears, cams, bearings etc. and particularly where moving nylon parts may often be operated without lubrication and are silent running.

The second largest consumer of polyamides is the electrical industry. In telecommunications nylon resins are used for circuits and relays, for contact makers, for drive units, for cable sheaths, for plug and socket devices, etc. Nylons 11 and 12 are used as coatings for underground cables in hot continents where the resistance to termite attack and formic acid are particularly important.

In building and sanitary industries they are used as water heater tanks, heat exchangers, thermostat casings, fans, chairs and other furniture, sterilizable moldings for medical uses, hair combs, etc.

Nylon films, particularly from nylon 11, are used in packaging pharmaceutical products and foodstuffs especially for those applications that need boil-in-the-bag treatment. Chemical and abrasion resistant coatings are obtained from nylon 11 by fluidized bed dipping and powder spraying techniques.

Nylon 6 is used for the production of large objects such as propellers for marine craft by the polymerization casting of caprolactam, as was already mentioned.

Composite glass reinforced nylon materials have replaced metals in many applications due to their outstanding mechanical properties. Carbon fiber reinforced nylon found use in aerospace and tennis racquet applications [13]. Blends of nylon 6,6 and thermoplastic elastomers such as ethylene–propylene–diene terpolymer and super-tough materials that offer new options for hard–soft composites [20].

Completely novel applications have been opened by blending copolyamides and amylose-containing starch which may be processed into biologically degradable molded articles and films [20].

5.2 ALIPHATIC POLYAMIDES FROM DIMER ACIDS

Drying vegetable oils (tung, soybean and linseed oils) contain a mixture of esters of polyunsaturated acids such as linoleic acid, $C_{18}H_{32}O_2$. The mixture of free acids obtained by the saponification of oils, is subsequently heated in the presence of a catalyst and the unsaturated acids are dimerized, trimerized and polymerized. This complex mixture of inexact structure ranging from dark viscous liquids to brittle resins is known as 'dimer fatty acids' as 60–75% of its content is composed of dimerized acids.

The dimer acid is subsequently used as the acid component for the preparation of polyamides with bulky side groups. Suitable amines are ethylenediamine or diethylenetriamine and the structure of the polyamide may be:

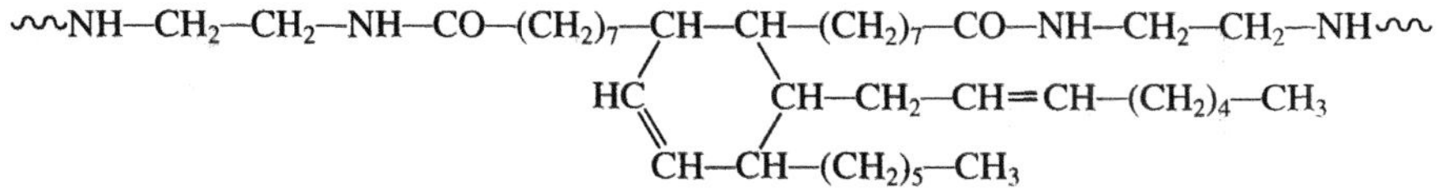

Some of the fatty polyamides are tough and flexible solid resins and find use as hot melt adhesives and as coatings for flexible substrates such as plastic films, paper, etc. Those fatty polyamides prepared with amines that have a functionality of higher than two, are highly branched and liquid. They have free amine groups and are mainly used as curing agents and flexibilizers for epoxy resins intended for adhesives and thixotropic paints.

5.3 POLYAMIDE THERMOPLASTIC ELASTOMERS

Polyamide thermoplastic elastomers are block copolymers that exhibit the characteristics of vulcanized rubber at normal temperatures but behave as thermoplastics on heating. Similar to other thermoplastic elastomers (polyurethanes and polyesters), the chains of polyamide thermoplastic elastomers are composed of two different alternating blocks, a hard thermoplastic block with a T_g well above room temperature and a soft elastomeric block which has a T_g below room temperature.

The hard segment may be any of the nylon-type polyamides described previously, most frequently polycaprolactam and polylaurinlactam. The soft segments are composed of polyether diols such as poly(oxyethylene) glycol, poly(oxypropylene) glycol or poly(oxytetramethylene) glycol. A typical structure of a polyamide thermoplastic elastomer based on polycaprolactam and poly(oxytetramethylene) glycol is:

$$\left[NH-(CH_2)_5-CO \right]_x O \left[(CH_2)_4-O \right]_y$$

Hard segment Soft segment

The hard segments of different chains are linked to each other by strong hydrogen bondings characteristic to polyamide polymers, and constitute hard thermoplastic semicrystalline domains interspersed in the continuous soft matrix. They act as physical crosslinking points and are responsible for the high tensile strength, abrasion resistance and high service temperature (up to 150°C). The flexibility properties of the material are imparted by the amorphous soft elastomeric matrix. Upon heating, the hard phases melt and the material is fluid and can be easily processed. On cooling, reformation of the physical crosslinks and the separation of phases occur and a rubbery material is obtained.

Polyamide block copolymers may be prepared either by the classical polycondensation process or by the newer reaction injection molding process (RIM) [7,15,22].

By the first route, a polyamide with MW = 300–15 000 and carboxyl end groups is first prepared, and it is subsequently condensed with the polymeric diol (MW = 100–6000) to give the block copolymer.

In the RIM process for nylons, as in the case of PU, the polymer is formed very rapidly from the reaction components. Two components are used in the nylon RIM process: the first is a solution of the catalyst in caprolactam and the second is the mixture of caprolactam with the initiators. Preferred catalysts are caprolactam magnesium bromide made from caprolactam and alkyl magnesium bromide and sodium caprolactam, made from caprolactam and sodium hydride [7]. The most frequently used initiators are bis-acyl-lactams such as adipyl-bis-caprolactam, the reaction products of diisocyanates with caprolactam or of isocyanate terminated prepolymers with caprolactam [7].

Contrary to PU, nylon–RIM reactions are endothermic and the process is conducted at 130–140°C with cycle times of 2–3 minutes. In the RIM process the injection pressures are low, requiring much lower locking forces of the mold than in conventional injection molding. This represents an advantage for the processing of large moldings.

One recent development has been the preparation of polyamide elastomers by a new route based on the reaction of dicarboxylic acids with diisocyanates [23–25]. The reaction implies the intermediate formation of unstable carbamic carboxylic anhydride which decomposes to amide with the evolution of carbon dioxide. The reaction is represented schematically below:

$$\text{R-N=C=O} + \text{R}'\text{-COOH} \longrightarrow [\text{R-NH-CO-O-CO-R}'] \longrightarrow \text{R-NH-CO-R}' + \text{CO}_2$$

It was shown that the temperature and catalysis affect the distribution of products obtained, as undesirable side reactions can take place [25]. Amide formation is base catalyzed but higher conversions of monomers to amide and consequently higher MW polyamides are obtained using as catalyst a mixture of methylphospholine oxide and dimethylphospholine oxide [26]:

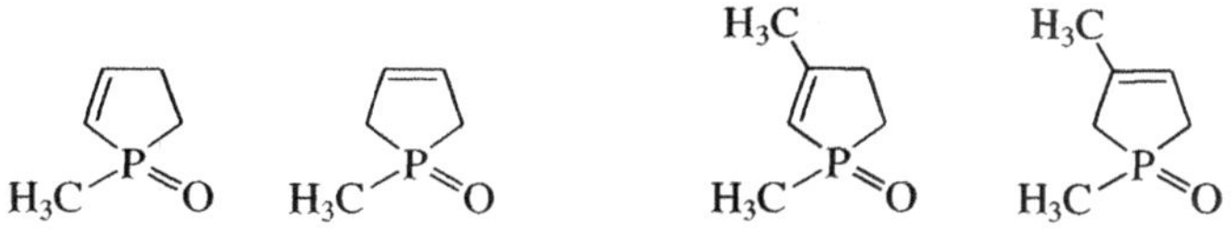

Methylphospholine oxide Dimethylphospholine oxide

The phospholine oxide catalyst participates in the reaction but is regenerated and recuperated at the end and can be reused. The solvents that gave better results at high polycondensation temperatures were tetramethylene sulfone, diphenyl sulfone and butyrolactone [27].

The polyamide elastomers are based on 4,4′-methylene bis(phenylisocyanate) (MDI), a small molecule dicarboxylic acid and an aliphatic prepolymer of polyester-, polyether- or polycarbonate-type with $M_n = 500$–5000 and carboxylic end groups [23,25]:

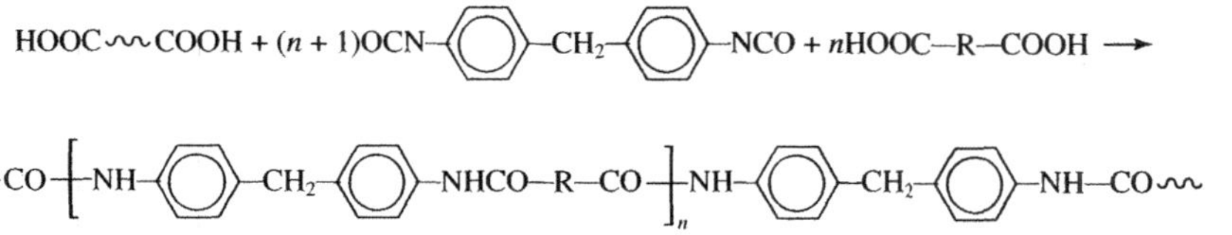

Soft segment Hard segment Soft segment

The polycondensation can be carried out in two ways. In the solution variant (Figure 5.4), the acid and acid prepolymer are dissolved in the inert solvent and mixed with the catalyst. MDI is then added under stirring to the solution at elevated temperature. Evolved carbon dioxide is vented out. At the end the polyamide is separated from the solvent and catalyst in film evaporators, solid polymer is pelletized and stored while the recovered solvent and catalyst are recycled to the reactor. An alternative and more

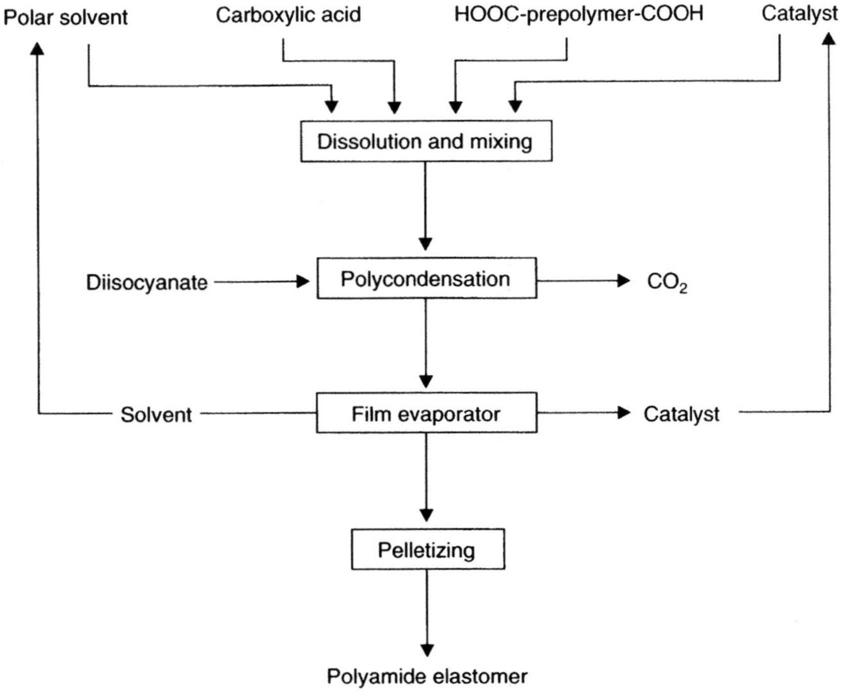

Figure 5.4 Preparation of polyamide thermoplastic elastomers from diisocyanates.

economic way is the melt polycondensation conducted in a properly vented extruder.

The properties of polyamide thermoplastic elastomers are controlled by the appropriate selection of the type, the length and the ratio of the hard and soft blocks. The hardness, the flexibility and the impact strength are maintained over the temperature range $-40°$ to $+80°C$. Polyester based elastomeric polyamides are susceptible to hydrolysis in humid conditions and hydrolytic stabilizers such as polymeric carbodiimides are used. Abrasion resistance is as good as in the case of PU. UV stability is excellent even without stabilizers. The general range of properties of polyamide thermoplastic elastomers are presented in Table 5.3.

The properties of thermoplastic polyamide elastomers place them on top of all other thermoplastic and thermoset elastomers but they are also the most expensive [28]. They share 2.5% of the market of all thermoplastic elastomers. Thermoplastic elastomers have the major environmental advantage over thermoset rubbers that they are completely recyclable at the end of their service life.

Polyamide elastomers can be favorably blended with nylons, PET and PBT. They accelerate the crystallization rates of PET and PBT [25]. Usual fillers are glass fibers and minerals.

Table 5.3 Some properties of polyamide thermoplastic elastomers

Property	*Value*
Density (g/cm^3)	1.01–1.15
Melting temperatures (°C)	120–270
Maximum service temperature (°C)	150
Tensile strength (MPa)	18–42
Tensile modulus at 300% (MPa)	18–33
Elongation (%)	270–500
Hardness (Shore)	60A–63D
Flexural modulus (MPa)	20–410
Water absorption (% vol.)	1.2–120

Applications include automobile bumpers and fascia, housings for business machines, gaskets and seals, watch straps, high performance athletic goods, military shoes, etc. The worldwide consumption of polyamide thermoplastic elastomers in 1992 was estimated at 7000 tons [20].

5.4 PARTIALLY AROMATIC POLYAMIDES

Partially aromatic polyamides include those polyamides in which at least one of the components is aromatic. This group of polyamides constitutes high performance engineering thermoplastics with excellent temperature stability. They comprise a large variety of materials from completely amorphous and transparent to partially crystalline [13,20,24]. Aliphatic and cycloaliphatic amines are preferred to aromatic amines which have poor reactivities and high melt viscosities. The acids generally used are terephthalic and isophthalic acids. Polyamides based on aromatic amines have also been prepared, but by using an alternative route based on commercially available aromatic diisocyanates and various dicarboxylic acids [25].

1. Poly(trimethylhexamethylene terephthalamide) is an amorphous copolyamide prepared by the polycondensation of terephthalic acid with a 1/1 mixture of 2,2,4- and 2,4,4-trimethylhexamethylene diamine:

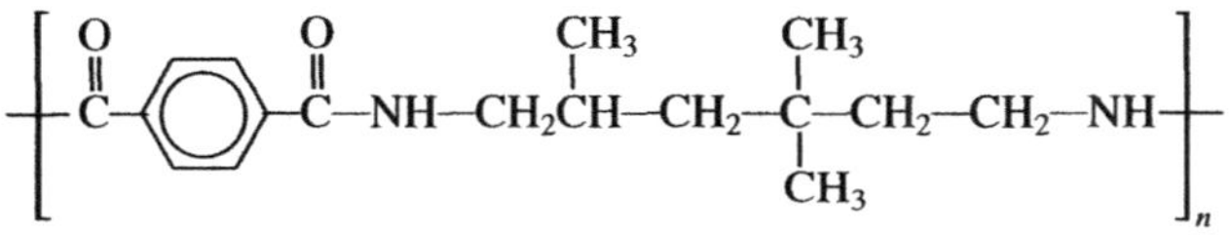

The presence of substituent methyl groups on both amine molecules as well as the irregular structure of the copolymer prevent the crystallization of the amorphous material, which is consequently highly transparent and more soluble than nylons. The aromatic structures in the polymer chain account

Table 5.4 Some properties of poly(trimethyl hexamethylene) terephthalamide

Property	*Value*
Density (g/cm^3)	1.12
Tensile strength (MPa)	68
Tensile modulus (MPa)	3000
Elongation (%)	70
T_g (°C)	150
Deflection temperature at 1.8 MPa (°C)	130
Molding shrinkage (cm/cm)	0.007
Refractive index	1.566

for a T_g as high as 150°C, good heat deformation resistance, rigidity and hardness. Compared to nylon, the material has lower water absorption and better electrical insulation properties and very low molding shrinkage. Some properties are given in Table 5.4.

2. Another amorphous and transparent copolyamide with similar properties is prepared from isophthalic acid (**4**), bis-(4-amino-3-methylcyclohexyl) methane (**5**) and laurinlactam (**6**):

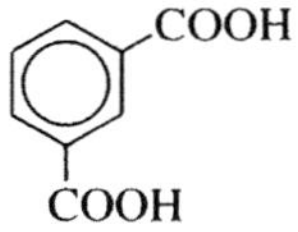

(**4**)

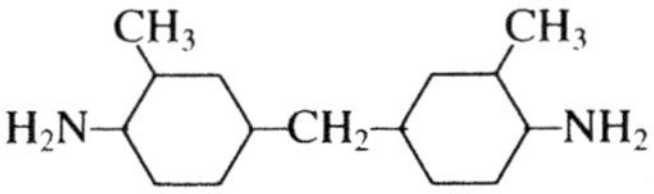

(**5**)

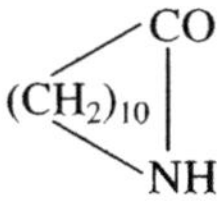

(**6**)

This copolyamide has a very low water absorption and the lowest density (1.06 g/cm^3) among polyamides.

3. Various transparent copolyamides have been prepared by the condensation of 4,4′-diphenylmethane diisocyanate with (a) a 30/70 mixture of adipic (C6) and azelaic (C9) acids and (b) a 50/50 mixture of isophthalic and azelaic acids and 2,4-tolylene diisocyanate. The general structure of type (a) polyamide is given below:

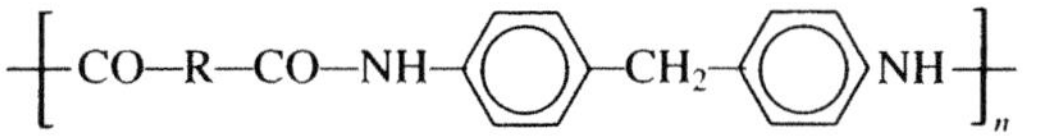

where R = $(CH_2)_4/(CH_2)_7 = 30/70$.

High MW polyamides are obtained when the condensation is conducted by adding MDI to a solution that contains the mixture of carboxylic acid and phospholine oxide catalysts at temperatures above 150°C [26]. The best solvents are tetramethylene sulfone, diphenyl sulfone or butyrolactone [27]. At the end of the polycondensation, the solution is fed into evaporators where the solvent and the catalyst are stripped and the polymer

Table 5.5 Some properties of MDI based polyamides

Property	*Value*	
	Polyamide 70/30	*Polyamide 50/50*
Density (g/cm^3)	1.17	1.16
T_g (°C)	140	180
Heat deflection temperature at 1.8 MPa (°C)	127	160
Tensile strength (MPa)	83	82
Tensile modulus (MPa)	1869	2289
Elongation at break (%)	80	30
Mold shrinkage (cm/cm)	0.005	0.0035

is pelletized and stored. The solvent and the catalyst are recycled and reused.

An alternative polycondensation process may be run in the melt by reactive extrusion with adequate venting of carbon dioxide by-product [26].

Mechanical properties compare favorably with those of polycarbonate and polysulfone plastics. Electrical properties are much less influenced by moisture because of the high aromatic content and their relatively high T_g. Some properties of MDI based polyamides are given in Table 5.5.

All these amorphous and transparent polyamides compete with poly(methyl methacrylate) for their transparency but they are as tough as polycarbonate and polysulfone. Their aging and chemical resistance are also good.

Partially aromatic polyamides are used for transparent flow meter parts, sight glasses and spectacle frames, counters and containers for solvents, transparent housings for different equipment, X-ray apparatus windows and sanitary fittings.

4. Partially crystalline polyamides are produced by the polycondensation of adipic acid with *m*-xylylene diamine:

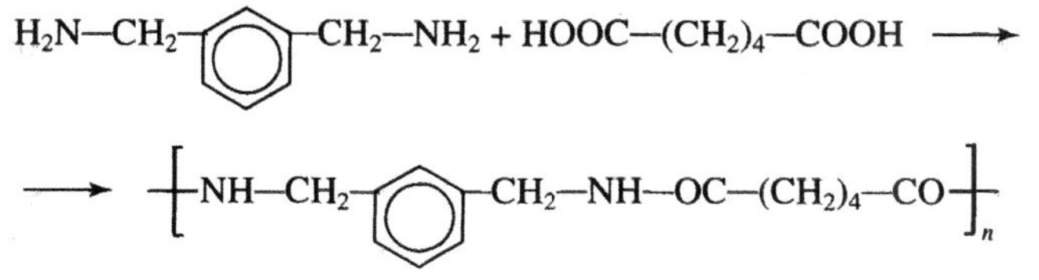

These polyamides have T_m values in the range 235–240°C, T_g = 100°C and low heat deflection temperatures of about 96°C. Therefore these polyamides are not suited for high temperature applications and they are used only in composite materials which have improved heat deflection temperature and mechanical characteristics. Glass fibers (30–50%), carbon fibers

Table 5.6 Some properties of glass filled composites

Property	*Value*
Tensile strength (MPa)	185
Flexural modulus (GPa)	10
Heat deflection temperature at 1.8 MPa (°C)	220–230
Molding shrinkage (cm/cm)	0.003–0.006

and particulates can be chosen as reinforcing fillers. Other additives such as impact modifiers and fire retardants are used specifically.

Some characteristics of glass filled composites are presented in Table 5.6.

Composite polyamides may replace metals in some applications. They are used as shafts and gears, electrical plugs and sockets, machine components and other mechanical parts.

5. Another semicrystalline polyamide is prepared by the condensation of MDI and azelaic acid in a similar way to that described in 3 above:

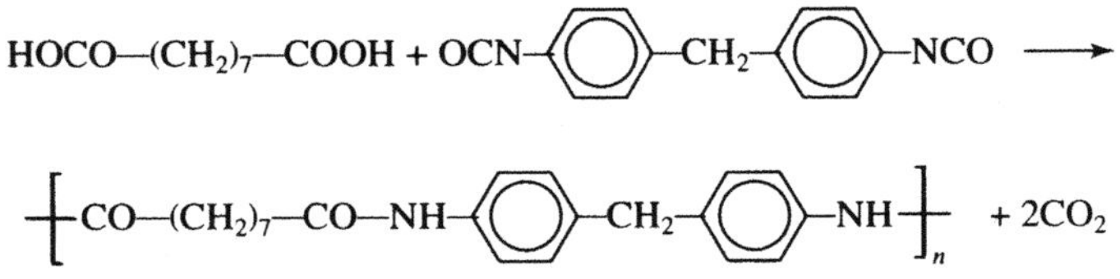

This polyamide has a relatively high T_m at 290°C and T_g at 135°C. Combined with suitable reinforcing and nucleating agents, the polyamide is a valuable engineering thermoplastic with high modulus and tensile strength, high heat resistance, dimensional stability, good electrical properties and excellent chemical resistance. It is used in automotive under-the-hood and other harsh industrial applications. Blends with 15–30% bottle grade PET or PBT are engineering resins with good properties and lower cost.

Table 5.7 Some properties of unfilled and glass reinforced semicrystalline polyamides

		Glass reinforced	
Property	*Unfilled*	*30% glass*	*40% glass*
Density (g/cm^3)	1.17	1.40	1.49
Tensile strength (MPa)	71	148	195
Tensile modulus (MPa)	1681	5757	7193
Elongation (%)	5.0	4.0	4.1
Flexural modulus (MPa)	2459	7648	10 018
Heat deflection temperature at 1.8 MPa (°C)	130	247	247

Some properties of unfilled and reinforced polyamides of this group are presented in Table 5.7.

5.5 AROMATIC POLYAMIDES

Aromatic polyamides or aramid fibers, according to the definition of the US Federal Trade Commission, are 'manufactured fibers in which the fiber-forming substance is a long-chain synthetic polyamide in which at least 85% of the amide linkages are attached directly to two aromatic rings'.

The synthesis of aromatic polyamides is not possible by the direct condensation of acids and amines as aromatic amines are less reactive than aliphatic and would not lead to macromolecular products. On the other hand, the very high melting temperatures and the discoloration and decomposition in the molten state make the bulk polycondensation technique impracticable. For these reasons, aromatic polyamides are obtained by the more costly solution and interfacial polycondensation methods of diacid chlorides and diamines [30–35].

Few details are made available on the synthesis of aramid fibers. A flow chart of the general method of interfacial polycondensation applied to aramid fibers is presented in Figure 5.5.

The acid chloride is dissolved in an organic solvent immiscible with water such as methylene chloride, carbon tetrachloride or cyclohexanone. The diamine is brought into an aqueous solution containing sodium carbonate as acid acceptor. As the monomers are very reactive to each other, the polycondensation is effected by vigorous mixing of the two immiscible solutions. The reaction takes place at the liquid interface or just inside the solvent boundary. The material is supplied by the manufacturers as fibers, obtained by spinning into hot air the solution of polymer in dimethylformamide containing lithium chloride.

Several aramid fibers are available on the market.

1. The first fiber, poly(*m*-phenylene isophthalamide) was produced at the end of 1960 by the polycondensation of isophthaloyl chloride with *m*-phenylenediamine:

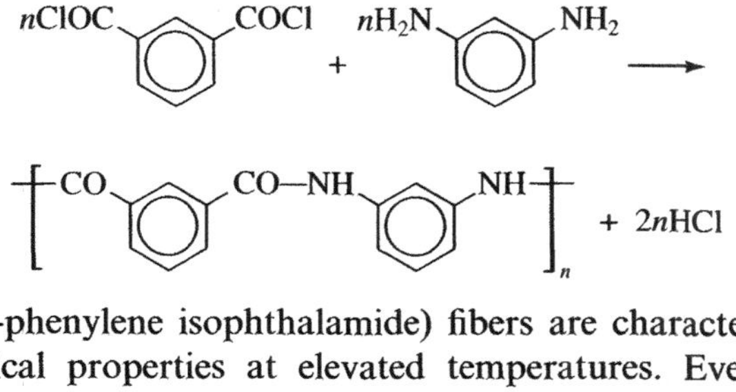

The poly(*m*-phenylene isophthalamide) fibers are characterized by their good mechanical properties at elevated temperatures. Even at ambient temperature, their mechanical, chemical and electrical insulation properties

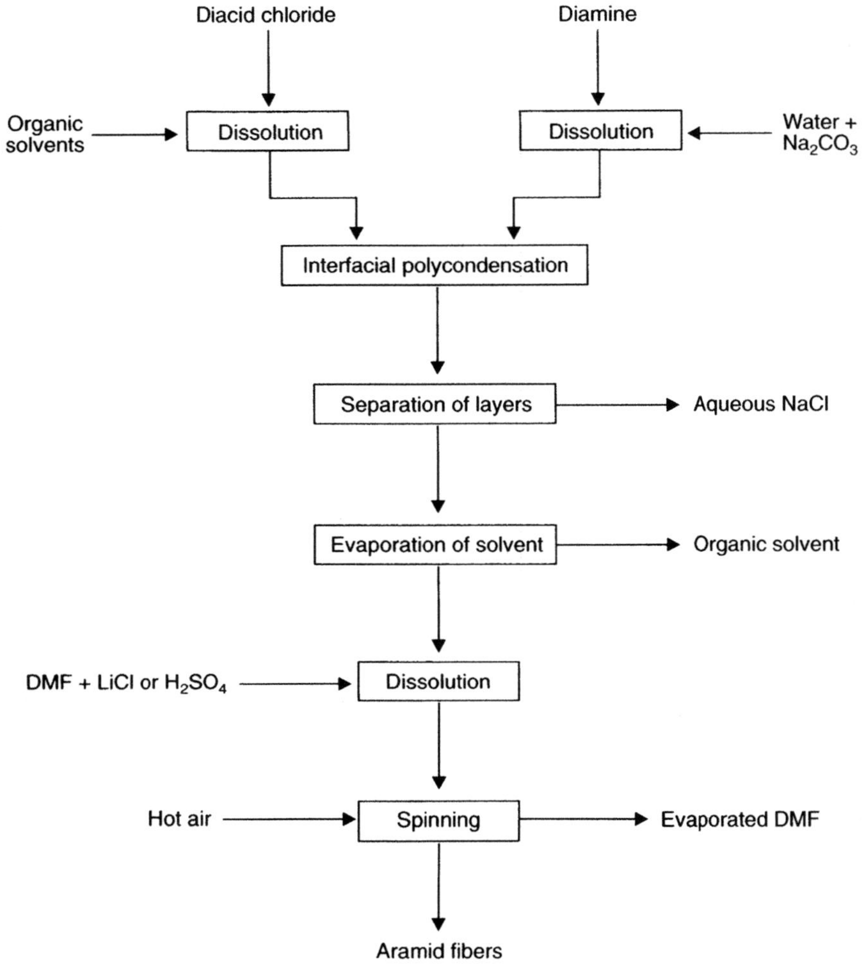

Figure 5.5 Preparation of aramid fibers by interfacial polycondensation.

are similar to those of nylons 6 and 6,6; at 260°C they still retain half of these properties. The crystalline melting point is 368–390°C; the material resists ignition but shrinks considerably.

The fibers are used for heat protective clothing and for electrical insulation at high temperatures.

2. Poly(*p*-phenylene terephthalamide) has the highest strength/density properties among commercially available fibers. It is another aramid fiber and was commercialized in 1972. The polymer is prepared by low temperature solution polycondensation of terephthaloyl chloride with *p*-phenylenediamine in a solvent composed of a 2:1 mixture of hexamethylphosphoramide and *N*-methylpyrrolidone at −10°C:

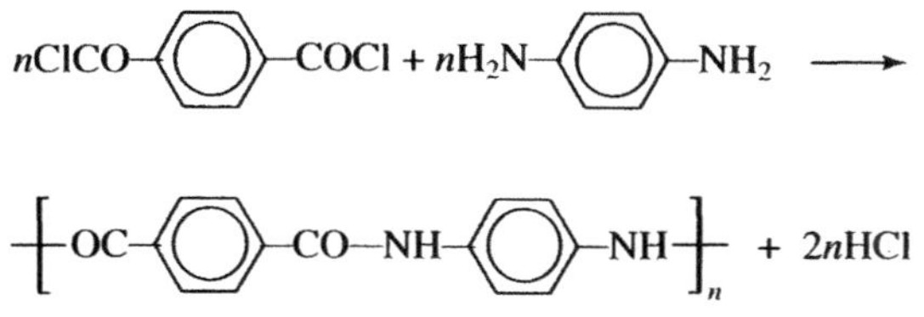

The polymer is dissolved in sulfuric acid and spun into fibers. In sulfuric acid solution the polymer is lyotropic as it exhibits a liquid crystal phase formed under the influence of the solvent. Extrusion of this solution into fibers and removal of the sulfuric acid results in further alignment of the molecules, with the result that they are even more uniform than could be obtained by drawing. As a general rule, *p*-linked benzene rings give rise to molecules that are rigid along their main axis and are typical for liquid crystal structures [36–41].

Poly(*p*-phenylene terephthalamide) fibers are competitive with glass, carbon and steel fibers. They are advantageously used as fillers in composite materials as they have lighter weight, greater stiffness and higher tensile strength than composites based on glass fiber reinforcement [42]. The heat resistance of the polymer is high as it retains about half of the tensile strength after 100 hours at 250°C. Some comparative properties of poly(*p*-phenylene terephthalamide) and glass fibers are presented in Table 5.8.

The polymer does not melt but begins to carbonize at 425°C. Thermal and electrical insulating properties of aramid based composite materials are better than those of glass fiber counterparts.

The volume of poly(*p*-phenylene terephthalamide) used in 1987 was about 10 tons and continues to grow as new applications are found. It is used as fiber and fabric reinforcement for composite materials, particularly with epoxy resins, in many applications. These advanced composites have replaced conventional aluminum for floors, ceilings, bulkheads, side walls and other parts in the aircraft and aerospace industry, reducing weight, manufacturing costs and maintenance demands [42].

Other uses for poly(*p*-phenylene terephthalamide) reinforced composites are for boat hulls, military protection helmets and filament-wound rocket

Table 5.8 Comparative tensile properties of aramid fibers and glass fibers

Property	*Aramid fiber**	*Glass fiber*
Density (g/cm^3)	1.44–1.47	2.49–2.55
Tensile strength (GPa)	3.45–3.62	3.10–4.14
Tensile modulus (GPa)	124–186	72–85
Elongation at break (%)	2.0–2.5	4.3–4.8

* Aramid = poly(*p*-phenylene terephthalamide)

motors. The fibers are also used as reinforcement in radial tires and mechanical rubber goods and for specialty clothing. Modern lightweight bullet-proof vests contain up to 18 layers of woven aramid cloth.

3. Aromatic polyether–amide copolymer fibers that may compete with poly(*p*-phenylene terephthalamide) fibers are prepared by the polycondensation of terephthaloyl chloride with a mixture of *p*-phenylenediamine (**7**) and 4,4′-diaminodiphenylether (**8**). The polymer melts at 515°C.

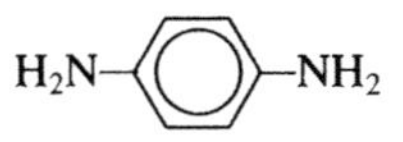

(**7**)

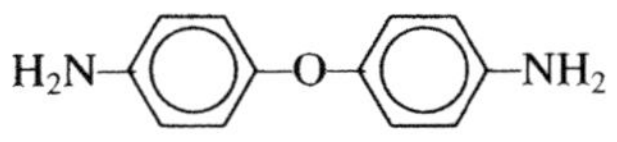

(**8**)

5.6 POLYIMIDES

Polyimides are high temperature engineering materials characterized by the presence in the polymer chains of cyclic structures that contain the imide group:

```
    CO–
   /
–N
   \
    CO–
```

The first polyimide was produced in the early 1960s but only in recent years has there been a rapid development for high strength composites, thermally stable films, molding compounds and adhesives [43–46].

Different types of polyimide polymers are marketed. Some of them contain the imide group as the main functional group and are known as unmodified polyimides. Modified polyimides that include polyamide-imides, polyesterimides, polyetherimides and polybismaleimides were produced in order to provide materials with the same superior physico-mechanical characteristics as unmodified polyimides but which were easier to process.

5.6.1 Unmodified polyimides

Unmodified polyimides are prepared by the polycondensation of aromatic dianhydrides with aromatic diamines. Examples of dianhydrides are pyromellitic dianhydride (**9**), 3,3′,4,4′-biphenyltetracarboxylic dianhydride (**10**), benzophenone-3,3′,4,4′-tetracarboxylic acid dianhydride (**11**) and the fluorinated dianhydride (**12**) which has found occasional use in moisture resistant polyimides. The most used aromatic diamine is 4,4′-diaminodiphenyl ether (**8**). Both dianhydride and diamine are required in high purity.

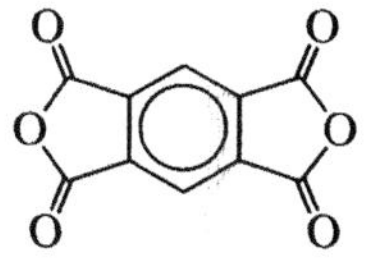

(9)

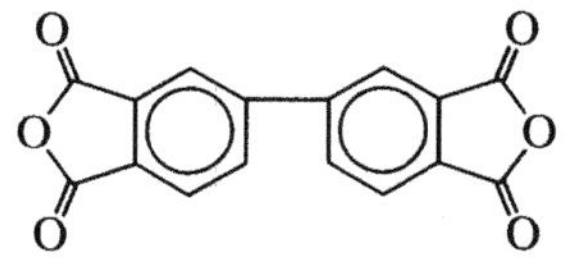

(10)

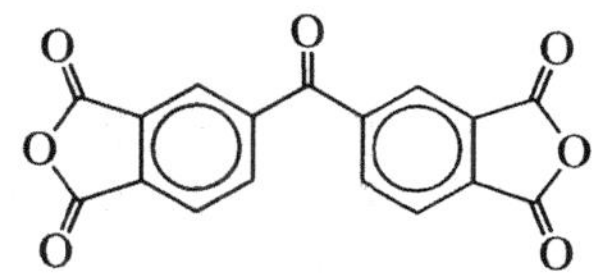

(11)

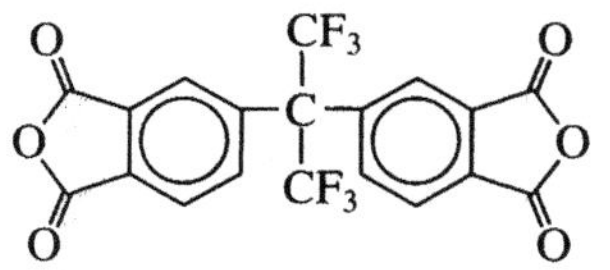

(12)

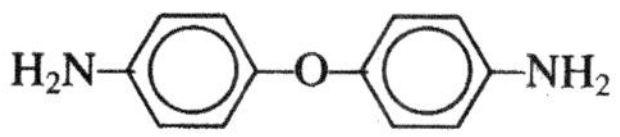

(8)

The synthesis of polyimides is carried out in two stages. The first step is effected by dissolving the diamine in polar solvents such as dimethylformamide, dimethylacetamide or dimethyl sulfoxide and by adding the dianhydride to the solution which is maintained at mild temperatures (under 50°C) in order to avoid cyclization (imidization). The reaction of the diamine with the dianhydride is fast and exothermic and it results in the formation of a viscous solution of polyamic acid prepolymer (**13**). The second step is the cyclization of the polyamic acid to the polyimide by intramolecular condensation.

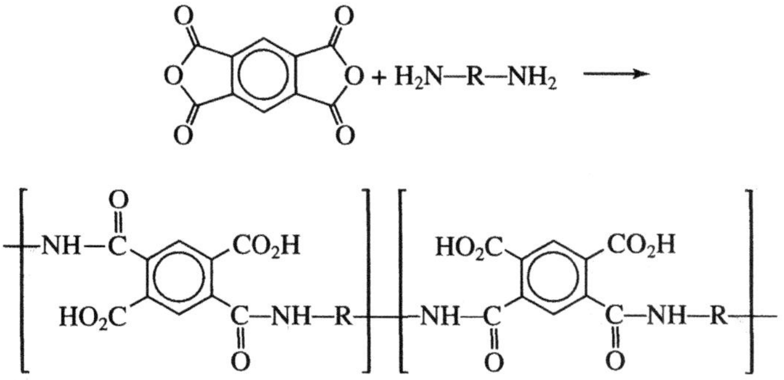

Polyamic acid (**13**)

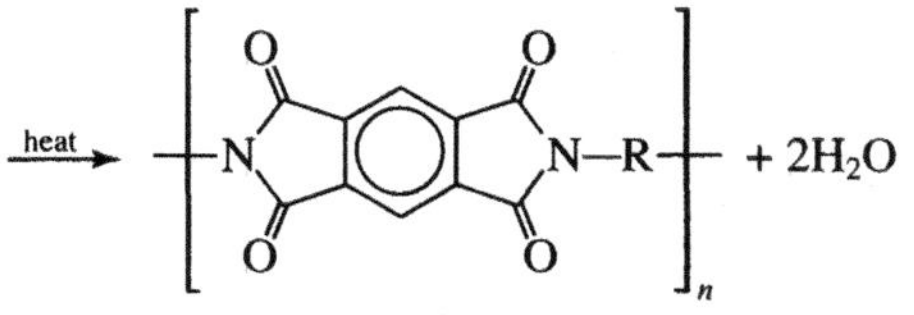

As the resultant cyclic polyimide is infusible and insoluble, the imidization step must be carried out in the solid state, i.e. in the physical form of the final product. The solution of polyamic acid obtained in the first step is cast to film or used for the impregnation of glass or carbon fibers for composite materials. After the evaporation of the solvent by heating for 1–2 hours at 150–200°C, the imidization is completed by an additional heating at 300°C. Careful processing is necessary as water formed as a by-product during the cyclization as well as residual solvent may cause voids that structurally weaken the finished product. Addition of acetic anhydride and pyridine to the polyamic acid solution contribute to easier removal of water.

The formation of five-membered imide rings by intramolecular polycondensation is the favored path of reaction. Nevertheless, some intermolecular condensation may also occur resulting in crosslinking and premature gelation which will bring difficulties in processing by conventional thermoplastic methods.

A simpler one-step process is based on the utilization of diisocyanates instead of diamines, which react with dianhydrides yielding the polyimide directly, with simultaneous evolution of carbon dioxide [47–49]. The polyimide prepared from benzophenone-3,3′,4,4′-tetracarboxylic acid dianhydride (**11**) and a 20/80 mixture of 4,4′-diphenylmethane diisocyanate (**14**) and 2,4-toluene diisocyanate is available commercially and may be processed either from dimethylformamide or from *N*-methyl pyrrolidone as fibers and films or as compression moldings [25]. The evolution of carbon dioxide from the condensation was turned to profit in the manufacture of polyimide foams used for acoustic insulation of jet engines.

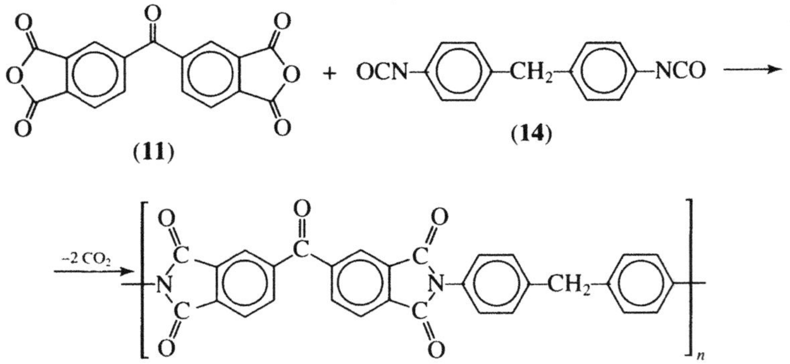

Polyimide polymers, especially in film form, have outstanding heat resistance with heat deflection temperatures as high as 357°C. After 1000 hours of exposure to air at 300°C the polymer retains 90% of its tensile and dielectric strength.

Polyimide film has excellent abrasion resistance, flame, solvent and oxidative degradation resistance as well as high energy radiation resistance. Chemical resistance of polyimides is very good, apart from with aqueous acids and alkalis which degrade the polymer.

Polyimides have found use as insulating enamels and varnishes for electric motors and cables and for coating glass cloth. Polyimides can be compounded with graphite, molybdenum disulfide, PTFE and metal powders such as aluminum for use as a wear surface in the manufacture of gaskets and piston rings for jet engines, and in soldering and welding equipment. Compounded with abrasives, polyimide composites are used in grinding wheels [25].

5.7 MODIFIED POLYIMIDES

5.7.1 Poly(amide-imides) (PAIs)

PAIs are produced by the condensation of high purity raw materials, namely the acid chloride of trimellitic anhydride (**15**) with 4,4′-diaminodiphenylmethane (**16**) in a solution of *N*-methylpyrrolidone. Hydrochloric acid and water evolve from the condensation.

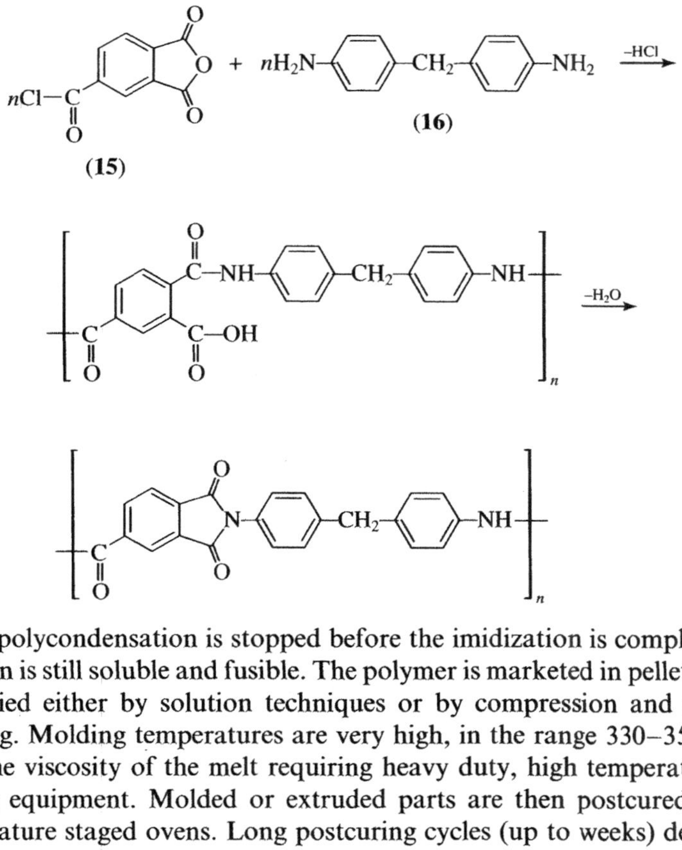

The polycondensation is stopped before the imidization is completed and the resin is still soluble and fusible. The polymer is marketed in pellet form. It is applied either by solution techniques or by compression and injection molding. Molding temperatures are very high, in the range 330–350°C and so is the viscosity of the melt requiring heavy duty, high temperature processing equipment. Molded or extruded parts are then postcured in high temperature staged ovens. Long postcuring cycles (up to weeks) depend on the part thickness since the reaction by-product (water) must diffuse through the part. Postcuring is thought to increase chain length, to fully imidize the chain and possibly to cause a small amount of crosslinking; this all contributes to high quality products [50].

Table 5.9 Some properties of poly(amide-imides)

Property	*At ambient temp.*	*at 260°C*
Density (g/cm^3)	1.14	–
Glass transition temperature, T_g (°C)	275	–
Heat deflection temperature at 1.82 MPa (°C)	270	–
Continuous use temperature (°C)	200	–
Tensile strength (MPa)	185	50
Flexural modulus (GPa)	4.55	3.00
Elongation (%)	12	22

High MW PAIs have been prepared starting from diisocyanates, particularly MDI (**14**), trimellitic anhydride (**17**) or an 80/20 mixture of trimellitic anhydride/isophthalic acid [25]. PAI ignition resistant foams can be prepared based on diisocyanates and the evolution of carbon dioxide from their reaction with dianhydrides [51].

PAI block copolymers have been prepared which are soluble in polar aprotic solvents, have reduced viscosities and T_g at 290°C [52]. A polyamide with amine end groups is prepared in the first step and is subsequently reacted in solution with a mixture of amine and dianhydride; a block PAI is formed by the condensation of the polyamide with the polyamic acid formed *in situ*, followed by cyclization in solution.

Requirements of high purity raw materials, demanding processing conditions and long curing cycles, are the reasons that make PAI products very expensive. As a result, only those applications that require their specific combination of characteristics use PAI materials. The commercial production of PAI was about 250 tons at the end of 1980s [50].

PAIs are hard, tough and creep resistant polymers with exceptionally high temperature stability [53,54]. Some properties of PAIs are given in Table 5.9. Creep modulus under 70 MPa load and 150°C is 1700 MPa in 100 hours. PAIs are resistant to aromatic and halogenated hydrocarbons, to other organic solvents and diluted acids but are attacked by alkalis. They have an excellent resistance to radiation.

PAIs are used as thermoplastic matrix in high performance composites. Up to 30% of graphite fibers give composites which have a very low coefficient of friction. Glass fibers (30–40%) and PTFE are also used as reinforcing materials.

PAIs are extensively used in the aerospace and automotive industries. They have lower weight and higher strength per unit weight than many high performance metal alloys.

5.7.2 Polyesterimides

Polyesterimides are another type of modified polyimides which contain ester and imide groups in the chain. They are synthesized in three steps. In the first step, trimellitic anhydride (**17**) is reacted with bis(2-hydroxyethyl) terephthalate (**18**) to give an intermediate product with ester groups in the chain and capped with anhydride groups (**19**):

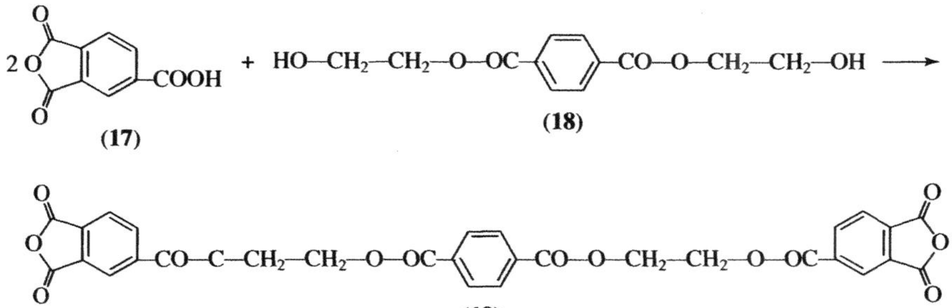

(**17**) (**18**) (**19**)

In the second step, this intermediate anhydride is treated with 4,4′-diaminodiphenyl methane (**16**) or other diamine and the resulting polyamic acid is then imidized in the third step in the same way that was described earlier for polyimides. Their general formula is:

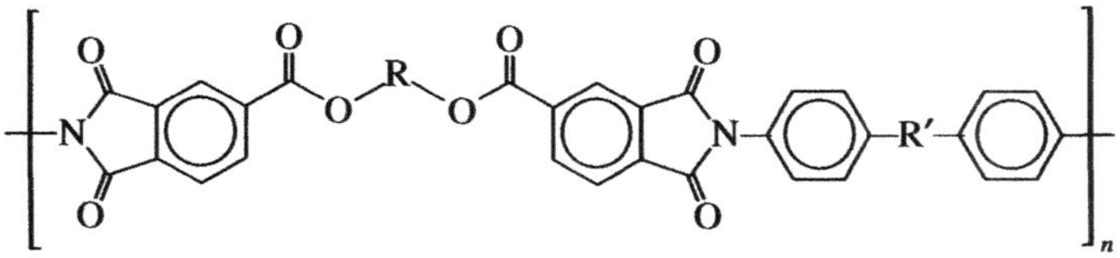

Polyesteramides are used for heat resistant wire enamels.

5.7.3 Polyetherimides (PEIs)

PEIs are high temperature high performance engineering thermoplastics materials of the polyimide family and constitute the last entry on the market in 1982. The presence of imide and aromatic units provides materials with the excellent properties characteristic to imide polymers, while the ether groups impart improved processability and melt flow. The synthesis of PEI involves three stages [55–57].

1. Firstly, a diamine is end capped with two molecules of nitrophthalic anhydride (**20**) leading to a low MW bisimide intermediate (**22**) with preformed imide groups. *m*-Phenylenediamine (**21**) is preferred commercially but other diamines may be used in order to obtain polyetherimides with different structures and properties.

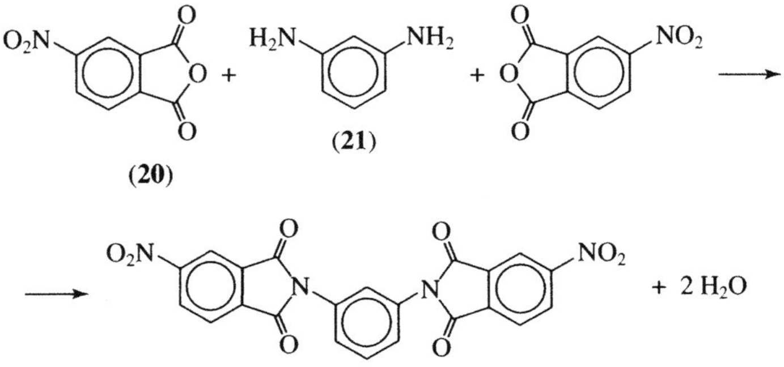

(20) (21) (22)

2. Preparation of the sodium salt of bis-phenol A (**23**):

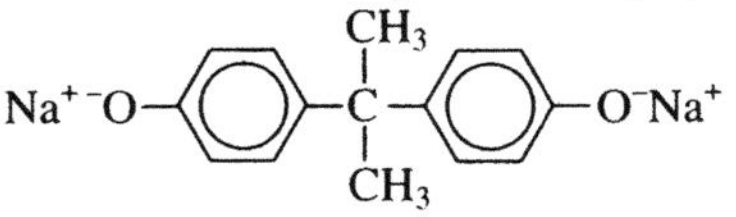

(23)

3. The polycondensation of bis-phenol A dianion with the bisimide intermediate in aprotic dipolar solvents and mild conditions leads to the formation of polyetherimide chains containing ether linkages (**24**).

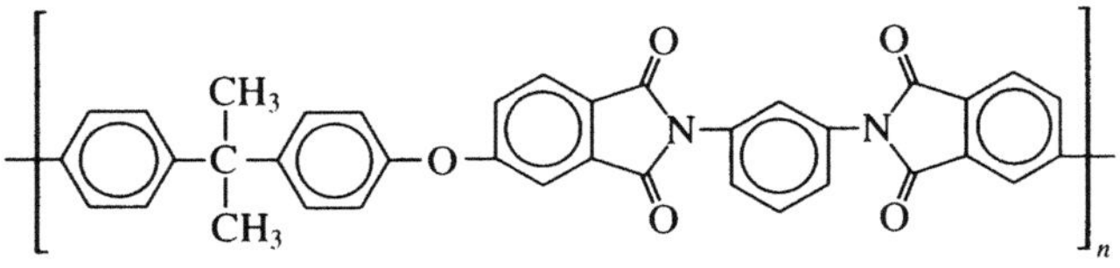

(24)

PEIs are amorphous and transparent materials. They have outstanding tensile properties even without reinforcement, an excellent balance of electrical properties and an outstanding thermal stability, retaining their strength at elevated temperatures. Table 5.10 illustrates some properties of PEIs.

PEIs are resistant to hydrocarbons, fully halogenated solvents and mineral acids and have an excellent hydrolytic stability. They are attacked by alkalis and dissolved by partially halogenated solvents. Even without the addition of stabilizers, PEIs are resistant to UV and ionizing radiations.

High flame resistance accompanied by very low smoke generation make PEIs superior to many engineering materials and are in agreement with stricter regulations on the flammability of plastics [13].

PEI resins are processed by injection molding, blow molding and by extrusion. Reinforcing materials such as carbon fiber or glass fiber (up to 40%) are used resulting in a substantial increase in tensile properties due to the excellent adhesion of PEI to the glass fiber [55].

PEIs have found various applications as car under-the-hood components and other mechanical parts, three-dimensional printed circuit boards,

Table 5.10 Some characteristics of polyetherimide resins

Property	*Value*
Density (g/cm^3)	1.27
Glass transition temperature, T_g (°C)	217
Heat deflection temperature at 1.84 MPa (°C)	200
Continuous use temperature (°C)	170
Tensile strength (MPa)	105
Tensile modulus (MPa)	3000
Elongation at break (%)	60
Flexural modulus (MPa)	3309
Molding shrinkage (cm/cm)	0.005–0.007

microwave equipment and different components for the aerospace industry [58]. PEI fibers and fabrics have been used for protective clothing for fire fighters and race drivers, for aircraft upholstery and carpets [55]. Film applications include insulating tapes, flexible circuitry and wire insulation.

5.7.4 Polybismaleimides

A different route to modified polyimide polymers is used in the preparation of polybismaleimides. They are synthesized from monomers or low MW oligomers which contain preformed imide groups in the chain and double bonds at the end of the chains; they are transformed to final crosslinked polymers by rearrangement reactions and by addition polymerization [59–61]. This avoids the production of volatile by-products during the final stage.

Commercial polymers are based either on maleic anhydride or on nadic anhydride, which are reacted with a diamine to form bismaleimides (**25**) and bisnadimides (**26**), respectively:

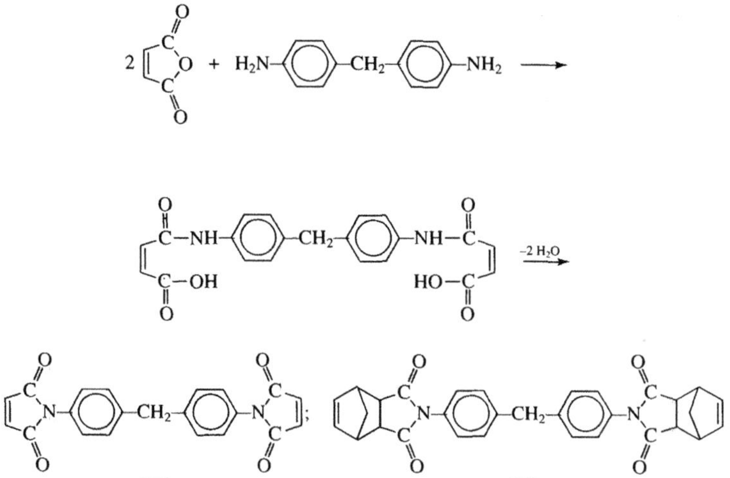

(**25**) (**26**)

These bisimides can be polymerized by thermal free radical polymerization through the end double bonds, but the resultant homopolymers have a high degree of crosslinking and are too brittle. Commercial products are based on chain extended bisimides which lead to reduced crosslink density and brittleness of the finished product. The extension of the chain can be done by the addition of a variety of bifunctional compounds such as diamines, disulfides and aldoxymes. Most commonly used is 4,4′-diaminodiphenyl methane:

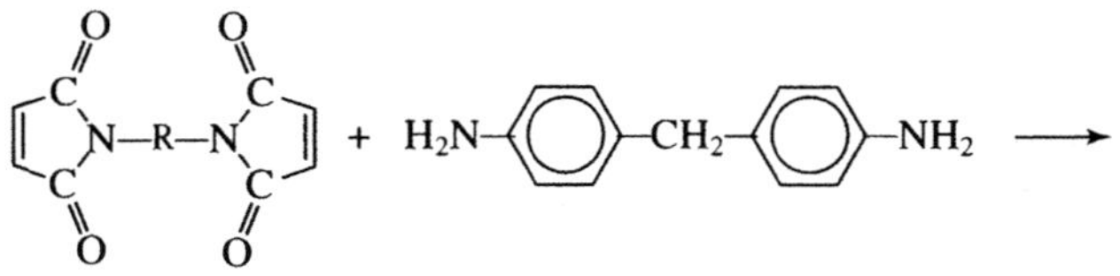

(25)

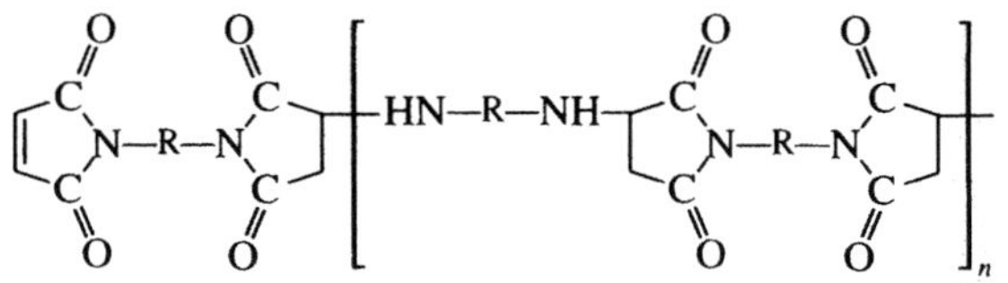

(27)

The reaction is effected by heating the mixture of bismaleimide with a molar deficiency of diamine at temperatures of 200–260°C. Two different reactions take place simultaneously without the formation of any volatile product: (a) the addition of the amino groups to maleic double bonds which extends the chains to form structure **(27)** which has maleic end groups and (b) thermal free radical polymerization of the remaining maleic end groups to crosslinked structures. Both these reactions take place during the molding process.

Unfilled and filled grades are commercially available. Unfilled polybismaleimides are applied for compression molded printed circuit boards. Fillers such as glass fibers, carbon fibers, asbestos, graphite and PTFE are used in composite materials which have found use in weapon and aerospace industries.

REFERENCES

1. Reisch, M.S. (1994) *Chem. Eng. News*, 7 November, 10.
2. Moncrieff, R.M. (1975) *Man-Made Fibers*, Butterworths, London, Chapters 19–22.
3. Kohan, M.I. (ed.) (1973) *Nylon Plastics*, Wiley-Interscience, New York.
4. Seymour, R.B. and Kirshenbaum, G.S. (eds) (1986) *History of High Performance Polymers*, Elsevier, New York: (a) p. 19; (b) p. 39; (c) p. 55.

5. Fredlina, L.K. and Karapetian, S.A. (1961) *Telomerization and New Synthetic Materials*, Pergamon, Oxford.
6. Roffia, P. *et al.* (1987) European Patent 208,311, Montedipe.
7. H. Ulrich (1992) in *Reaction Polymers* (eds W.F. Gum, W. Riese and H. Ulrich), Hanser Publishers, Munich, p. 217.
8. Reimschuessel, H.K. (1977) *J. Polym. Sci. Macromol. Rev.*, **12**, 65.
9. Millich, F. and Seshadri, K.V. (1972) in *Cyclic Monomers* (ed. K.C. Frisch), Wiley-Interscience, New York, Chapter 3.
10. Reimschuessel, H.K. (1969) in *Ring-Opening Polymerization* (eds K.C. Frisch and S.L. Reegen), Marcel Dekker, New York, Chápter 7.
11. Stevens, M.P. (1990) *Polymer Chemistry*, 2nd edn, Oxford University Press, New York.
12. Dai, S.A. *et al.* (1985) US Patent 4,550,157 to Upjohn; (1986) US Patent 4,604,450 to Upjohn.
13. Brysdon, J.A. (1995) *Plastic Materials*, 6th edn, Butterworths, London.
14. Iobst, S.A. (1985) *Polym. Eng. Sci.*, **25**, 425.
15. Sweeney,F.M. (1990) in *Handbook of Plastic Materials and Technology* (ed. I.I. Rubin), Wiley-Interscience, New York, p. 1273.
16. Jacobs, D.B. and Zimmerman, J. (1977) in *Polymerization Processes* (ed. C.E. Schildknecht), Wiley-Interscience, New York, p. 424.
17. Griehl, W. and Ruestem, D. (1970) *Ind. Eng. Chem.*, **62**(3), 16.
18. Damm, P.O. and Matthies, P. (1990) in *Handbook of Plastic Materials and Technology* (ed. I.I. Rubin), Wiley-Interscience, New York, p.161.
19. Roerdink, E. and Warnier, J.M.M. (1985) *Polymer*, **26**, 1582.
20. Maskus, P. and Capaul, M. (1993) *Kunststoffe*, **10**, 20.
21. Ellis, T.S. (1988) *J. Appl. Polym. Sci.*, **36**, 451.
22. Mumcu, S., Burzin, K., Feldmann, R. and Feinauer, R. (1978) *Angew. Makromol. Chem.*, **74**, 49.
23. Chen, A.T., Farrissey, W.J. and Nelb, R.G. (1978) US Patent 4,129,715 to Upjohn.
24. Nelb, R.G., Onder, K. and Chen, A.T. (1987) *Thermoplastic Elastomers*, Hanser Publishers, Munich, p. 197.
25. Onder K. (1992) in *Reaction Polymers* (eds. W.F. Gum, W. Riese and H. Ulrich), Hanser Publishers, Munich, p. 405.
26. Onder, K. (1979) US Patent 4,156,065 to Upjohn.
27. Onder, K., Andrews, P.S., Farrissey Jr, W.J. and Tilley, J.N. (1980) *Polymer Preprints*, **21**(2), 130.
28. Rader, C.P. (1993) *Kunststoffe*, **10**, 36.
29. Nelb, G.R., Onder, K., Raush, K.W. and Vanderlip, J.A. (1987) US Patent 4,672,094 to The Dow Chemical Co.
30. Morgan, P.W. (1965) *Condensation Polymers by Interfacial and Solution Methods*, Wiley-Interscience, New York, Chapter 5.
31. Nikonov, V.Z. and Savinov, V.M. (1977) in *Interfacial Synthesis* (eds F Millich and C.E. Carraher), Vol. 2, Marcel Dekker, New York, Chapter 15.
32. Korshak, V.V. (1972) *Heat-Resistant Polymers*, Halsted Press, New York.
33. Critchley, J.P., Knight, G.J. and Wright, W.W. (1983) *Heat-Resistant Polymers*, Plenum Press, New York.
34. Lenk, R.S. (1978) *Macromol. Rev.*, **13**, 355.

35. Jaffe, M. and Jones, R.S. (1985) in *High Technology Fibers*, Part A (eds M. Lewin and J. Preston), Marcel Dekker, New York, Chapter 9.
36. Gordon, M. (ed.) (1984) *Adv. Polym. Sci.*, **59**, **60–61**, Springer-Verlag, New York.
37. Blumstein, A. (ed.) (1978) *Liquid Crystalline Order in Polymers*, Academic Press, New York.
38. Ciferri, A., Krigbaum, W.R. and Meyer, R.B. (eds) (1982) *Polymer Liquid Crystals*, Academic Press, New York.
39. Varshney, S.K. (1986) *J. Macromol. Sci., Rev. Macromol. Chem. Phys.*, **C26**, 551.
40. Lenz, R.W. (1985) *Polym. J.*, **17**, 105.
41. Finkelmann, H. (1987) *Angew. Chem. Int. Ed.* (English), **26**, 816.
42. Langston, P. and Zahr, G.E. (1990) in *Handbook of Plastic Materials and Technology* (ed. I.I. Rubin), Wiley-Interscience, New York, p. 931.
43. Mittal, K.L. (ed.) (1984) *Polyimides*, Vols 1 and 2, Plenum Press, New York.
44. Beesonov, M.I., Koton, M.M., Kudryavtsev, V.V. and Laius, L.A. (1987) *Polyimides – Thermally Stable Polymers*, Plenum Press, New York.
45. Ranney, M.W. (1971) *Polyimides Manufacture*, Noyes Data Corp., Park Ridge, New Jersey.
46. Cassidy, P.E. (1980) *Thermally Stable Polymers*, Marcel Dekker, New York, Chapter 5.
47. Carleton, P.S., Farissey Jr, W.J. and Rose, J.S. (1972) *J. Appl. Polym. Sci.*, **16**, 2983.
48. Ghatge, N.D. and Mulik, U.P. (1980) *J. Polym. Sci., Polym. Chem. Ed.* **18**, 1905.
49. Alvino, W.M. and Edelman, L.E. (1975) *J. Appl. Polym. Sci.*, **19**, 2961.
50. Throne, J.L. (1990) in *Handbook of Plastic Materials and Technology* (ed. I.I. Rubin), Wiley-Interscience, New York, 225.
51. Nelb, R.G. and Saunders, K.G. (1988) US Patent 4,738,990 to Dow Chemical Co.
52. Phaendner, R. (1992) Ger. Offen. DE 4,332,248 to Ciba-Geigy A.-G.; *CA* (1994), **121**(10), 231660.
53. Billerbeck, C.J. and Henke, S.J. (1985) in *Engineering Thermoplastics: Properties and Applications* (ed. J.M. Margolis), Marcel Dekker, New York, p. 373.
54. Cekis, G.V. (1987) in *Modern Plastic Encyclopedia*, Vol. 63, 10A, McGraw-Hill, New York, p. 37.
55. Serfaty, I.W. (1990) in *Handbook of Plastic Materials and Technology* (ed. I.I. Rubin), Wiley-Interscience, New York, p. 263.
56. Johnson, R.O. and Burlhis, H.S. (1983) *J. Polym. Sci., Polym. Symp.*, **70**, 129.
57. White, D.M. (1981) *J. Polym. Sci., Polym. Chem. Ed.*, **19**, 1635.
58. Bailly, C., Sederel, W. and Stahl, P.O. (1993) *Kunststoffe*, **10**, 48.
59. Stenzenberger, H.D., Herzog, H., Romer, W. *et al.* (1983) *Br. Polym. J.*, **15**, 2.
60. Trivedi, B.C. and Culbertson, B.M. (1982) *Maleic Anhydride*, Plenum Press, New York.
61. Varma, I.K. and Sharma, S. (1984) *Eur. Polym. J.*, **20**, 1101.

Silicone polymers

6

6.1 INTRODUCTION

Like carbon, silicon has the possibility of forming covalent compounds. The siloxane bond –Si–O–Si is more stable than the Si–Si one. It is the basic bond in the silicone polymers, and the same linkage is present in glass and quartz. Silicon multiple bonds polymers are uncommon; as a result, silicone polymers are obtained only by polycondensation. The combination of inorganic polymer networks with organic groups leads to inorganic–organic polymers. A convenient method for the introduction of organic radicals into a mineral backbone is the use of organo substituted silicoesters in a polycondensation process [1]. In the 1980s, the USA's silicone polymers annual consumption exceeded 60 000 tons. Since 1990 worldwide production of silicones has risen by 8–10% annually on average. In 1991/1992 the world market for silicones, including chip material, was roughly worth 4.5 billion US $ [2].

Depending on the substituents at the oxygen atom position of the Si–O– bond, various type of compounds have to be distinguished [3].

- Silanols and their condensation products, the siloxanes: Si–OH and Si–O–Si
- Organyloxisilanes: –Si–OR, R = alkyl, aryl
- Acyloxysilanes (also named silyl esters)

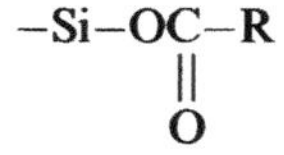

- Silylenol ethers

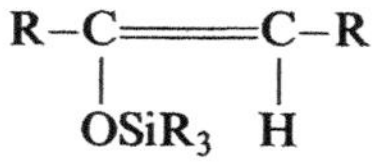

- Cyclic Si–O compounds (e.g. siloxa-heterocycles)

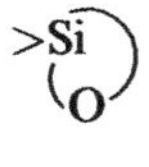

Linear siloxanes are primarily formed by either condensation of silanols or by hydrolysis of organyloxysilanes with functional (e.g. halogens, amino, etc.) substitutents. If a chain is produced with two substituted organic groups on each silicon atom a linear silicone polymer known as polyorganosiloxane would result:

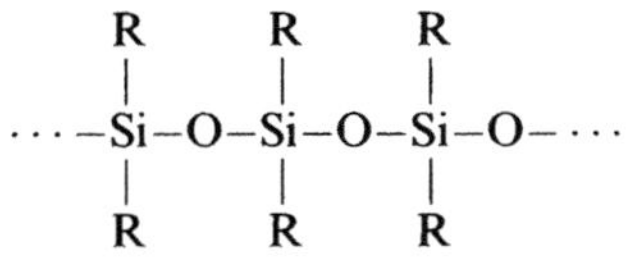

Polysilicic acid is produced when silicone tetrachloride is hydrolyzed [4]:

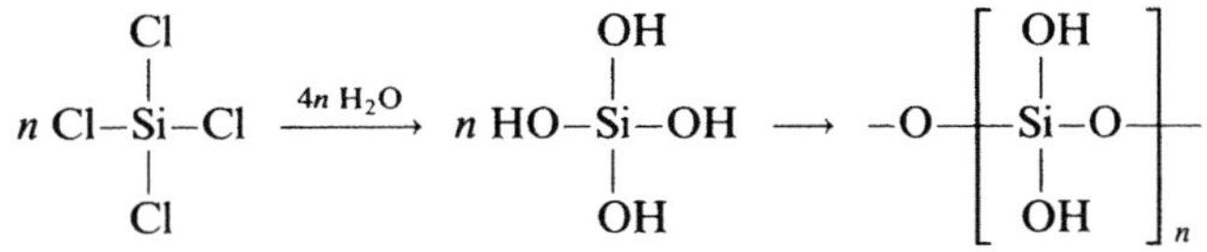

Silicic acid

Many silicone polymers, especially those with elastomeric characteristics, are organomineral polymers of predominantly dimethyl siloxane [5]:

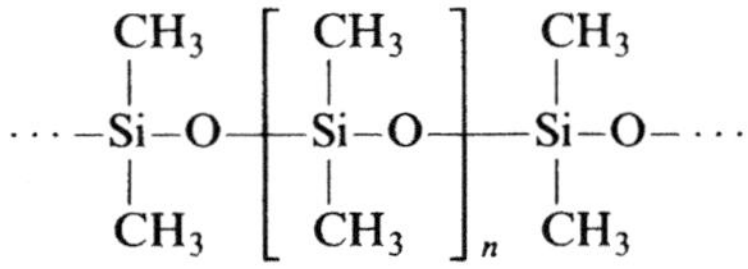

Modifications may be made by substituting small quantities of other groups (e.g. vinyl, phenyl, etc.) for methyl ones along the macromolecular chain. The properties of silicone polymers vary with the nature of the substituent groups, R/Si ratios and MW.

Dimethyldichlorosilane, $(CH_3)_2SiCl_2$, is the basic raw material for most silicone polymers; modifications produced by varying reaction conditions and adding controlled amounts of other silanes have a great influence on the final product. If a mixture of mono- and dichlorosilane is hydrolyzed, the resulting product is a linear macromolecular compound in which the macromolecules of dimethylsiloxane units are terminated by monofunctional trimethyl units as follows:

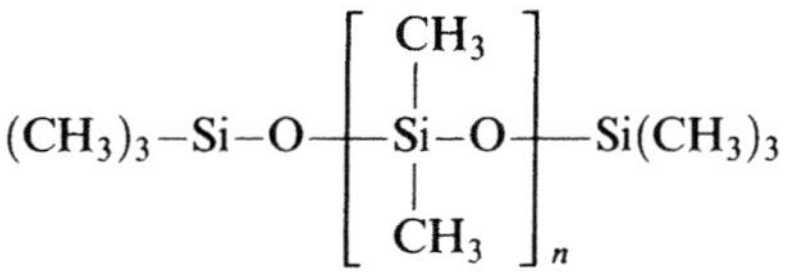

The end trimethyl siloxane units prevent further condensation and impart chemical stability.

Polydimethylsiloxane is the basic polymer for a series of different products varying from oils to elastomers. The oils are based on low MW linear macromolecules; resins possess a crosslinked microstructure and elastomers are based on linear high MW chains.

6.2 SILICONE OILS

Silicone oils consist of linear polydimethylsiloxane macromolecules with a MW in the range of 4000–25 000. The synthesis starts with dimethyldichlorosilane or a mixture of methane and phenylchlorosiloxanes which is continuously hydrolyzed by mixing with dilute hydrochloric acid. The produced oil consists of a mixture of long and short polydimethylsiloxanes and various cyclic siloxanes, especially the tetramer, octamethylcyclotetrasiloxane. This mixture is separated and heated at 150°C with a small amount of aqueous sulfuric acid or sodium hydroxide, and a chain stopper (hexamethyldisiloxane) is added to obtain a certain MW and, respectively, viscosity. Hexamethyldisiloxane produces trimethylsiloxil capped ends. The viscosity of silicone oils is controlled by the ratio of hexamethyldisiloxane to dimethyldimethoxysilane (or other suitable silane) employed [6]. The oily fluid is separated from the aqueous layer, filtered, dried and heated to remove volatiles. In the case of silicon oils it must be noted that a linear relation exists between the logarithm of the average chain length and the logarithm of the viscosity (Figure 6.1). The relationship is valid for a wide range of viscosity and shows that polydimethylsiloxanes, with MW over 100 000, are still capable of flowing while organic polymers are already solid at much lower MW. The reason for the low viscosities at relatively high MW is the low intermolecular force in the polydimethylsiloxanes. This also explains the small temperature dependence of physical and mechanical properties of silicon oils and elastomers which is important for many applications [7].

Such polydiorganosiloxane fluids containing both methyl and phenyl groups are produced.

6.3 SILICONE ELASTOMERS

In contrast with the oils, silicone resins contain Si atoms with no or only one organic substituent. They are, therefore, crosslinkable to harder and stiffer compounds than the elastomers, but many must be handled in solution to prevent premature cure. They are used as insulating varnishes, impregnating and encapsulating agents, and in industrial paints [8].

General purpose silicone elastomers are based on polydimethylsiloxanes, but special purpose products containing a small proportion of groups other than methyl are also available [9]. In order to obtain elastic properties, it is necessary to crosslink the linear siliconic macromolecules. The crosslinking (vulcanization) can be achieved in different ways. The most important

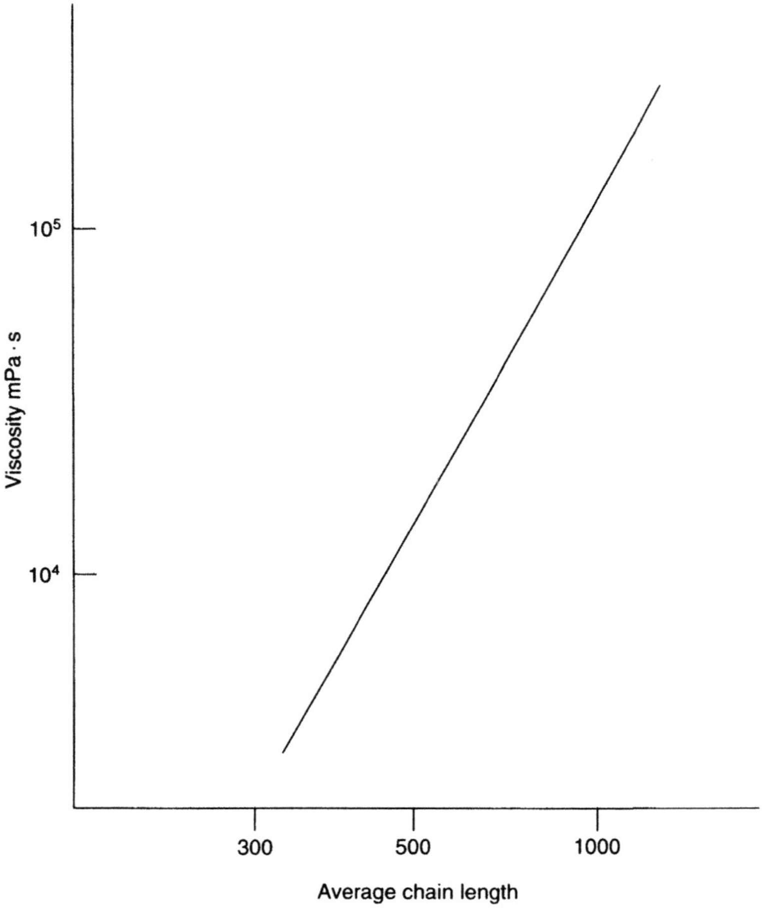

Figure 6.1 Polydimethylsiloxane; average chain length–viscosity [7]. (Reprinted by permission of Vulkan Verlag GMBH from *Silicones – Chemistry and Technology* by A.G. Bayer *et al.*, 1991.)

technical methods are based on organic peroxides at high temperatures and crosslinking with organometallic and organic silicon compounds at room temperature or at only slightly elevated temperatures.

The production of silicone polymers involves the preparation of organosilicon halides or esters, their isolation and purification, the hydrolysis of an appropriate mixture of these intermediates and the condensation to achieve the polymers with the desired microstructure. The chlorosilanes may be produced by a Grignard reaction or by the 'direct' process.

Highly defined telechelic star polymers of poly(dimethylsiloxane) of controlled MW, low polydispersity, controlled arm number and quantitative end reactivity can be produced [10].

Various silicone elastomers are manufactured [11].

1. Dimethyl silicone elastomer: in the technology of this elastomer, dimethyldichlorosilane is dissolved in ether and the solution is mixed with an excess of water. Cyclic oligomers make up about 50% of the product; octamethyltetrasiloxane, the tetramer, is the main cyclic component.

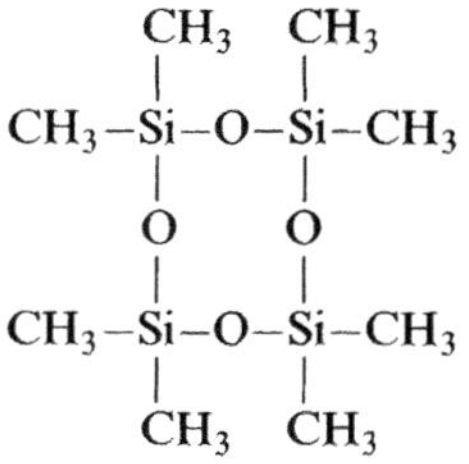

The polymer, which has no elastic characteristics, is synthesized by heating at 150–200°C in the presence of NaOH and a monofunctional additive to control MW. The obtained polymer with a MW in the range of 300 000–800 000 is transformed into a true elastomer after crosslinking with organic peroxides [12].

2. Vinyl silicone elastomers: unlike the former dimethyl gums, the vinyl silicone gums used for the vinyl silicone elastomer synthesis contain about 0.1% molar methyl-vinyl-siloxane groups which permit crosslinking with less organic peroxide.

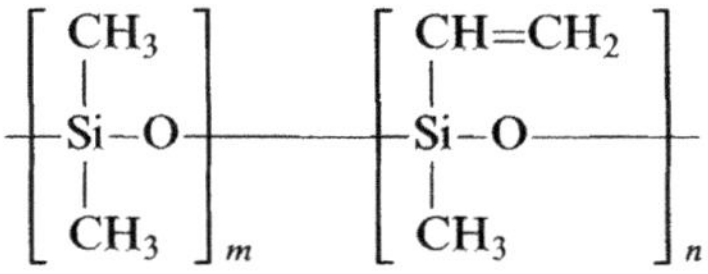

3. Phenyl silicone elastomers: in this product 5–15% of the methyl groups are replaced with phenyl ones which provide good low temperature properties.

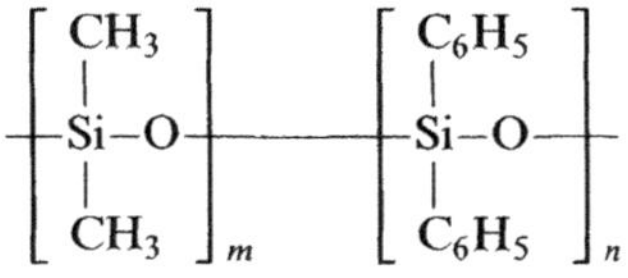

4. Nitrile silicone elastomers contain cyanoalkyl groups.

5. Fluorosilicone elastomers have γ-trifluoropropylmethyl-siloxane units which bring better solvent resistance:

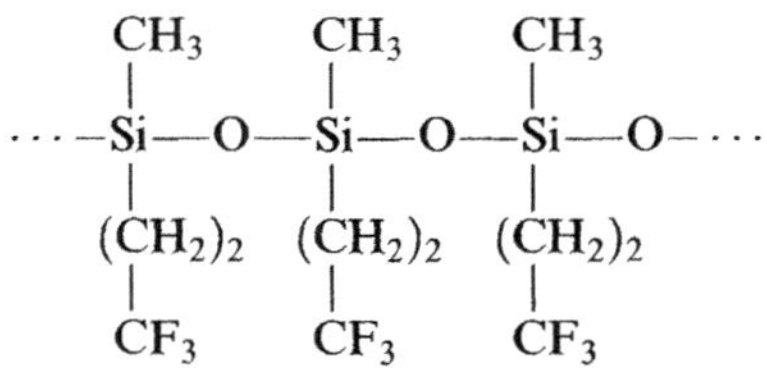

6. Room temperature vulcanizing (RTV) silicone elastomers: these polydimethylsiloxane-type elastomers are terminated by silanol groups which participate in vulcanization. The MWs, in the range of 10 000 to 100 000, are much lower than the MWs of silicone elastomers cured at elevated temperature. The technological process of their synthesis is based on dimethyldichlorosilane hydrolysis in the presence of calcium carbonate or another base which is added to neutralize the HCl formed, and thus to avoid the further condensation of silanol groups. As vulcanizing agents for such elastomers, tri- or tetralkoxysilane or polyalkoxysiloxane are used.
7. Liquid silicone elastomers are produced from polydimethylsiloxanes which contain a small amount of vinyl groups. These elastomers can be processed by injection molding and vulcanized in the mold at 100–200°C.
8. Borosilicone elastomers are produced by treating RTV silicone elastomers with low MW (1000–10 000) with boric acid; as a result, gums with 300 000–700 000 MWs are obtained. Peroxides are used for their crosslinking.

Polysiloxane thermoplastic interpenetrating polymer networks (IPNs) are a group of blends that significantly extend the process and mechanical property range of silicone elastomers and allow modification of thermoplastics. Such composites are characterized by reactive processing consistent with traditional thermoplastic technologies. These semi-IPNs are obtained from polysiloxanes which are crosslinked in a variety of thermoplastic matrices, including polyurethanes, polyamides and polyolefins [13].

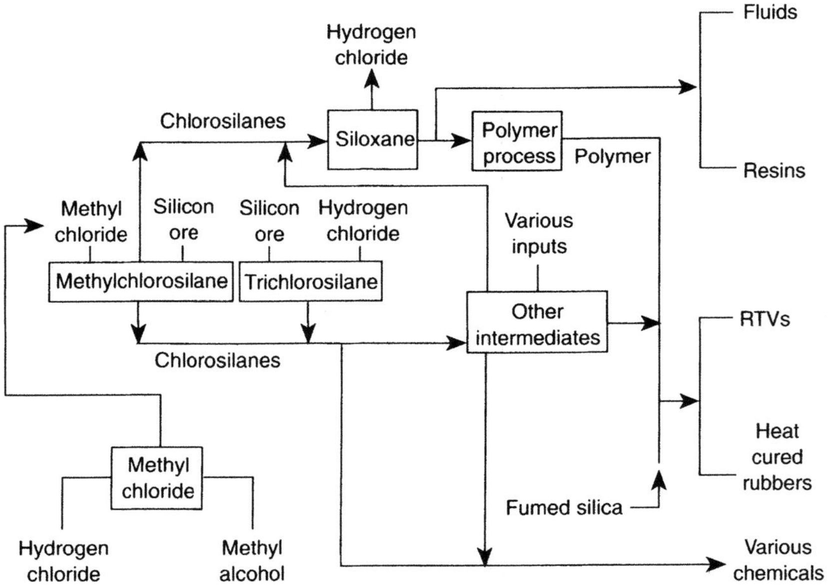

Figure 6.2 Silicone manufacturing. (Reprinted by permission of John Wiley & Sons Inc. from *Encyclopedia of Polymer Science and Engineering*, 2nd edn, by H. Mark, N.M. Bikales, C.G. Overberger and G. Menges (eds), Vol. 15, 1989.)

Silicone rubber compounds are normally heat cured in the presence of dibenzoil peroxide, dicumyl peroxide, ditertiary-butyl peroxide or other peroxides. It is apparent that the curing reactions could produce no more than one mole of chemical crosslinks per mole of peroxide [12].

Peroxides such as α,α'-di(tert-butylperoxy)-*m*/*p*-diisopropyl benzene (DBPIB) and 2,5-dimethyl-2,5-di(tert-butylperoxy) hexane (DBPH) are preferred for heat curing vinylmethylsilicone elastomers, because the free radical initiated crosslink does not reduce the inherent stability of the polymer [14].

A flow chart of a silicone manufacturing facility is presented in Figure 6.2. [15].

6.4 PROPERTIES

Depending on the microstructure, it is possible to obtain silicone elastomers with similar basic properties but which differ in certain special characteristics. The heat vulcanizable types are characterized by high MWs (300 000–800 000), while the room temperature vulcanizable types suitable for dipping and spreading processes have MWs in the range of 10 000–100 000 and consequently lower viscosities. The strength of these elastomers improves with increasing MW. The branching of polydimethylsiloxanes improves low temperature flexibility [16].

Vinyl silicone elastomers have much reduced compression set. Unlike polydimethylsiloxane elastomers which are filled with silica, vinyl silicone ones can be filled with carbon black. Elastomers with methylphenylsiloxane units have a T_g around −115°C compared to −50°C for plain polydimethylsiloxane. The presence of phenyl groups decreases the vulcanization yield as shown by reduced elasticity, higher tensile strength and higher tear resistance of the vulcanizates.

Besides vulcanization, the presence of phenyl groups may reduce compression set, resistance to swelling in polar or aromatic and naphthenic solvents. Silicone rubbers with high phenyl group content are particularly resistant to β and γ radiation [16].

The nitrile silicone elastomers and fluorosilicone ones have enhanced oil resistance and good low temperature properties but have not achieved much commercial importance.

RTV silicone elastomers are supplied as two separate components which are mixed together prior to the use. The vulcanizing agent might be with the part which contains the silanol terminated component or with the catalyst. Liquid silicone elastomers which cure only on heating should not be confused with two-part RTV silicone elastomers.

Polyethylmethylsiloxanes are distinguished from polydimethylsiloxanes by the fact that at a content of 95 mol% diethylsiloxane units, the T_g is

−136°C, their rate of crystallization is also lower, and as in the case of the presence of vinyl groups, the degree of crosslinking is higher.

The general characteristics of silicone elastomers are [17]

- very wide service temperature range (−100 to +310°C);
- excellent resistance to attack by oxygen, ozone and sunlight;
- excellent non-stick, non-adhesive properties;
- low toxicity;
- possibility of optical transparency;
- good electrical insulation properties;
- low chemical reactivity, and excellent chemical and weathering resistance.

Improvements in thermal stability and flame retardance were achieved by replacement of the methyl groups in poly(dimethylsiloxane) with phenyl or vinyl groups or by incorporation of carborane units [17,18].

Possible limitations of the material are

- the low tensile strength of vulcanized rubbers;
- poor hydrocarbon, oil and solvent resistance (except with silicone elastomers that have fluorine, vinyl, nitrile and methyl substituent groups on the macromolecular chain);
- high gas permeability;
- somewhat high cost.

The solubility of carbon dioxide, methane and propane in five different silicone polymers was studied by V.M. Shah *et al.* [19].

The silicone polymers exhibit low interchain attractions and have a highly flexible backbone. These characteristics, as well as their high molar volume, low solubility parameter, low surface tension and high compressibility, contribute to the ability of silicone products (rubber, sealant, etc.) to remain elastic down to very low temperatures.

The high thermal stability of polydimethylsiloxane rubber is a reflection of the good stability of the bonds involved (which comprise −Si−O−, −Si−C−, −Si−H−) and lack of weak bonds, rather than any special outstanding feature [20].

The reason for the greater heat resistance of silicone rubber can be found in the bond energy of the Si−O bond. Values for this bond range from 88–117 kcal/mol, while values for the C−C single bond range from 83–85 kcal/mol. While the bond energy is helpful in understanding the heat resistance of silicone rubber, a more thorough understanding of the chemical behavior of the bond would require a look at the three components of the bond, i.e. covalent, polar and doublebond contribution [12]. Unlike the polysiloxanes which show low values of surface tension, the polysilanes exhibit surface tension which varies widely depending on the substituents [4,19,20].

A typical silicone thermoplastic elastomer has a tensile strength of 5 MPa and a hardness of (Shore A) 40–100 [21].

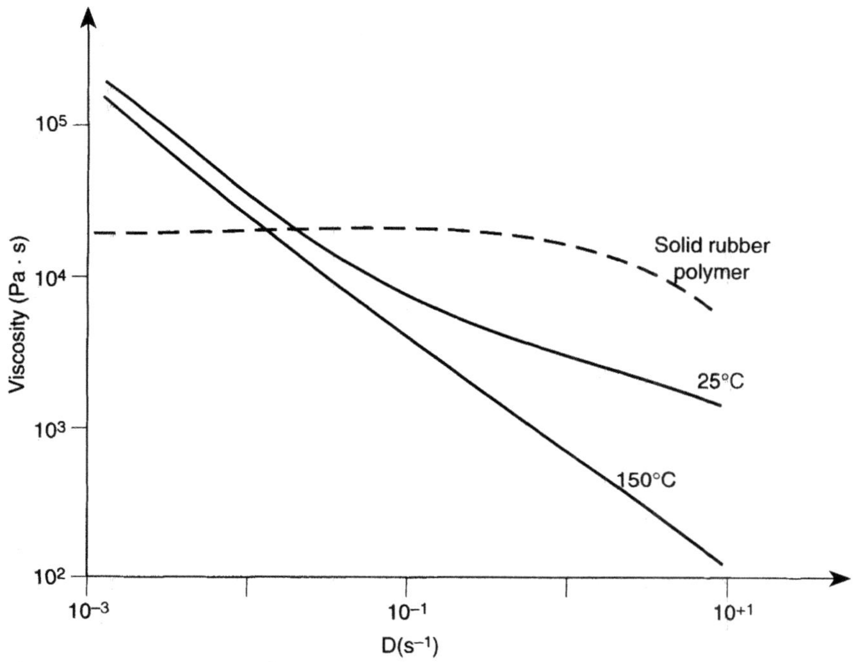

Figure 6.3 Viscosity variations of liquid silicone rubbers [7]. (Reprinted with the permission of Vulkan Verlag GMBH from *Silicones – Chemistry and Technology*, 1991.)

Processing of silicone rubber mixtures involves a variety of shear stresses. The characterization of the material therefore requires finding the shear rate dependence of the viscosity. Because of the high content of filler, silicone rubbers exhibit an important pseudo-plastic behavior even at low shear rates. This applies not only to solid rubbers but also to liquid silicone rubbers (Figure 6.3). For comparison, the viscosity behavior of a solid rubber is included in the figure. Below a shear rate of $D = 10^{-1}\,s^{-1}$, the polymer behaves as a Newtonian liquid in contrast to liquid silicone rubbers [7].

6.5 APPLICATIONS

The silicone products were first developed for high temperature insulation and for encapsulation of electrical parts, but due to their relative inertness, new applications emerged in other fields (e.g. biomedical). Today silicone products range from liquids to reinforced plastics and elastomers. The liquid silicones have the advantage of oxidative stability, low freezing point, and a small temperature coefficient of viscosity.

Liquid silicone elastomer is a popular product with broad industrial use. These products are pumpable liquids that can be injected into low pressure molds for curing in a few seconds to several minutes at 100–200°C. Network formation is by a platinum catalyzed coupling of a siloxane hydride with a vinyl siloxane. The complete system contains reinforcing fillers and pigments to provide an elastomer with properties similar to those of materials prepared from high MW polymers.

The range of silicones for foam-in-place applications continues to grow. A variety of silicone foam is particularly useful in fire resistant penetration seals. A promising new silicone is a latex that deposits a coherent and fully cured elastomer on loss of water. The new types offer the convenience of silicones, form water based systems and are appearing as easy-to-use water based silicone caulks and as high performance exterior coating systems [22].

Bouncing putty is based on polydimethylsiloxane polymer modified with boric acid, additives, fillers and plasticizer to give a product that on shock behaves like an elastic material but flows as a viscous liquid on slow application of pressure [23].

Silicone sealants are high quality, high recovery elastomers which are stable in both the cured and uncured states. They are easy to work with and provide a very neat joint. Their outstanding property is high recovery, and after one year of service the sealant may show as much as 98% recovery. The drawback of high recovery is the inherent low tear resistance, which is about half the value of the tear resistance of polysulfides or polymercaptans [24]. Once the deformed sealant is cut or punctured, the existing internal stresses will extend a tear rapidly, leading to sealant failure. Silicones require no solvent for workability, thus curing shrinkage is negligible due to almost 100% solids.

Sealants are classified as RTV because they cure at room temperature and do not require heat to cure, as in the case for heat cured rubbers [25]. The polymer chain used in heat cured rubber has a degree of polymerization (DP) of 7000, while the DP of RTV elastomers is 1000. The shorter chain length is necessary because lower viscosity is required for RTV applications, but the tensile strength and elongation are lower than for heat cured rubber. Varying the organic groups attached to the silicon atoms can expand the capabilities of silicone elastomers. Substituting methyl with phenyl groups, for example, provides greater low temperature flexibility, which is required for sealants used in the aerospace industry. Fillers can be used to enhanced desirable properties, reduce cost, add color or to increase resistance to degradation.

One-component silicone sealants are condensation cured, which means that they cure by reacting with moisture present in the atmosphere. A trifunctional organosilane is the crosslinking system, while a metal soap acts as a catalyst for the condensation reaction. The sealant is stable until it is exposed to the atmosphere, and when exposed it undergoes hydrolysis with atmospheric moisture. The crosslinked network between polysiloxane chains is formed in a reaction which is catalyzed by the metal soap. This reaction liberates an acid such as acetic, which gives off a slight odor. If the sealant is allowed to develop its entire crosslinked network through complete cure, all of its inherent characteristics and properties can be achieved.

Curing is accelerated by hot humid weather and retarded by cold weather. If cured in cold weather, the sealant remains more flexible at low temperatures (although it may harden and become more brittle than organic elastomers). The range of color tints is virtually unlimited and color stability is good; thus substrates are not stained. Silicone sealants exhibit excellent resistance to weathering, moisture and ultraviolet radiation, and are not affected by ozone [25]. After 8000 hours of accelerated weathering, there is no discoloration, appreciable hardening, loss of flexibility or staining of substrates. This makes them prime candidates for exterior one-stage sealing. Then tensile adhesive strength is low in relation to the tensile cohesive strength. This necessitates the use of a primer for working joints, which is one of the primary applications of silicone sealants. Installation requires a very clean joint; thus silicone is a more labour intensive sealant than others.

Room temperature curing silicone sealants, according to Greenlee [26], will adhere well to umprimed, unanodized aluminum and steel if they contain alkoxysilyl groups and if the substrate has been brought into contact with certain anions either before or during cure.

Silicone elastomers consist of a polyorganosiloxane (usually polydimethylsiloxane, due to its cost and wide range of attainable properties), additives providing strength and other desirable properties and the vulcanization system.

Vulcanized elastomers maintain their physical properties over a large temperature range as other silicone polymers do. When elastomers are exposed to air at temperatures above 200°C, crosslinking and embrittlement slowly occur; in the absence of air, chain scission and softening predominate.

The mechanical properties of silicone elastomers at room temperature are inferior to those of other elastomers. For example, natural rubber vulcanizates normally have tensile strengths in the range 21–28 MPa (3000–4000 lbf/in^2) whilst the corresponding range for general purpose heat cured silicones is only 3.5–7.0 MPa (500–1000 lbf/in^2); room temperature cured material is usually inferior. This low strength results from the small intermolecular attraction between silicone chains and from the highly coiled nature of the chains which results in a small end-to-end length and, consequently, limited chain entanglement. However, the mechanical properties of silicone elastomers at elevated temperatures are greatly superior to those of other elastomers (except fluoro elastomers). For example, in a heat aging test at 250°C a silicone elastomer had useful tensile strength and elongation after two weeks whereas a natural rubber vulcanizate had lost all rubber properties after four hours.

Silicone elastomers have good electrical insulation properties, especially if the decomposition products of the vulcanizing agent are removed by heating the vulcanizate after cure [26].

Silicone elastomers are used also as single ply synthetic roof membranes applied to concrete [27], coatings, printing inks, cellular plastics, textiles, consumer products [28–30], and materials for corrosion protection in adverse environments [31]. The development of flowable silicone rubbers with low and extremely low viscosities and outstanding quality structure makes possible the work without solvents in practically all the areas of coated fabric manufacture [32].

REFERENCES

1. Schmidt, J. and Seiferling, B. (1986) *Nat. Res. Symp. Pr 73*, Materials Research Society, Pittsburgh, p. 739.
2. Langer, M. (1993) *Kunststoffe*, **83**(10), 805.
3. Brikofer, L. and Stuhl, O. (1989) in *The Chemistry of Organic Silicon Compounds*. Vol. 1 (eds S. Patai and Z. Rappaport), J. Wiley & Sons, New York, p. 655.
4. Fujisaka, T., West, R. and Murray, C. (1993). *J. Organometallic Chem.*, **449**, 105.
5. Elias, H.G. (1993) *An Introduction to Plastics*, VCH, Weinheim.
6. Seymour, R.B. and Carracher, Jr, C.E. (1992) *Polymer Chemistry*, 3rd edn, Marcel Dekker, New York.
7. Bayer AG, Leverskusen, Th. Goldschmidt AG, Essen, Waeker-Chemie GMBH-Munich, Hans der Technik e.v., Essen (1991) *Silicones – Chemistry and Technology*, CRC Press, Boca Raton, and Vulcan Verlag, Essen.
8. Billmeyer Jr, F.W. (1988) *Textbook of Polymer Science*, 3rd edn, J. Wiley & Sons, New York.
9. Lee, K.S. (1990) in *Engineering Polymers* (ed. R.W. Dyson), Blackie, Glasgow, p. 38.
10. Dickstein, W.H. (1990) *Rigid Rod Star-block Copolymers*, Technomic, Lancaster, p. 51.
11. Saunders, K.J. (1988) *Organic Polymer Chemistry*, Chapman & Hall, London.
12. Caprion, J.C. and Macander, R.F. (1987) in *Rubber Technology* (ed. M. Morton), Van Nostrand Reinhold, New York, p. 375.
13. Ali, S.A.M. and Hourston, D.J. (1994) in *Advances in Interpenetrating Polymer Networks*, Vol. IV (eds D. Klempner and K.C. Frisch), Technomic, Lancaster, p. 17.
14. Class, J.B. and Grasso, R.P. (1993) *Rubber Chem. Technol.*, **66**(4), 605.
15. Mark, H.F., Bikales, N.M., Overberger, C.G. and Menges, G. (1989) *Encyclopaedia of Polymer Science and Engineering*, Vol. 15, J. Wiley & Sons, New York.
16. Noll, W. (1968) *Chemistry and Technology of Silicones*, Academic Press, New York.
17. Brydson, J.A. (1988) *Rubbery Materials and Their Compounds*, Elsevier Applied Science, London.
18. Brydson, J.A. and Cheremisinoff, N.P. (1989) *Handbook of Polymer Science and Technology*, Vol. 2, Marcel Dekker, New York.
19. Shah, V.M., Hardy, B.J. and Stern, S.A. (1986) *J. Polym. Sci. Part B Polymer Physics*, **24**, 2033.
20. Brydson, J.A. (1978) *Rubber Chemistry*, Applied Science Publishers, London, p. 396.

21. Grady, B.P. and Cooper, S.L. (1994) in *Science and Technology of Rubber*, 2nd edn (eds J.E. Mark, B. Erman and F.R. Eirich), Academic Press, New York, p. 601.
22. Weyenberg, D.R. and Lane, T.H. (1990) in *Silicone-Based Polymer Science; A Comprehensive Resource* (eds J.M. Zeigler and G.F.W. Fearon), American Chemical Society, Washington, p. 753.
23. Ulrich, J. (1982) *Introduction to Industrial Polymers*, Hanser Publishers, Munich.
24. Beznaczuk, L.M. (1985) Master's Thesis, Concordia University, Montreal.
25. Klowsowski, J.M. and Gant, G.A.L. (1979) in *Plastics, Mortars, Sealants and Caulking Compounds* (ed. R.B. Seymour), ACS Symposium Series 113, American Chemical Society, Washington, p. 133.
26. Greenlee, T.W. (1975) in *Adhesion Science and Technology* (ed. L.H. Lee), Plenum Press, New York, p. 339.
27. Feldman, D. (1989) *Polymeric Building Materials*, Elsevier Applied Science, London.
28. Ranney, M.W. (1977) *Silicones*, Vol. II, Noyes Data Co., Park Ridge.
29. Bennington, L.D. (1990) *Rad. Tech 90 North Am., Conf. Proc.*, **1**, p. 233.
30. Reti, R. (1993) *Em. Polym. Print Colour J.*, **183**(4335), 414.
31. Ryntz, R. and Elms, R. (1991) *Adv. Org. Coat. Sci. Technol. Ser.*, **13**, 519.
32. Mueller, J (1993) *J. Coated Fabr.*, **22**, 308.

7 Polyurethanes (PUs)

7.1 INTRODUCTION

PUs, as commercial polymers, constitute one of the most complete and versatile class of polymeric materials known today. The term polyurethane is extended to all complex reaction products of isocyanates and polyhydroxy compounds. From the chemical point of view they contain a diversity of functional groups along with the urethane group (–NH–CO–O–), which sometimes is not even the principal group present. Many other groups such as ether, ester, amide, urea, biuret, allophanate, uretidione, carbodiimide and isocyanurate enter the chemical composition.

PUs are synthesized by a step growth polymerization process in which the chain length of the polymer increases steadily as the reaction progresses.

The polymer chains may be linear or slightly branched as in thermoplastic PUs or they may have a crosslinked network leading to thermosetting products.

A large diversity of raw materials has been employed to produce PUs. They may be classified in three principal groups: isocyanates, polyols and chain extenders. The synthesis of the polymer is always based on the utilization of at least one component of each principal group. Complementary materials, such as water, blowing agents, catalysts, structure regulators and solvents, have been used to manufacture specific products (elastomers, foams, coatings). Commercial formulations may also include other additives such as anti-aging compounds, fillers, colorants, flame retardants and others.

A large spectrum of properties that can be tailored by the selection of raw materials, formulations and manufacturing processes, has led to a broad variety of end products that are in continuous diversification. The following major types of commercial PU are produced: flexible foams, rigid foams, integral (or self-skinning) foams, elastomers, fibers, surface coatings, adhesives and sealants.

PUs date from 1937 when fiber forming polymers were produced by the German chemist, O. Bayer (I.G. Farbenindustries), by reacting diisocyanates with aliphatic diols. No extensive commercial importance was reached until after the Second World War when elastomers (1950) and foams (1952) were developed at Bayer. But the expansion of the market started after 1957 when American companies introduced polyether–polyol based foams at a lower price and with a wider range of properties. The production volume of PUs has been growing continuously since then as new and valuable applications have been found [1].

PUs are among the top ten global polymers with a volume of 5.1 million tons per year in 1990 and an estimated demand of 8.5 million tons in 2000 [2]. The principal consumers of PUs are Western Europe with 2 million tons, the USA with 1.5 million tons and the Far Eastern countries with 1.2 million tons [2]. Flexible foams accounted for about 50% of the market, rigid foams for 30% and the remainder is covered by elastomers, adhesives, coatings and sealants.

7.2 RAW MATERIALS

7.2.1 Isocyanates and their chemistry

The specific raw material for polyurethane production are isocyanates, more precisely di- and polyisocyanates. Many preparative reactions are known to synthesize isocyanates [3,4] but they are used exclusively on a laboratory scale. The only methods that have acquired commercial importance consist of the phosgenation of the appropriate amines or amine hydrochlorides.

The phosgenation is usually carried out in a solvent inert to isocyanate, i.e. does not contain active hydrogen atoms (usually *o*-dichlorobenzene). The process is run continuously or in batch in two stages known as the cold and the hot stages. In the cold stage (25–100°C) a 20% solution of the appropriate amine is reacted with a solution containing a 20–50% excess of phosgene. Intermediate carbamoyl chloride is formed:

$$\mathrm{R{-}NH_2 + COCl_2 \longrightarrow R{-}NHCOCl + HCl}$$

Then the carbamoyl chloride is thermally decomposed to isocyanate in the hot stage (150–160°C) and the hydrogen chloride and excess phosgene are removed with nitrogen:

$$\mathrm{R{-}NHCOCl \longrightarrow R{-}N{=}C{=}O + HCl}$$

At the end, pure isocyanate is separated from the solvent by fractional distillation, by crystallization or by sublimation.

As isocyanates are very reactive compounds, side reactions may occur, especially if unreacted amine is present, leading to the formation of a urea:

$$\mathrm{R{-}N{=}C{=}O + R{-}NH_2 \longrightarrow R{-}NH{-}CO{-}NH{-}R}$$

To minimize this reaction, amine hydrochloride can be used instead of amine and excess phosgene is maintained during the process.

The most important commercial isocyanates are given below.

(a) The group of toluene diisocyanate (TDI) which comprises

- 80:20 TDI, a mixture of two isomers, 80% 2,4-toluene- and 20% 2,6-toluene diisocyanates;
- 65:35 TDI, a mixture of 65% 2,4- and 35% 2,6- isomers;
- 2,4-TDI which is pure 2,4- isomer.

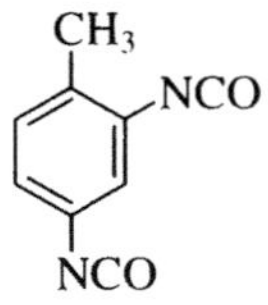

2,4-Toluene diisocyanate (2,4-TDI)

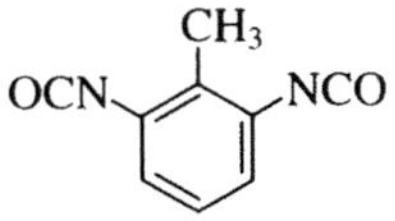

2,6-Toluene diisocyanate (2,6-TDI)

These different mixtures of isomers are obtained by various ways of conducting the dinitration of toluene. The mixtures of 2,4- and 2,6-dinitrotoluenes are subsequently reduced to diaminotoluenes without any fractionation and then phosgenated to diisocyanates.

TDI mixtures are liquids at ambient temperatures that have the melting range at 5–15°C; while pure 2,4-TDI melts at 22°C. The 4-isocyanato group is more reactive than either 2- or 6- position groups at ambient temperature and in the absence of catalysts.

TDIs are respiratory toxic materials and the toxicity is closely related to their vapor pressure. Safety measures should be taken in handling these materials.

TDI mixtures are extensively used for the production of polyurethane flexible foams, elastomers and adhesives.

(b) The diphenylmethane diisocyanate group is composed of

- pure 4,4′-diphenylmethane diisocyanate (MDI);
- 'polymeric' diphenylmethane diisocyanate (PMDI);
- modified diphenylmethane isocyanates.

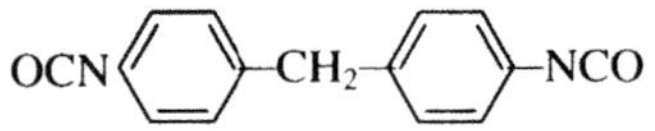

4,4′-Diphenylmethane diisocyanate (MDI)

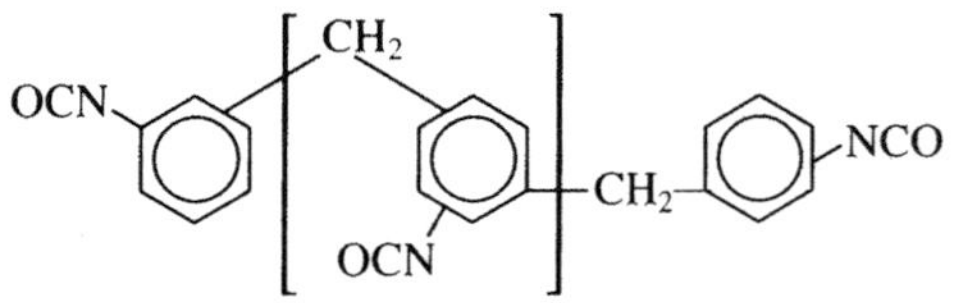

Polymeric diphenylmethane diisocyanate (PMDI)

As the melting point of pure 4,4′-MDI is 37–38°C, some liquid derivatives have been prepared. The so called 'polymeric' diphenylmethane diisocyanate

(PMDI) is a crude liquid mixture with the functionality 2.3–3.1, composed of around 55% MDI (4,4′- and 2,4′- isomers), 25% triisocyanates and 20% higher polyisocyanates. The mixture results at the step of amine preparation which is realized by the acid catalyzed condensation of aniline with formaldehyde and its composition depends on the aniline to formaldehyde ratio. By the phosgenation of the whole mixture, polymeric PMDI is obtained in quantitative yield. Pure 4,4′-MDI is separated from PMDI by continuous thin film distillation.

Using various chemical modifications of pure MDI, stable liquid derivatives with free isocyanate groups have been prepared. They are prepolymers with polyesters or polyethers, carbodiimide modified or isocyanurate modified MDI derivatives [5].

All these products are less toxic than TDI isocyanates as their vapor pressures are lower.

Crude PMDI oligomers are mainly used in the production of rigid insulation foams. Pure MDI is used in the manufacture of elastomers.

(c) Naphthalene, 1,5-diisocyanate (NDI) is a solid compound melting at 128°C. It was first prepared in Germany and is used almost exclusively in Europe for the production of Vulkollan type elastomers:

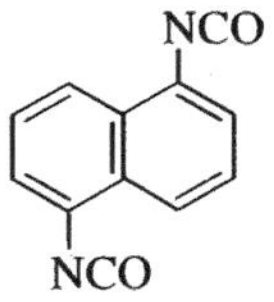

1,5-Naphthalene diisocyanate (NDI)

(d) Hexamethylene diisocyanate (HDI) is a volatile and high vapor pressure liquid; consequently, it is a toxic product:

$$O{=}C{=}N{-}(CH_2)_6{-}N{=}C{=}O$$

In order to reduce its volatility, hexamethylene diisocyanate is reacted with a controlled amount of water resulting in a triisocyanate with biuret structure:

$$H_2O + 3OCN(CH_2)_6NCO \longrightarrow \begin{array}{c} NH(CH_2)_6NCO \\ | \\ CO \\ | \\ N(CH_2)_6NCO \\ | \\ CO \\ | \\ NH(CH_2)_6NCO \end{array} \quad \text{Biuret triisocyanate}$$

Light stable non-yellowing PU are obtained from HDI and its derivatives due to the absence of aromatic structures and therefore aliphatic isocyanates are used for coatings.

(e) Another diisocyanate for the production of polyurethanes with improved resistance to discoloration and to thermal and hydrolytic attack is 4,4′-dicyclohexylmethane diisocyanate ($H_{12}MDI$), a mixture of stereoisomers obtained on the hydrogenation of 4,4′-diaminodiphenylmethane:

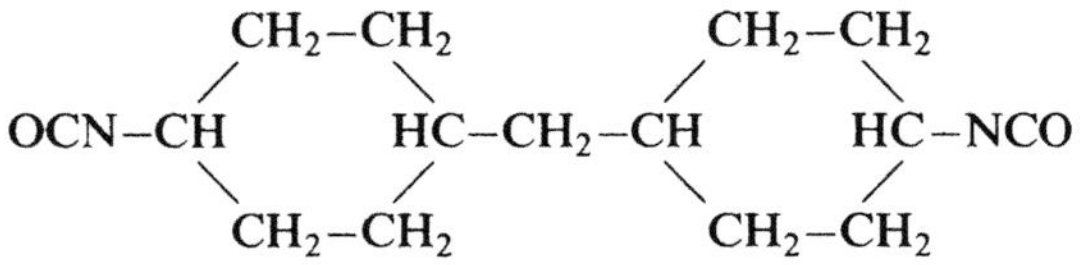

4,4′-Dicyclohexylmethane diisocyanate ($H_{12}MDI$)

(f) Isophorone diisocyanate (IPDI), 3-isocyanatomethyl-3,5,5-trimethylcyclohexyl isocyanate, has both isocyanate groups attached to saturated organic groups:

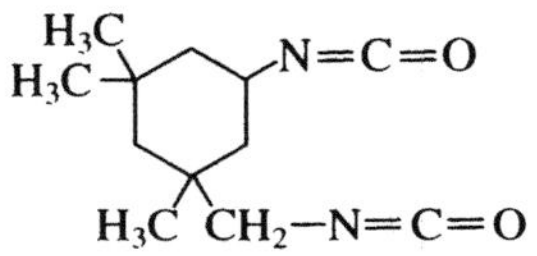

3-Isocyanatomethyl-3,5,5-trimethylcyclohexyl isocyanate
(IPDI)

It is a liquid less volatile than TDI and is used for light stable flexible coatings.

(g) 1,3-bis(isocyanatomethyl) cyclohexane (H_6XDI) is another commercial isocyanate suitable for non-yellowing PU applications:

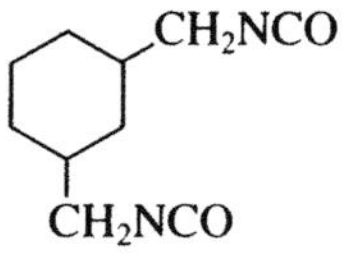

1,3-Bis(isocyanatomethyl) cyclohexane
(H_6XDI)

(h) A more recent technological development is the preparation of polymeric aliphatic isocyanates obtained by the copolymerization of styrene with 2-isocyanatoethyl methacrylate and other suitable acrylic acid esters [6].

The isocyanate group is very reactive and gives rise to a large number of reactions. Only those reactions employed in different polyurethane technologies are presented below. They may be classified as follows:

- addition reactions involving active hydrogen compounds (1–7);
- self-cycloaddition reactions (8 and 9);
- self-condensation reactions (10).

1. With alcohols and phenols (R′ is aliphatic or aromatic):

$$R{-}N{=}C{=}O + HO{-}R' \longrightarrow R{-}NH{-}CO{-}O{-}R' \qquad \text{Urethane group}$$

2. With amines:

$$R{-}N{=}C{=}O + H_2N{-}R' \longrightarrow R{-}NH{-}CO{-}NH{-}R' \qquad \text{Urea group}$$

3. With water:

$$R{-}N{=}C{=}O + H_2O \longrightarrow \underset{\text{Unstable carbamic acid}}{(R{-}NH{-}CO{-}OH)} \xrightarrow{-CO_2} R{-}NH_2 \qquad \text{Amine group}$$

4. With carboxylic acids:

$$R{-}N{=}C{=}O + HOCO{-}R' \longrightarrow \underset{\text{Unstable anhydride}}{(R{-}NH{-}CO{-}O{-}CO{-}R')} \xrightarrow{-CO_2}$$

$$R{-}NH{-}CO{-}R' \qquad \text{Amide group}$$

5. With urethanes:

$$R{-}N{=}C{=}O + R'{-}NH{-}CO{-}O{-}R'' \longrightarrow R{-}NH{-}CO{-}\overset{\displaystyle R'}{\overset{|}{N}}{-}CO{-}O{-}R''$$

Allophanate group

6. With ureas:

$$R{-}N{=}C{=}O + R'{-}NH{-}CO{-}NH{-}R'' \longrightarrow R{-}NH{-}CO{-}\overset{\displaystyle R'}{\overset{|}{N}}{-}CO{-}NH{-}R''$$

Biuret group

7. With amides:

$$R{-}N{=}C{=}O + R'{-}NH{-}CO{-}R'' \longrightarrow R{-}NH{-}CO{-}\overset{\displaystyle R'}{\overset{|}{N}}{-}CO{-}R''$$

Acylurea group

As can be seen, reactions (1–4) lead to linear chains whereas reactions (5–7) result in branched chains.

8. Dimerization:

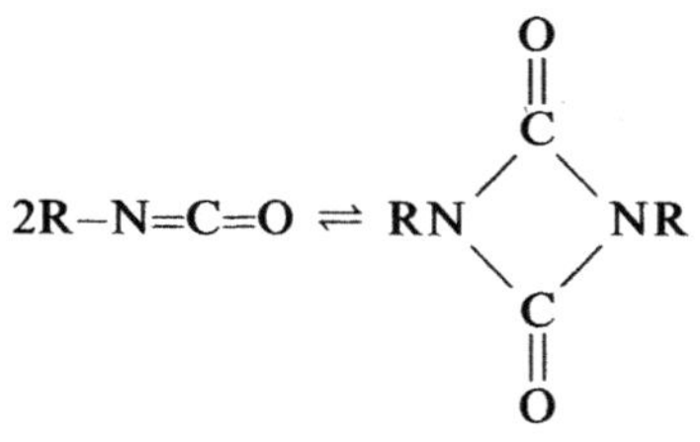

Uretidione group

9. Trimerization:

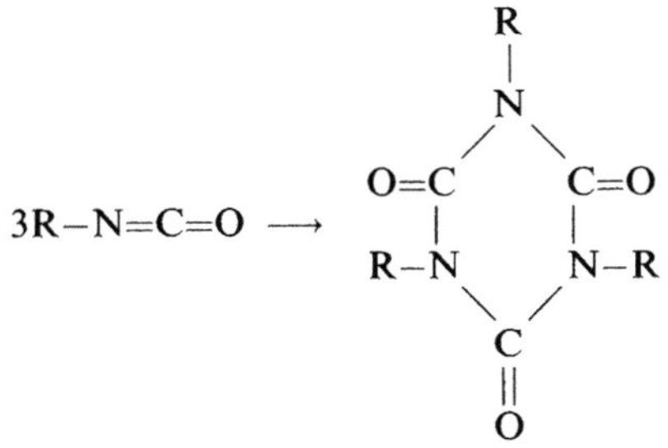

Isocyanurate group

10. Self-condensation:

$$2R{-}N{=}C{=}O \longrightarrow R{-}N{=}C{=}N{-}R + CO_2 \qquad \text{Carbodiimide group}$$

These reactions may also be used to modify the isocyanates in order to make them more versatile for various technologies or to introduce certain groups such as carbodiimide or isocyanurate into the end products.

It should be noted that the reaction rates depend on the structures of the involved compounds. The reactivity of the isocyanate group is enhanced by the presence of electron withdrawing groups on the isocyanate molecule whereas the electron donating substituents reduce the reactivity [7]. Aromatic isocyanates are much more reactive than aliphatic ones. Steric hindrance also influences the rate of isocyanate reactions. Contrary to isocyanates, the active hydrogen compounds are activated by the presence of electron donating groups on their molecules. The active hydrogen compounds can be arranged in decreasing order of their nucleophilicities which are proportional to their reactivities toward isocyanates:

aliphatic amines > aromatic amines > primary alcohols > water > secondary alcohols > phenols > acids > ureas > amides > urethanes.

The above considerations are true only in the absence of catalysts; their presence can substantially modify the rate of each individual reaction as well as their relative rates.

7.2.2 Polyols

Besides isocyanates, the other major components of all types of PU products are hydroxyl-containing compounds known as 'polyols'. Polyols can also be used as intermediates or modifiers for other polymers such as epoxy resins or unsaturated polyesters, as well as for non-polymer domains as surfactants and functional fluids.

Polyols are oligomeric products with MW in the range 250–8000 and a hydroxyl functionality from 1–8. The composition and the structure of the polyol polymeric backbone, the hydroxyl functionality and its distribution, and the hydroxyl equivalent weight are all essential characteristics of polyols that are reflected in the properties of the resulting PU polymers.

A large variety of polyols is used in PU manufacture. They are homo- or copolyethers, homo- or copolyesters, polyester–polyethers or polyisobutylene polyols. Another category of polyols are the so called 'polymer polyols' that are polyether polyols containing dispersions of organic fillers as an integral part of the polyol. Polyether amines have been introduced more recently for RIM technologies. The choice of polyol is imposed by the end application of the product. For example, PU that require hydrolytic stability and thermal resistance are based on polyether polyols, whereas a better resistance to solvents is imparted by the use of polyester polyols.

As a general guideline, linear or slightly branched polyols are used for the manufacture of elastomers and flexible foams, whereas moderately and highly branched polyols are used for coatings and rigid foams.

The technology of polyols for PU applications described particularly in patents has been reviewed extensively [8–12].

(a) Polyether polyols (PEPs)

PEP represent around 90% of the total polyols used for PU. They consist in a variety of linear or branched compounds with polyether backbone and hydroxyl functional groups. PU based on PEP have better hydrolytic resistance than those prepared from polyester polyols [13].

(a) The bulk of PEPs most widely used for PU are poly(propylene oxide) polyols (PPOs) and propylene oxide/ethylene oxide random or block copolymers. Two main categories of these polyols are prepared. For low modulus PU such as flexible slabs, elastomers (including RIM) and coatings, linear or slightly branched PEPs are prepared, with a 2–3 hydroxyl functionality and MW in the range of 1000–8000. High modulus highly crosslinked PU such as rigid foams, hard coatings, microcellular or solid plastics require highly branched, low MW polyols (250–1000), with an average hydroxyl functionality of 3–7.

PEP can be synthesized by anionic, cationic or coordinate catalytic mechanisms [14]. Commercially PPOs are produced by anionic polymerization of propylene oxide initiated by different glycols (starters). The functionality of the initiator determines the hydroxyl functionality of the polyol.

With propylene glycol as initiator linear PPOs are obtained:

$$HO{-}\underset{\displaystyle CH_3}{\underset{|}{CH}}{-}CH_2{-}O{-}\left[CH_2{-}\underset{\displaystyle CH_3}{\underset{|}{CH}}{-}O\right]_n{-}CH_2{-}\underset{\displaystyle CH_3}{\underset{|}{CH}}{-}OH$$

Secondary hydroxyls which are less reactive toward the isocyanates can be replaced by more reactive primary hydroxyls by copolymerizing propylene oxide with ethylene oxide so that the chains are tipped with ethylene oxide units:

$$HO\!-\!\!\left[CH_2-CH_2-O\right]_x\!\!\left[CH_2-\underset{\displaystyle CH_3}{\underset{|}{CH}}-O\right]_y\!\!\left[CH_2-CH_2-O\right]_z\!\!-H$$

PEPs based exclusively on ethylene oxide are too hydrophilic to be satisfactory for PU.

To prepare branched PPO, it is common practice to initiate the polymerization of propylene oxide by triols such as glycerol, 1,1,1-trimethylolpropane or 1,2,6-hexane triol. When glycerol is used as the initiator, the resulted PPO are triols such as:

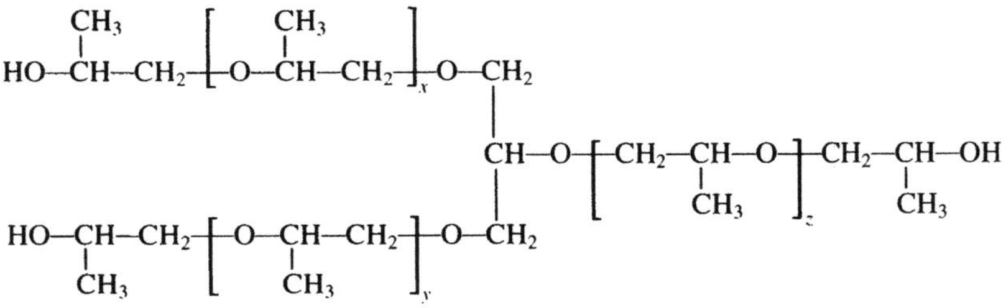

For even more branched structures, polymerization is initiated by polyfunctional initiators such as pentaerythritol or sorbitol with four and six hydroxyl groups, respectively:

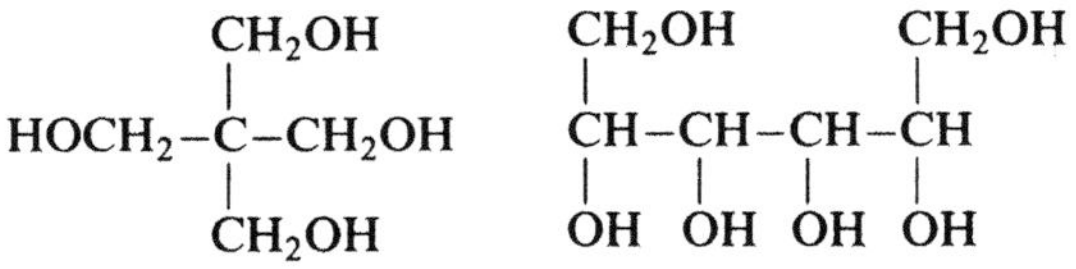

Pentaerythritol Sorbitol

Amines and other active hydrogen compounds can act as initiators.

The initiation step of the anionic polymerization is the formation of the alkoxide ion (alcoholate) of the initiator in the presence of a basic catalyst, most frequently potassium hydroxide:

$$M^{\oplus}OH^{\ominus} + InH \longrightarrow In^{\ominus}M^{\oplus} + H_2O$$

Propagation takes place by the addition of alkylene oxide units to the desired MW:

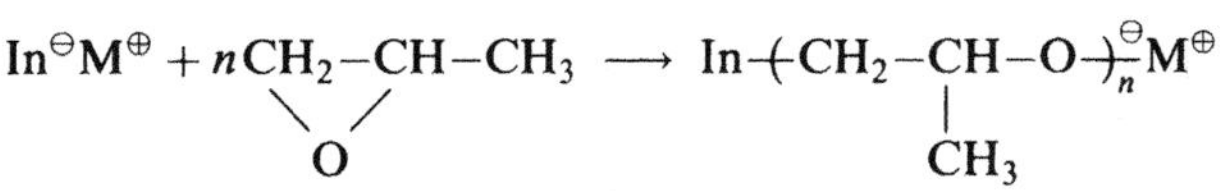

Chain growing is concluded by chain transfer and termination:

$$In\!\left(CH_2-\underset{\displaystyle CH_3}{\underset{|}{CH}}-O\right)_n^{\ominus}M^{\oplus} + ROH \longrightarrow In\!\left(CH_2-\underset{\displaystyle CH_3}{\underset{|}{CH}}-O\right)_n\!H + RO^{\ominus}M^{\oplus}$$

The manufacturing process can be run continuously or in semibatch operations and it is controlled extensively by instrumentation in order to obtain consistent products. Reaction temperatures are in the range 80–150°C. Polyols for PU must have a very high level of purity in order to ensure the control of subsequent reactions with isocyanates. The removal of catalyst traces has to go as far as 5 ppm for the majority of PPO applications, and abundant patent literature describes a variety of purification processes [12]. PPOs are stabilized with proprietary inhibitor antioxidants (hindered phenols, aromatic amines, phenothiazene and other compounds), saturated with nitrogen and stored hot. A flow chart of the preparation of PEP is presented in Figure 7.1.

(b) Another PEP is poly(oxytetramethylene) glycol (PTMG) or poly(tetrahydrofuran) (PTHF) introduced on the market in 1955:

$$HO{-}\left[(CH_2)_4{-}O\right]_n{-}(CH_2)_4{-}OH$$

It is prepared by acid catalyzed, cationic, ring opening polymerization of tetrahydrofuran (oxacyclopentane) [15,16]. The most commonly used catalyst for the commercial polymerization is fluorosulfuric acid [17]. The initiating species is the oxonium ion and the propagation involves nucleophilic attack of monomer with simultaneous ring opening.

Initiation:

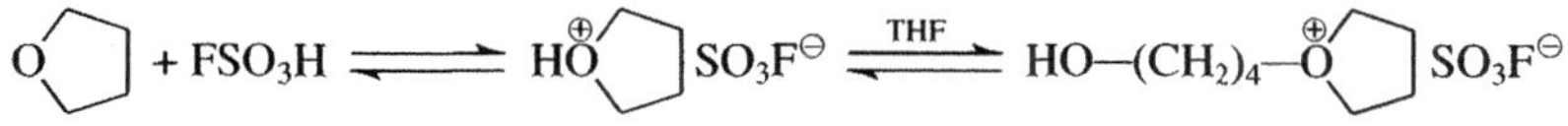

Propagation:

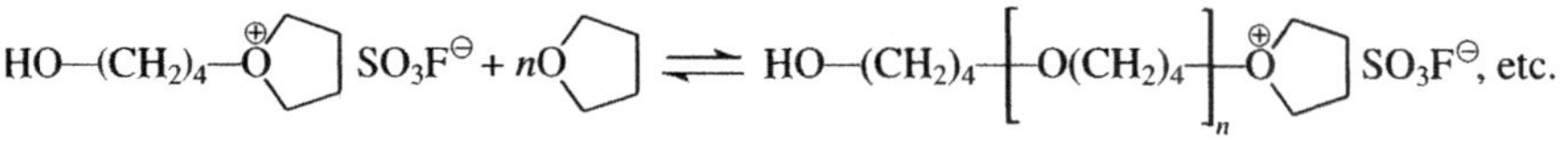

Commercial PTMGs have MW in the range 600–3000, a functionality of two and only primary hydroxyls. It is the most frequently used polyol for the preparation of PU thermoplastic elastomers and fibers due to its low T_g (−84°C), the regularity of its chain structure and its ability to crystallize upon extension.

(c) Filled PEPs have become major polyol raw materials for PU foams. This group of products consists of dispersions of different reinforcing polymers which are not additives but integral parts of the polyol [18]. They are made *in situ* in the liquid polyol through either free radical polymerization or step polyaddition reactions. Many commercial compositions of these particulate PEPs are prepared. Styrene–acrylonitrile or other copolymers grafted on PEP are known as 'polymer polyols'. Other filled PEPs contain particulate dispersed phases based on polyureas, polyurethanes or epoxy resins.

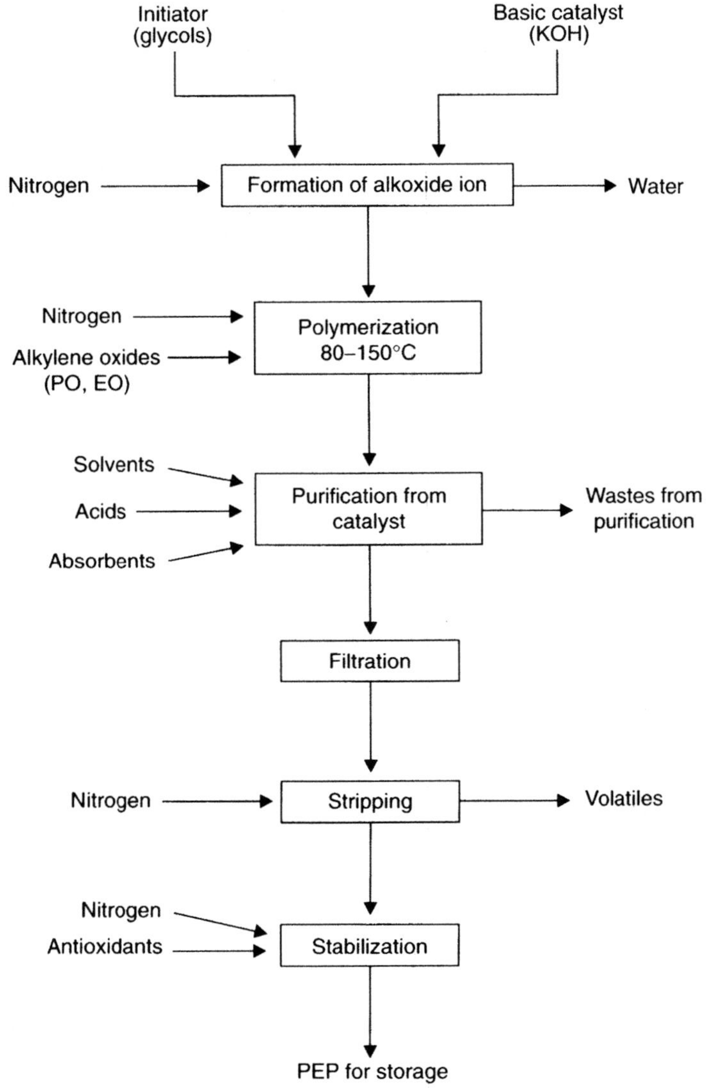

Figure 7.1 Preparation of polyether polyols.

Similar to the filler effects in composite materials, filled polyols enhance the load bearing characteristics of high resilience flexible PU foams and the strength properties of PU elastomers.

(d) Amine terminated PEPs have been prepared more recently and used in different polyurethane/polyurea technologies such as RIM applications, foam and cast elastomers, in epoxy resins as well as in many other non-polymeric fields [19]. Primary and secondary hydroxyl groups can both be aminated with ammonia in the presence of catalysts. These polyetheramines have increased reactivities toward isocyanates allowing shorter manufacturing cycles and, in many applications, the use of catalysts can be avoided [20,21]. The resulting polyurethane/polyurea products exhibit better mechanical properties.

(b) Polyester polyols (PESPs)

Chronologically PESPs are the first polyols used for the preparation of PUs [11]. Besides this main application, they are also used as components of epoxy, alkyd or unsaturated polyester resins.

Two different categories of PESPs are manufactured and intended for specific applications [22]. High purity PESPs with MWs in the range 2000–4000 are prepared by the condensation of dibasic acids such as adipic, sebacic, phthalic and aliphatic diols: ethylene glycol, diethylene glycol, propylene glycol, 1,4-butanediol, 1,5-pentanediol and 1,6-hexanediol. Slightly to highly branched PESPs are prepared from the same acids with polyhydric alcohols such as glycerol, trimethylolpropane or pentaerythritol. A typical polyester is poly(ethyleneadipate) diol:

$$HO{-}CH_2{-}CH_2{-}O{-}\left[CO{-}(CH_2)_4{-}CO{-}O{-}CH_2{-}CH_2{-}O\right]_n{-}H$$

Polycaprolactone diol is another PESP prepared by the ring opening polymerization of ϵ-caprolactone and diol or triol initiators:

$$H{-}\left[O{-}(CH_2)_5{-}CO\right]_x{-}ORO{-}\left[CO{-}(CH_2)_5{-}O\right]_y{-}H$$

Castor oil, a natural compound of ester type largely composed of glyceroltriricinoleate, is a triol specifically used in the manufacture of PU adhesives and coatings.

The composition and MW of PESP determine their physical properties, such as the ability for hydrogen bonding, the presence of crystalline regions, the melting temperatures and the viscosity of the melt. The hydrolytic stability of PESP depends on the concentration of the ester groups. The PESPs are used for PU elastomers, coatings and fibers with high tensile properties, excellent abrasion resistance and flex fatigue.

The manufacture of PESP is operated in batch or continuous reactors. The polyesterification of acids and glycols is conducted in two stages. In the first stage, the temperature is raised from 135°C to above 200°C in order to remove 90% of water from the condensation reaction. The second stage is conducted at a reduced pressure of 0.4–2 kPa which removes water and drives the polyesterification to the desired degree of conversion [11]. The preparation of PESP is presented in Figure 7.2.

The second category of PESPs consists of low MW aromatic PESP prepared by the transesterification of wastes of poly(ethylene terephthalate) or

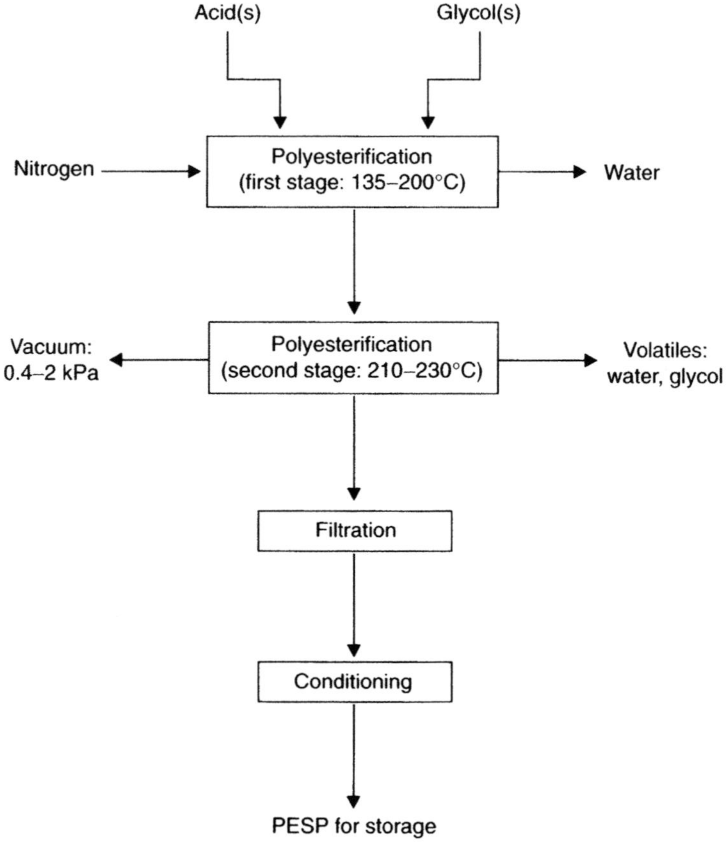

Figure 7.2 Preparation of polyester polyols.

dimethylterephthalate distillation bottoms [22]. Poly(ethylene terephthalate) (from recycled film or bottles) or dimethylterephthalate are transesterified with a glycol in a ratio determined by the desired MW. In addition to the low cost, the flammability of these PESPs is reduced due to their aromatic content. They are successfully used for rigid PU foams with improved burn characteristics.

(c) Polyisobutylene glycols

Polyisobutylene glycols are linear or branched hydroxyl terminated hydrocarbon chains:

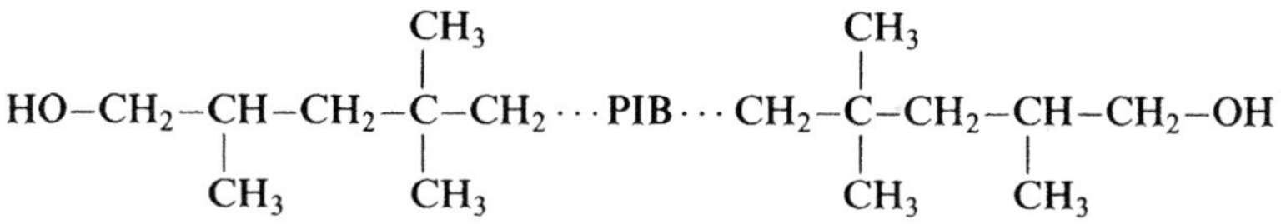

They are prepared by the polymerization of isobutylene in the presence of bi- or trifunctional initiating transfer agents (called inifers) in conjunction with Friedel–Crafts acids [23]. They have recently become commercially available and are intended for polyurethanes for medical and engineering applications [24]. Polyisobutylene based polyurethanes exhibit improved hydrolytic and oxidative stabilities and excellent gas barrier properties due to low water absorption and low moisture permeability of hydrophobic hydrocarbon domains [25].

7.2.3 Chain extenders

Chain extenders play a major role in determining the final properties of PU even though they are relatively minor components in the formulations. They contribute to the nature and the density of chemical crosslinks as well as to the nature and size of the so called 'hard segments' that are responsible for physical crosslinks in linear thermoplastic polyurethanes.

Aromatic diamines and aliphatic or aromatic hydroxy compounds are used as chain extenders.

The selection of the chain extender depends on the formulation (diisocyanate and polyol), the ultimate properties and cost intended for the PU, as well as the processing characteristics. Aromatic diamines are very reactive toward the isocyanate groups and, as a general rule, they are used with prepolymers or systems that are less reactive, such as TDI–polyol prepolymers or for those technologies where very short gel times are required. Less reactive hydroxy extenders are used with more reactive MDI–polyol prepolymers [26]. Aliphatic diamines are too reactive to be used in practice.

In diamine extended PU, the rigid aromatic rings and the highly polar urea groups formed in the reaction with the isocyanates, impart high mechanical strength to the material. Similar properties are obtained when aromatic hydroxy extenders are used. On the other hand, aliphatic short chain glycols are used as extenders when softer and medium strength products are desired [27]. Table 7.1 illustrates the general characteristics of urethanes cured by the two principal types of curatives.

Table 7.1 Influence of glycol and amine chain extenders on PU properties

Glycols	*Property*	*Amines*
————	Reactivity of the system	————→*
————	Tensile strength	————→
————	Modulus	————→
————	Hardness	————→
————	Abrasion resistance	————→
————	Compression set	————→
←————	Resilience	————

* low ——→ high

The most extensively used diamine is 4,4′-methylene bis(2-chloroaniline) known under the trade name MOCA:

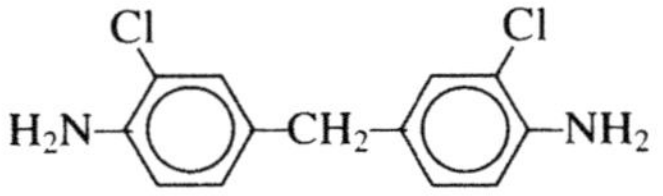

Because of concerns over the potential health hazards to humans, MOCA has to be handled with special care and its use is avoided whenever possible. In 1985 MOCA was labeled a 'probable human carcinogen' by the US Environmental Agency [28].

During the past few years, several new aromatic substituted diamine extenders became available commercially as a result of the intensive research work done to replace MOCA [27,28]. Electron withdrawing substituents on the aromatic ring in addition to their steric hindrance further reduce the reactivity of the amino groups and also lower the toxicity of the amines. Chemical formulas and trade names of the most common amine extenders are presented below:

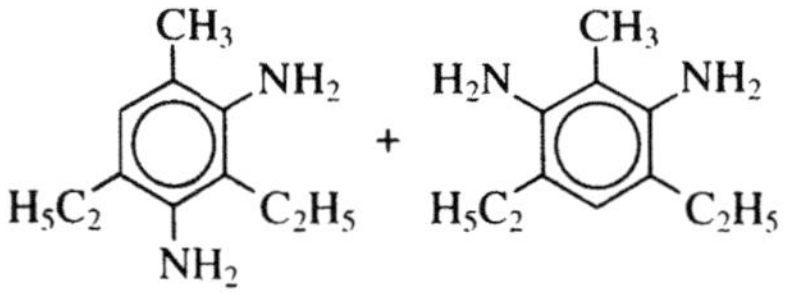

DETDA
3,5-Diethyl-2,4(2,6)-diamino toluene

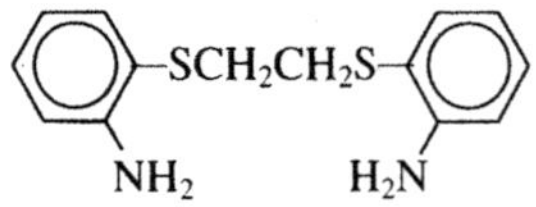

CYANACURE
1,2-Bis(2-aminophenylthio) ethane

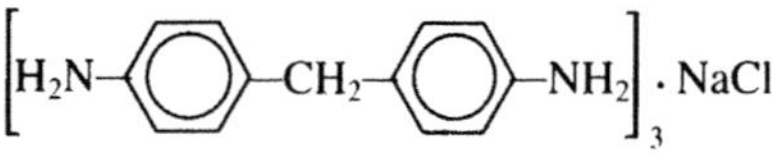

CAYTUR 21
4,4′-Diaminodiphenylmethane complex with NaCl

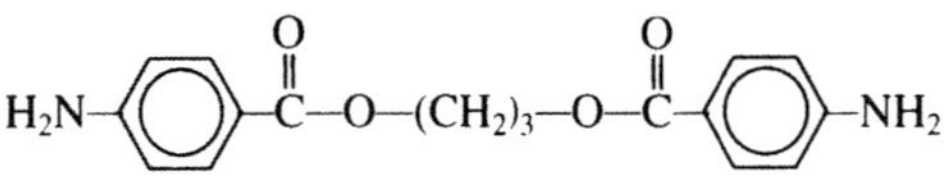

POLACURE 740M
Trimethyleneglycol-bis(4-aminobenzoate)

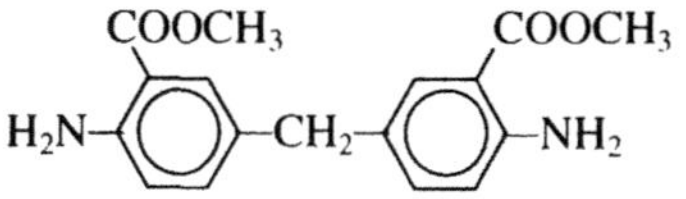

POLACURE 1000
4,4′-Methylene-bis(2-carbomethoxyaniline)

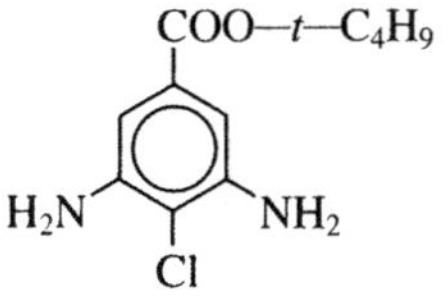

BAYTEC 1604
4-Chloro-3,5-diamino-isobutylbenzoate

Even if they have been found to be safer than MOCA, they are more expensive, require narrower processing parameters and are designed to more specific applications. In RIM formulations amines and particularly DETDA are the chain extenders of choice because they provide not only processing advantages over the glycol extenders, but also yield products with superior physical properties [29]. Gel times as short as 0.8–1.4 s have been practiced in PU/polyurea RIM technologies. For more processing latitude, mixtures of amines or of amines/glycols can be used.

Some of the more common hydroxy chain extenders are presented below:

$HO{-}(CH_2)_2{-}OH$
Ethylene glycol(EG)

$HO{-}(CH_2)_4{-}OH$
1,4-Butanediol (BDO)

$HO{-}(CH_2)_2{-}O{-}(CH_2)_2{-}OH$
Diethylene glycol

$HO{-}(CH_2)_2{-}O{-}(CH_2)_2{-}O{-}(CH_2)_2{-}OH$
Triethylene glycol

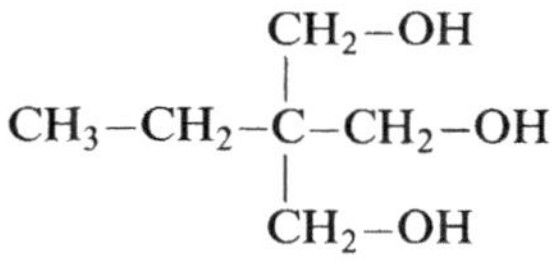
1,1,1-Trimethylolpropane (TMP)

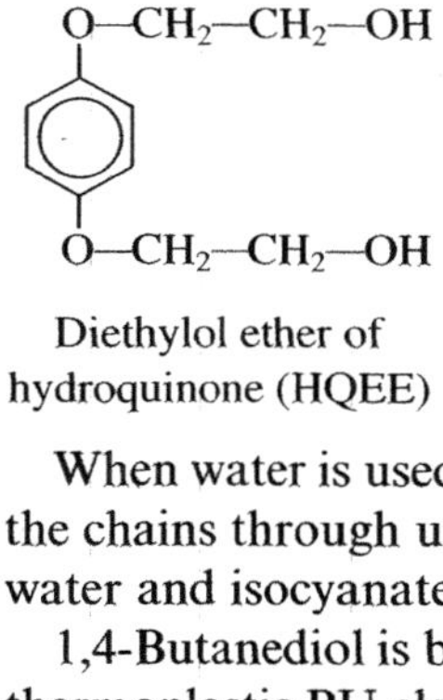
Diethylol ether of hydroquinone (HQEE)

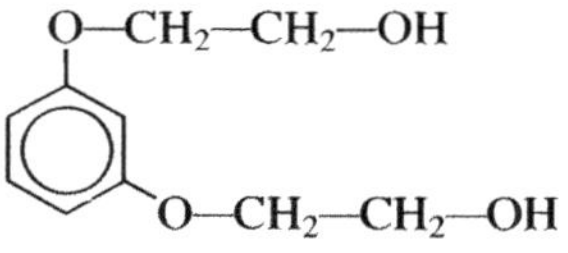
Diethylol ether of resorcinol (HER)

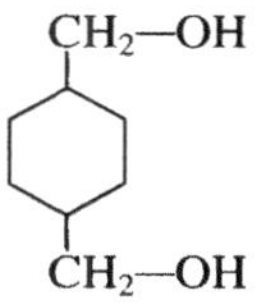
1,4-Dimethylol cyclohexane

When water is used as in flexible PU foams, in addition to the extension of the chains through urea linkages, the evolution of CO_2 from the reaction of water and isocyanate provides the gas that raises the foam.

1,4-Butanediol is by far the most used aliphatic diol for the preparation of thermoplastic PU elastomers [27,28]. Low hardness elastomers are obtained with TDI based prepolymers whereas high hardness products result with MDI prepolymers.

The two aromatic diols, HQEE and HER, yield products with increased thermal resistance and better retention of properties at elevated temperatures, and exceptionally good compression set.

Trimethylolpropane and glycerol used either alone or mixed with diols introduce additional chemical crosslinking in polyurethanes.

7.2.4 Catalysts

The utilization of catalysts is essential in many commercial procedures in order to achieve the desired processing parameters and physical properties.

During the preparation of PU, more than one of the previously mentioned isocyanate reactions take place simultaneously; therefore, the relative rates of all reactions should be controlled very carefully as they have an important bearing in determining the final structure and the properties of the resulting PU. It is thus evident the importance of the utilization of the appropriate catalyst or mixture of catalysts.

Many classes of compounds have been investigated as catalysts but only tertiary amines and some tin, mercury and bismuth compounds have acquired practical importance [11].

The catalytic activity of amines depends not only on their base strength but even more on steric hindrance at the nitrogen atom [30]. Amines with a less hindered lone pair of electrons on the nitrogen (with smaller or no substituents) are stronger catalysts. Thus, triethylenediamine (1,4-diazabicyclo-2,2,2-octane or DABCO) is a weak base but a powerful catalyst:

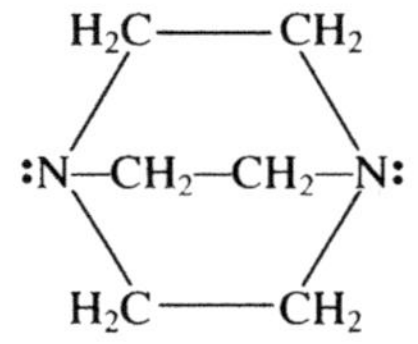

More specific tertiary amine catalysts with delayed initiation or heat activation have been developed recently [31].

The rates of various isocyanate reactions are influenced to a different extent by various catalysts, as can be seen from Table 7.2. The reaction (a) corresponds to OH/NCO reaction (growing of the polymer chains); (b) is a HOH/NCO reaction and results in the formation of amine groups with simultaneous evolution of CO_2 and the blowing of the foam, and (c) leads to crosslinking and to the formation of the final polymer network. A comparison between the relative rates of the three reactions for each individual catalyst is given in the fourth column. It is

Table 7.2 Effect of various catalysts on isocyanate reactions [32]

	Relative rates of isocyanate reaction with			
	n-*Butanol* (*a*)	*Water* (*b*)	*Diphenylurea* (*c*)	*a*:*b*:*c*
None	1.0	1.1	2.2	0.45: 0.5:1
Triethylamine	86	47	4	21.5:11.7:1
Tetramethyl-1,3-butanediamine	260	100	12	21.6: 8.3:1
Triethylenediamine	1200	380	90	13.3: 4.2:1
Tributyltin acetate	80 000	14 000	8000	10.0: 1.7:1
Dibutyltin diacetate	600 000	100 000	12 000	50.0: 8.3:1

evident that metallic compounds, and particularly dibutyltin diacetate, are the most powerful catalysts and they selectively speed up the formation of polymer chains while the amines promote more strongly the gelling reaction, the reaction with water and the subsequent blowing of the foam. When used in mixture, tertiary amines and tin compounds are synergistic, i.e. their catalytic activity is more than twice the sum of their separate activities.

Furthermore, various classes of isocyanates respond in a different way to the action of catalysts. The relative rates of reaction of aromatic and aliphatic isocyanates are not strongly modified by tertiary amines. Tin and lead compounds provide a more pronounced catalytic activity on aliphatic isocyanates making them as reactive as the aromatic ones whereas iron, cobalt and titanium compounds make aliphatic isocyanates even more reactive than the aromatic ones [33].

As for the polyols, certain catalysts manifest a selective promoting effect on the reactivity of either primary or of secondary hydroxyl groups.

Hexahydrotriazines, potassium carboxylates and quaternary ammonium hydroxides are the catalysts of choice to introduce isocyanurate structures in PU–isocyanurate foams, whereas phospholene oxides catalyze the formation of carbodiimide structures in low density poly(carbodiimide) and PU–carbodiimide–isocyanurate foams [34].

7.3 POLYURETHANE ELASTOMERS

On the market there is a large spectrum of PU elastomers from highly flexible and resilient to hard, plastic-like materials [26]. The application of analytical techniques during the last years has led to a comprehensive understanding of structure–property relationships in PU and of the kinetics of formation processes. These two types of information have resulted in a high level of control of formulations and manufacturing.

The variation in properties is based on the balance of two kinds of structures in the PU macromolecule: a 'soft segment' block and a 'hard segment' block. The soft segments are formed from the isocyanate and the high MW polyol; the hard segments are formed from the isocyanate and the chain extender. In linear or slightly crosslinked polymers these two blocks are phase separated, giving rise to the outstanding properties of PU elastomers. On the other hand, crosslinked PUs are single phase, amorphous and non-elastomeric materials.

Different types of PU elastomers include cast elastomers both non-cellular and microcellular, thermoplastic elastomers, RIM elastomers and millable gums. PU elastomers can also be formulated into coatings, sealants and adhesives [2]. The main categories of PU elastomers are presented below.

7.3.1 Cast urethane elastomers

Casting as a manufacturing technique represents one of the major methods of preparing PU elastomers and foams. Casting involves a precise metering of chosen components, their thorough mixing and then pouring the reaction mixture into a heated mold wherein the crosslinked tridimensional polymer structure is simultaneously formed and molded in the desired size and shape. The final requirements of the product properties as well as the economical aspects dictate the selection of basic raw materials (isocyanate, polyol and extender), other ingredients as well as the manufacturing technology.

Linear or branched polyether and polyester polyols are used in thermoset PU elastomers with MWs in the range 2000–6000 [26]. Pure, crude or polymeric MDI as well as TDI mixtures are the isocyanates used according to the application. As for the chain extenders, they consist of both amine and glycol types (ethylene glycol, diethylene glycol, 1,4-butanediol).

The casting procedure generally follows two basic techniques known as the prepolymer route and the one-shot technique [26]. The reaction injection molding (RIM) procedure has evolved from the one-shot casting technique and will be discussed separately.

The prepolymer technique is by far the most commonly used for elastomers. The predried polyol is reacted with a molar excess of isocyanate (20–40%) resulting in a prepolymer that consists of two to three polyol molecules joined together by diisocyanate molecules and urethane linkages and terminated with isocyanate end groups. This is termed 'full' prepolymer and it results in extremely high viscosities. A variation of this method consists of the preparation of a 'quasi' prepolymer using only a part of the polyol so that higher ratio isocyanate/polyol is obtained. After the reaction, the mixture is composed of isocyanate capped polyol plus free, unattached isocyanate and the viscosity is much lower. The remaining polyol is mixed up with the chain extender and the other ingredients so that the overall stoichiometry is not changed. Metering and mixing operations are easier in the quasi prepolymer technique due to better balanced volumes to be mixed and to lower viscosities of the quasi prepolymer compared to the full prepolymer [26].

Chemical reactions in the preparation of a full prepolymer are presented below:

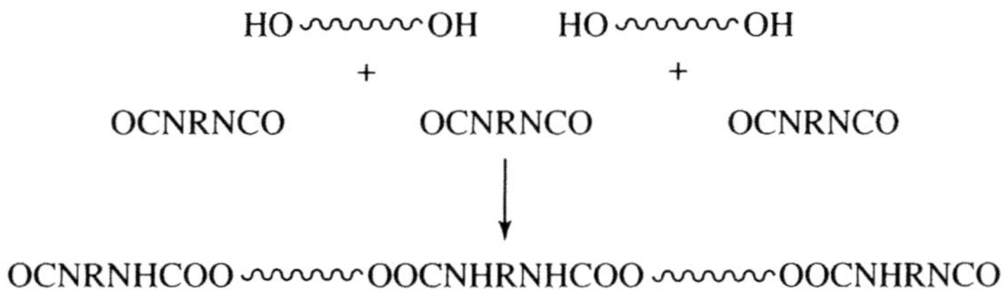

Storage stable prepolymers with isocyanate end groups are now available commercially [27].

In the second step of the process the liquid or molten prepolymer, which is in fact a 'macrodiisocyanate', is mixed and reacted with the previously heated and degassed chain extender. The time of mixing, known as 'pot life', depends on both the extender and the temperature and it generally takes some minutes. A very efficient mechanical mixing head in a mixing chamber is used to thoroughly mix the components. After each shot, the mixing chamber is cleaned with a solvent or an air blast. The mixture is poured into a heated open mold before the viscosity of the melt starts to grow significantly (gel time). During this stage the macrodiisocyanate reacts with the chain extender with the formation of new urethane or urea linkages that extend the polymer chains. The amount of chain extender is smaller than the equivalent of free isocyanate groups on the prepolymer so that free isocyanate groups still remain on the extended polymer chains. A schematic representation of glycol chain extension is illustrated below:

OCN∼∼∼∼NCO OCN∼∼∼∼NCO OCN∼∼∼∼NCO
\+ \+ \+
HOR′OH HOR′OH HOR′OH

↓

HOR′OOCNH∼∼∼∼NHCOOR′OOCNH∼∼∼∼NHCOOR′OOCNH∼∼∼∼NCO

Diamine chain extension is represented schematically below.

OCN∼∼∼∼NCO OCN∼∼∼∼NCO OCN∼∼∼∼NCO
\+ \+
$H_2NR'NH_2$ $H_2NR'NH_2$ $H_2NR'NH_2$

↓

$H_2NR'NHCONH$∼∼∼∼NHCONHR′NHCONH∼∼∼∼NHCONHR′NHCONH∼∼∼∼NCO

In the curing stage that takes place in the mold, these free isocyanate groups react with the urethane or urea groups present on the polymer chains resulting in a crosslinked tridimensional structure. After demolding, additional postcuring at elevated temperatures is necessary in order for the products to acquire their final physical properties.

The flow charts of the preparation of casting polyurethane elastomers by the full prepolymer and quasi prepolymer methods are presented in Figures 7.3 and 7.4, respectively.

In the one-shot system all the ingredients are mixed together simultaneously and the rate of different reactions is carefully controlled by the use of catalysts. A flow chart of the preparation of cast elastomers by the one-shot method is illustrated in Figure 7.5.

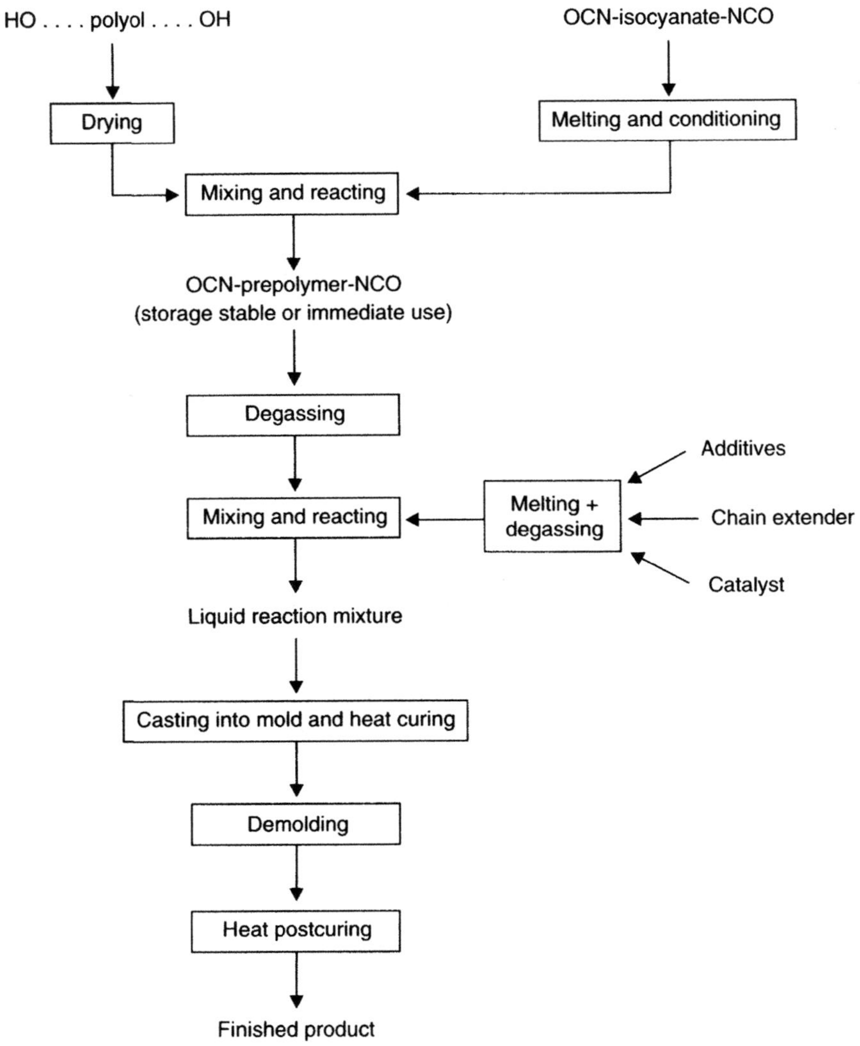

Figure 7.3 Preparation of PU cast elastomers by the full prepolymer method.

The overall stoichiometry of reactants in any PU system is expressed by the isocyanate index which is the molar ratio (NCO/all OH) × 100 and must be controlled accurately. When all the components are bifunctional and the index is 100, the resulted polymer is linear and thermoplastic. If the index is less than 100, the end groups are –OH. An index higher than 100 means that after the reaction with the chain extender, the polymer chains are –NCO terminated and crosslinking occurs on curing by allophanate or

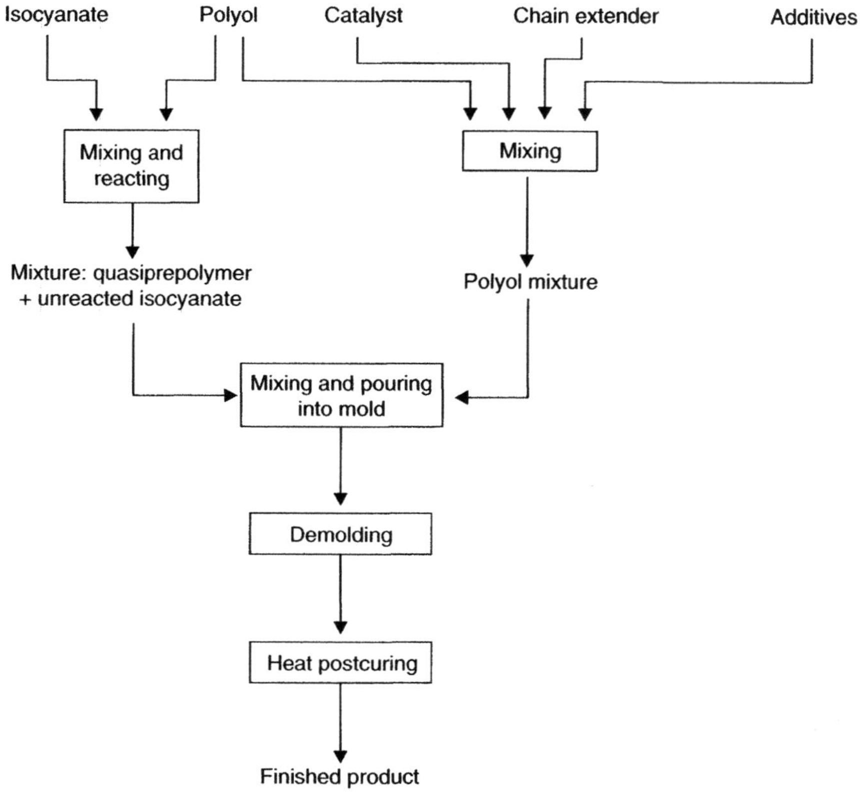

Figure 7.4 Preparation of PU cast elastomers by the quasi prepolymer method.

biuret structures. The degree of crosslinking is controlled by the isocyanate index [26].

Another way to introduce crosslinking is the utilization of polyols, isocyanates or extenders with functionalities higher than two (branched polyester or polyether polyols, polymeric or modified MDI compounds or trihydroxy extenders).

An example of typical formulations and of reaction conditions for casting polyurethane elastomers is given in Table 7.3.

A broad range of properties can be obtained by selecting appropriate formulations and reaction conditions. Among the most outstanding properties of polyurethane elastomers are high abrasion resistance, tear strength and tensile strength as well as excellent resistance to aliphatic hydrocarbon fuels, oils, oxygen and ozone. They are also good electrical and thermal insulators. They may be formulated from hard materials (50 Shore D) to very soft elastomers (10 Shore A). Hardness is controlled by the amount of hard segment that results from the reaction of the isocyanate and chain extender (Table 7.4) [35].

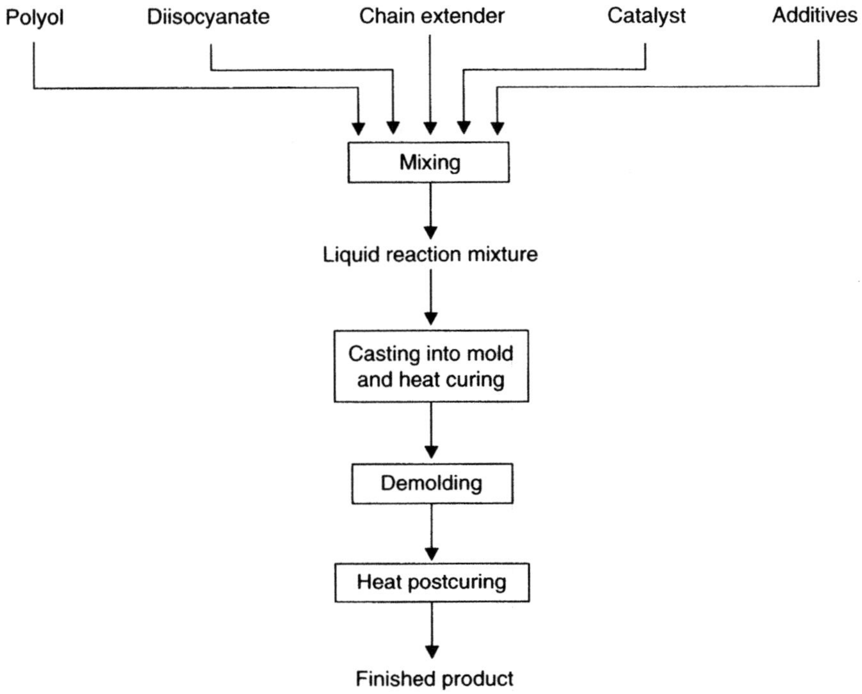

Figure 7.5 Preparation of PU cast elastomers by the one-shot method.

Some properties of a PU cast elastomer based on a 6000 MW polyether triol chain extended with 1,4-butanediol are given in Table 7.5.

Major shortcomings of PUs are their low resistance to steam, aromatic and chlorinated solvents and strong acids and bases. The maximum service temperature is 120°C.

Table 7.3 A typical formulation and reaction conditions for casting polyurethane elastomers

	Parts by weight	
Prepolymer (PTMG/TDI), 4.2% NCO	100	100
MOCA	12.5	–
1,4-Butanediol	–	4.0
Trimethylolpropane	–	0.3
	Conditions	
Prepolymer temperature (°C)	100	100
Extender temperature (°C)	115	100
Pot life (min)	12	45
Demolding time at 100°C (min)	30–45	90
Postcuring at 100°C (h)	16	20

Table 7.4 Effect of the amount of chain extender on some PU properties [35]

Hard segment (%)	*Hardness (Shore A/D)*	*Flexural modulus (MPa)*
25	82 A	4.3
50	60 D	484
75	78 D	1382

Common applications for cast PU elastomers are printing and industrial rollers, pipes and impellers for mining and quarrying industries, flexible tooling molds, recreational surfaces (in multisports halls, tennis courts and children's playgrounds), solid industrial tires for fork and lift trucks, skateboard and roller skate wheels, soft cast elastomer (tire filling material for heavy vehicles), sealing gaskets and oil field equipment, diaphragms and a variety of other mechanical applications [36].

Microcellular elastomers obtained by the incorporation of a blowing additive are prepared by different technologies and will be discussed with the RIM elastomers.

7.3.2 Thermoplastic polyurethane elastomers (TPUs)

TPUs represent about 9.8% of the total of thermoplastic elastomers [37]. TPUs have become the second most important class of urethane elastomers. They are linear block copolymers that can be processed on conventional plastic processing equipment by any standard method used for thermoplastics such as injection and compression molding, extrusion, calendering and film blow molding. They are supplied as dry granules [38].

TPUs are synthesized from the general types of PU components, namely a diisocyanate, a polyol and a chain extender. The preferred isocyanates

Table 7.5 Some properties of PU cast elastomer based on a 6000 MW polyether triol chain

Property	*Value*
Density	1.10 g/cm^3
Tensile strength	24 MPa
Elongation	275%
Modulus 100%	13.7 MPa
Hardness (Shore A)	95

are MDI for general applications and $H_{12}MDI$ for color-stable coatings. The most used polyols are poly(tetrabutylene adipate) glycols, poly(caprolactone) glycols and poly(oxytetramethylene) glycols. 1,4-Butanediol is the choice chain extender. Reaction stoichiometry is calculated based on the complete reaction of isocyanate groups with all hydroxyl groups, from the polyol and the chain extender (isocyanate index 100), so that linear chains are formed.

Full prepolymer, quasi prepolymer or one-shot techniques are all applied in the manufacture of TPUs. A typical formulation (by weight) based on 1/3/2 equivalent ratio of polyol/diisocyanate/extender are given below:

- Polycaprolactone (MW 2000), 1000;
- MDI, 375;
- 1,4-Butanediol, 90.

Using the two-step full prepolymer or quasi prepolymer routes, all or some of the predried polyol is reacted with the isocyanate at 110–130°C for about two hours and the resultant prepolymer is then mixed with the predried diol (chain extender or chain extender plus the remaining polyol) at 100°C. The mixture is poured continuously onto a heated endless belt and is removed after 20 minutes. After cooling, the product can be chopped or granulated and stored in air-tight containers. In the one-shot method, all the components are mixed together simultaneously and polymerization is allowed to proceed. The polymerization may alternatively be conducted in an extruder with intensive mixing.

The polyaddition reactions in the preparation of TPUs are illustrated schematically below:

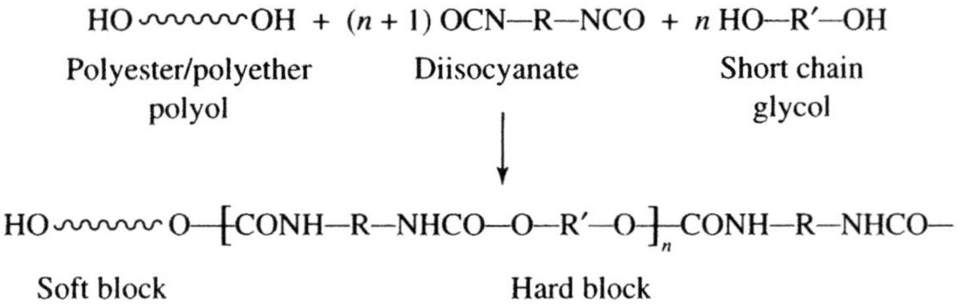

TPUs are $(AB)_n$ block copolymers. They are composed of 'soft segment' blocks and 'hard segment' blocks. The soft blocks are formed from the high MW polyols (polyester or polyether diols) and isocyanate and they are connected through urethane linkages to the hard blocks that are formed from the chain extender and isocyanate. TPUs have the properties of a vulcanized elastomer although they are thermoplastic. Many studies have been done to elucidate the morphology of thermoplastic elastomers and to explain their properties. During the polymerization, phase separation occurs due to the basic thermodynamic incompatibility of the soft and hard segments [39]. Recent studies on the morphology of various TPUs show that the hard segments tend to cluster or aggregate with each other into crystalline, hard and rigid domains through hydrogen bonding which is due to the polar nature of urethane or urea linkages [40,41]. These hard domains are dispersed within the soft rubbery matrix composed of soft segments and act as filler particles reinforcing the matrix. They function as 'virtual crosslinkers' of TPU preventing permanent deformations of the elastomeric soft segments.

DSC measurements also evidenced the two-phase structure of TPUs by the presence of two separate T_g values corresponding to the soft and hard domains and a melting endotherm for the crystalline hard segment [42].

As in thermoset PUs, the hard segments control the hardness, modulus and tear strength while the soft segments provide flexibility, low temperature properties and solvent, oil and hydrolysis resistances [43–45].

Chain extenders play a major role in the final elastomer properties as they determine the nature of the hard segments. The content of the hard blocks is determined by the free unattached isocyanate content in the prepolymer.

Some properties of TPUs based on polycaprolactone polyester with different hardnesses (hard block content) are given in Table 7.6 [41].

TPUs have quite similar properties to those of cast PU rubbers but the absence of true covalent crosslinks results in higher values of compression set and in a more pronounced loss of strength with increasing temperature.

TPUs are used for the preparation of polyurethane engineering thermoplastics that are polyblends of TPUs and a variety of other polymers such as nylon, PVC, ABS, SAN, PC, PET, PBT, acetals and those reinforced with glass or mineral fillers [46–48]. For medical purposes, siloxane modified TPUs are being investigated [5].

The variety of applications is extensive. Among the most important are wire and cable jacketing, sheet and film, household gadgets, gears, various backing pads in abrasive applications, components for mining and drilling machinery and for agricultural equipment, medical devices, rolls and diaphragms [27,38,48]. TPUs are used not only as elastomers, but also for the preparation of adhesives, coatings and films.

Table 7.6 Some properties of TPUs based on polycaprolactone polyester

Hardness	80 Shore A	55 Shore D
Tensile strength (MPa)	52	43
Elongation (%)	490	350
Modulus 100% (MPa)	6	20

7.3.3 Polyurethane fibers

PU fibers are known under the generic name of Spandex fibers, which are composed of at least 85% segmented block TPU [38]. Long chain polyols are polyethers such as PTMG or linear polyesters that are reacted with TDI and/or MDI to form the prepolymer. For chain extension aliphatic amines (ethylene diamine), aromatic amines (*m*-phenylendiamine) or hydrazine are usually used. Spandex fibers are prepared by either solution or reaction spinning [38].

In solution spinning, the NCO terminated prepolymer is dissolved in a highly polar solvent such as *N,N*-dimethylformamide or *N,N*-dimethylacetamide, and the chain is subsequently extended by the addition of aliphatic diamines or hydrazine. Additives and stabilizers are also added to the solution, which is then pumped through a spinnerette plate into a hot atmosphere where the solvent diffuses out to form the filaments.

In reaction spinning, the additives and stabilizers are added to liquid prepolymer, which is then extruded into a coagulating bath that contains the amine chain extender which extends the polymeric chains and forms the fiber.

Crosslinks are produced through biuret and allophanate structures at elevated temperatures during the curing of fibers.

The basic characteristics of Spandex fibers are softness, high modulus, elongation and instant recovery when the stress is released, excellent resistance to oxidation, body oils and dry cleaning solvents and good dyeability. Spandex fibers have surpassed the properties of any elastomeric fiber known and they have replaced natural rubber fibers in many applications.

7.3.4 Millable polyurethane elastomers

Vulcanizable PU gums are used in small volume. They are processed on conventional rubber machinery. Before vulcanization they are essentially linear rubber-like compounds. They are prepared as storage-stable prepolymers with hydroxyl terminal groups (isocyanate index lower than 100) from linear polyols such as poly(ethyleneglycol adipate) or PTMG and MDI or TDI. In the next step these prepolymers are compounded on two roll mills with the other ingredients. Crosslinking is achieved by one of the three ways of curing used with PU millable elastomers, namely by sulfur, peroxides or blocked isocyanates [26].

Better properties are obtained by conventional sulfur vulcanization in the presence of accelerators and activators. Special prepolymers with pendant carbon–carbon double bonds are prepared using unsaturated chain extenders such as the monoallyl ether of glycerol, $H_2C{=}CH{-}CH_2{-}O{-}R(OH)_2$. A schematic representation of sulfur vulcanization is presented below:

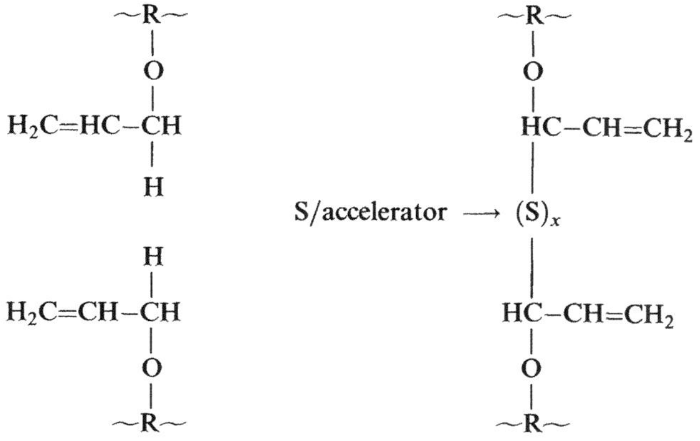

Millable prepolymer chains   Sulfur crosslinked urethane

Curing with peroxides (e.g. dicumyl peroxide) by a free radical mechanism is used with prepolymers based on MDI. Free radicals which are produced on the polymer chains at the site of the active methylene group in MDI, join together in a crosslinked network.

The third way of curing is done by incorporation on the rolls of a blocked or 'latent diisocyanate' which is a dimer of toluene-2,4-diisocyanate. During the molding at about 150°C the dimer dissociates into free isocyanate:

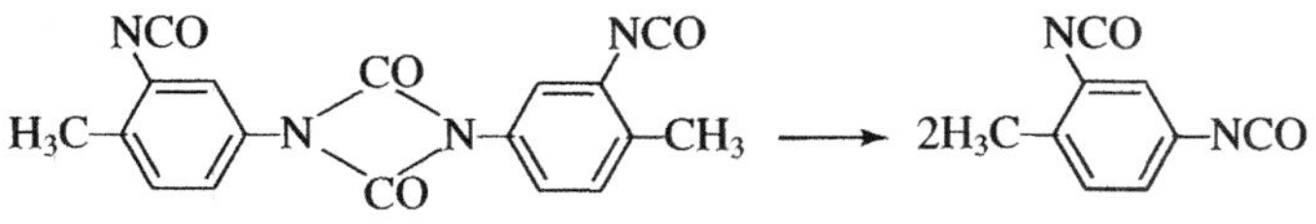

which reacts with the urethane or urea groups on the prepolymer backbone and crosslinking occurs via allophanate or biuret linkages.

Millable PUs are compounded with reinforcing fillers such as carbon blacks, silicas and clays, as well as with other additives, just as other rubber materials are. Their properties are not generally as good as those of cast elastomers and they are of limited importance [2].

7.4 POLYURETHANE FOAMS

Polyurethane foams are by far the most versatile of all the polymeric foams. They may be manufactured in forms from the very soft to the very rigid. Properties can be tailored by the selection of suitable raw materials and formulations. The requirements for particular applications dictate the density of the foams [49].

Flexible and rigid foams are manufactured with different equipment, using different techniques and entirely different raw materials and formulations. In flexible foams, the chains are long and the crosslinks are farther apart, permitting more movement. In rigid foams, the crosslinks are close together creating a tight network and restricting movement. As a result, flexible and rigid polyurethane foams are very different in properties and uses and they will be discussed separately.

7.4.1 Flexible foams

These foams can be made with a large range of densities, flexibilities, elongations and compressive strengths. They are slightly crosslinked or even thermoplastic materials. The flexible foams are open celled [49,50].

Two categories of flexible foams are manufactured, hot molded and cold molded, high resilience (HR) foams. Hot molded foams are based on TDI and less reactive polyether polyols containing secondary hydroxyls. Their molding temperatures are from 80–120°C and cycle times from 20–30 minutes. Cold molded, high resilience (HR) foams are based on more reactive MDI or PMDI and primary hydroxyl-containing polyethers. Enhanced reactivity of HR foam components leads to molding temperatures of only 30–50°C and 4–20 minutes cycle times. HR foams have a higher resiliency and comfort for cushioning applications. Special polyethers, which contain grafted organic polymers such as acrylonitrile/styrene copolymers or polyurea dispersions, are used in HR foams in order to enhance the load bearing capacity.

Saturated polyesters such as poly(ethylene adipate) glycols or poly(propylene succinate) glycols are used in the higher grade foams especially for cloth lining where higher temperature resistance and dry cleaning solvents resistance are needed.

Water is by far the most successful blowing agent for flexible foams as it plays a double role. It reacts with isocyanate groups with the evolution of CO_2 gas that blows the foam and introduces urea groups on the polymer chains. The latter act as chain extenders and subsequently lead to biuret crosslinking and to tridimensional elastomeric foams. In some formulations additional physical blowing agents such as fluorocarbons and methylene dichloride can be introduced for lower density foams. The heat of the water–isocyanate and polyol–isocyanate reactions help to expand the CO_2 and to evaporate the added blowing agent. The density of the foam can be evaluated by the blowing index, I_B, calculated with an empirical relationship based on the overall amount of blowing agent per 100 parts of polyol [49]:

$$I_B = [\text{water}] + [CCl_3F]/10 + [CH_2Cl_2]/9$$

Foam formulations include at least one catalyst; however, with polyether polyols, which contain in the majority less reactive secondary hydroxyl groups, a dual catalyst system is required. This is composed of a tertiary amine such as triethylene diamine, bis-(dimethylaminoethyl) ether or other similar amines and a metallic salt, usually dibutyltin dilaurate or stannous octoate [51]. The two catalysts are synergistic. Their ratio is very critical as each one preferentially promotes a different type of reaction. Tin compounds are particularly effective on the formation of urethane links, i.e. on polymer chain growth and gelation. Tertiary amines favor isocyanate–water reaction and CO_2 evolution and blowing. If the tin concentration is too high, gelation is too fast and the foam results tight, with closed cells and low air flow. At higher amine concentrations, the foam blows too fast and may collapse from lack of gel strength. Delayed action or heat activated catalysts, introduced more recently, are used for the manufacturing of large objects where mold filling is essential [31]. Latent catalyst systems have been created by the adsorption of conventional amines on molecular sieves (zeolite) [5].

Other important components in foam formulation are surface active agents, used at 1% level, which decrease the surface tension, facilitate the dispersion of water, and stabilize and control the structure of cells. The most used agents are water soluble polyether siloxane block or graft copolymers.

An example of formulation of a flexible polyurethane foam is given in Table 7.7.

Flexible polyurethane foams are manufactured almost exclusively by the one-shot process in which the ingredients are all mixed simultaneously or are previously preblended. The mixing head is the heart of the equipment. They are of two types: the recirculating type used in high pressure impingement RIM mixing machines and the non-recirculating, low pressure mechanical mixing machines. Both types are provided with sophisticated control system to monitor pressures, temperatures and flow rates and automatic flow control. Most slabstock foams are produced with non-recirculating heads with mixer speeds in the range 2000–6000 rpm. Their disadvantage is the need to flush the head with solvents after each shot. The reactive liquid mixture is poured continuously onto a conveyer system yielding foam slabstocks up to 2.2 m wide and 1.25 m high. The conveyer

Table 7.7 A formulation of a flexible polyurethane foam

	Parts by weight
Polyether triol (PPO MW 3000)	100
80/20 TDI, 110 index	41
Water	3
Triethylenediamine	0.3
Stannous octoate	0.15
Silicone block copolymer	1.0

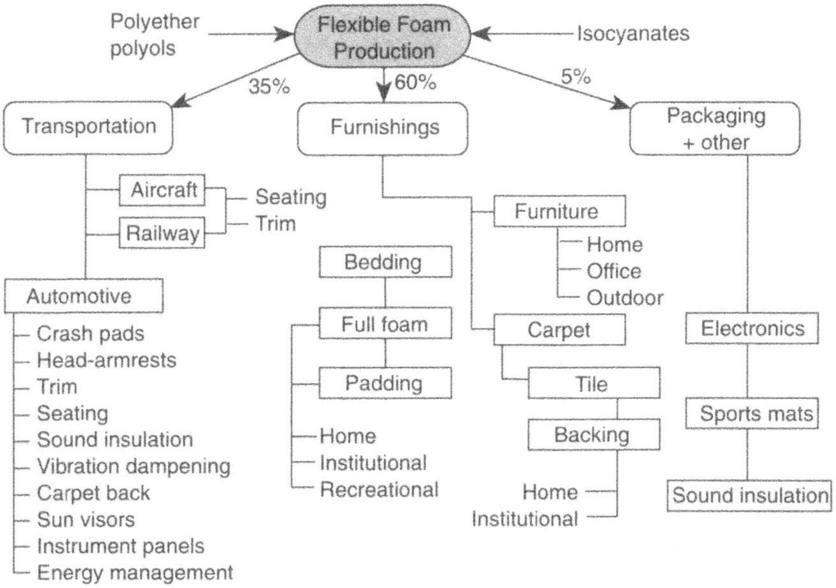

Figure 7.6 Application areas for PU flexible slabstock foam [52]. (Reprinted by permission of Carl Hanser Verlag from *Reaction Polymers* by W.F. Gum, W. Riese and H. Ulrich (eds), 1992, Carl Hanser Verlag Munich-Cincinnati.)

and the mixing head are located in a ventilated tunnel. After 5–10 minutes, the foam can be cut and handled. The slabstocks are then heat cured for 24 hours and then cut into different sized products and shapes.

When the product shape is more complex or contains inserts, the polymerizing liquid is poured into molds; the mixture foams and fills the mold. Another application is in foam carpet cushioning and is realized by pouring the mixture directly onto the carpet that has been previously coated with an intermediate layer.

A recent review of foam morphology studies has shown that the polymer that composes the foam is phase separated PU/urea that contains a soft segment phase, a hard segment or microdomain and a precipitated, aromatic polyurea discontinuous phase [52]. The polymer is a lightly crosslinked $(AB)_n$ block copolymer with excellent elastomeric properties similar to that of elastomers. Dynamic mechanical spectroscopy (DMS) and differential scanning calorimetry (DSC) clearly demonstrated the presence of a transition T_g at −60 to −40°C from the soft polyether block and a high T_g at about 250°C attributed to the polyurea hard phase. In addition to this physical crosslinking due to phase separation, some chemical crosslinking is present from branched polyols and from biuret and allophanate links.

The geometry of the cells is a very important factor in the foam's ultimate properties. The first step in cell formation is the nucleation, i.e. the initiation of the bubbles, which is done by the gases already present in reactants. The next step is the growth of the bubbles by thermal expansion and by evolution of CO_2 from the water–isocyanate reaction. Spherical bubbles grow and are transformed into pentagonal dodecahedral cells [49]. The process ends by the rupture of cell walls and the opening of the cells. In a good foam, a majority of cells have open pentagonal windows with tridimensional triangular cross-section struts in between.

The final mechanical properties of a flexible foam depend both on the macroscopic cell geometry and on the morphology of the polymer contained within the structural elements of the foam. In cushioning applications, a low hysteresis and a high resiliency are of major importance in compressive loading. Proper formulation will achieve low hysteresis values based on the advanced phase separation. Increasing the density of the foam increases the durability of the foam measured by tensile, tear, elongation and fatigue tests.

Flexible urethane foams are manufactured with densities in the range 0.016–0.048 g/cm^3; semiflexible foams range from 0.096–0.192 g/cm^3 and high resilience foams (HR) are in the range 0.028–0.048 g/cm^3.

Based on the broad range of load bearing capacities and resiliencies, the largest volume of foam is used for cushioning applications for furniture, automobiles, aircraft, bedding and carpet underlay. Polyether based foams are preferred as they have a higher resilience than polyester based foams. The applications of flexible PU foams are summarized in Figure 7.6. On the other hand, flexible foams based on polyesters perform better for textile applications as laminates for cold weather clothing; they are particularly resistant to dry cleaning solvents.

Other major uses include packaging of military hardware, electronic parts and delicate products, acoustic and thermal insulations and safety equipment. An important use for these open cell foams is as filters for oils, chemicals and various air pollution control devices.

7.4.2 Rigid polyurethane and polyisocyanurate foams

Rigid PU and polyisocyanurate foams are highly crosslinked materials in contrast with flexible foams, which are only lightly crosslinked. They are mainly used for insulation purposes on account of their high insulating efficiency and their high strength to weight ratio that provides significant structural integrity. The raw materials used in flexible and rigid foams are quite different.

The rigidity is provided by the high functionality of both the isocyanate and the polyol components. The majority of isocyanates used are crude and undistilled, such as PMDI or its mixtures with modified TDI, and mixtures of MDI with its oligomers [53,54]. The functionality of the isocyanate component is between 2.6 and 3.2. The isocyanate index in different formulations is of the order of 250 or even as high as 450–500 in PU/isocyanurate compositions.

Three basic types of polyols are generally used: polyester, polyether and halogen- or phosphorus-containing polyols. Polyester polyols have a better flammability performance due to an increased formation of char that serves as a heat barrier for inner foam. They are derived from aliphatic or aromatic dicarboxylic acids and their derivatives such as phthalic anhydride, dimethyl terephthalate residues and poly(ethylene terephthalate) scrap by transesterification with short chain diols [55]. For some applications polyethers are preferred to polyesters; they are branched and polyfunctional compounds with MW as low as 500. The short distance between the hydroxyl groups of the polyol accounts for an increased density of crosslinks in the final polyurethane foam. The functionality of polyols may be even as high as eight when they are synthesized in the presence of polyfunctional initiators such as pentaerythritol or sorbitol. Polyethers have better compatibility with chlorofluorocarbon blowing agents than polyesters.

Polysiloxane–polyether block copolymers are used as surfactants providing nucleating sites for bubble formation and regulating the size, shape and distribution of cells. Common catalysts for PU rigid foams are tertiary amines and only infrequently metallic salts. When PU/isocyanurate are prepared, specific isocyanate trimerization catalysts are used such as carboxylic metal salts (e.g. potassium acetate or octoate) [56].

The choice of the blowing agent for PU rigid foams is of major importance for the insulating efficiency. Rigid foams are closed cell and, at a density of 0.032 g/cm^3, 97% of their volume is occupied by gas. The overall thermal conductivity of a foam or the K-factor, expressed in watts per meter and Kelvin (W/m · K), depends on three contributions:

$$K = K_g + K_s + K_r$$

where K_g = the gas contribution, 45–50%; K_s = the solid polymer contribution, 25%; K_r = the radiative heat transfer contribution, 30–35%.

Solid polymer has relatively high thermal conductivities of 0.180–0.250 W/m · K but it contributes only a small fraction (about 25%) due to its small volume fraction in the foam matrix. K_s is only slightly affected by changes in polymer composition [57]. Radiative heat transfer K_r increases strongly with decreasing foam density owing to thinner cell walls. The contribution from gas thermal conductivity is by far the largest.

Thermal conductivities in W/m · K of some gases are

- air, 0.024;
- carbon dioxide, 0.014;
- fluorocarbons, 0.008–0.010.

For thermal insulating properties, trichlorofluoromethane (CFC-11 b.p. = 23.8°C) is the best blowing agent due to its low thermal conductivity.

It remains incorporated in the closed cells of the insulating foam. Concern for depletion of stratospheric ozone layer by CFC-11 imposed its partial to total replacement during the present decade [58]. To date, only two halocarbons, partly hydrogenated, have been found to be satisfactory blowing agents, namely 1,1-dichloro-2,2,2-trifluoroethane (HCFC-123) and 1,1-dichloro-1-fluoroethane (HCFC-141b) [59]. They are environmentally safer as they can be destroyed in the troposphere. More recently a technique has been developed for the cleavage of partly hydrogenated halocarbons at 2000°C in an oxyhydrogen flame to afford HF, HCl and CO_2 without formation of polychlorinated dioxins and furans [5].

Water is used in addition to halocarbons, and CO_2 gas produced in the water–isocyanate reaction acts as a blowing agent. The resulting polyurea structures contribute to increased mechanical strength of the foam.

The initial values of thermal conductivities of PU/isocyanurate rigid foams are ranging from $K = 0.015\text{–}0.020 \mathrm{W/m \cdot K}$, but these values deteriorate over time [53,59]. The cell gas consists mainly of CFC-11 and CO_2. The cell membranes are permeable to each of the gases present, although to a different degree. CO_2 diffuses completely out of the cells in a few days. The larger and heavier CFC-11 molecules diffuse much more slowly and remain in the foam for years. As time progresses, the concentration of CFC-11 in the cell diminishes as air permeates into the foam and its concentration increases until an equilibrium is reached at a 1/1 ratio of CFC-11/air in the cell. The net effect of the dilution of CFC-11 with air is a gradual increase in the thermal conductivity of the foam. Halocarbon blown foams have ultimate thermal conductivities significantly lower than those of CO_2 blown foams, as can be seen from the values in $\mathrm{W/m \cdot K}$ below:

- CFC-11 blown foam (final), 0.024;
- CO_2 blown foam (final), 0.035.

Strength characteristics of a foam depend mainly on the polymeric phase whereas its thermal conductivity depends on the gas phase [60]. When structural strength is required, the foam is blown by CO_2 from the isocyanate–water reaction that introduces PU/urea units and biuret additional crosslinks. In thermal insulation foams, where thermal conductivity is the most important characteristic, halocarbon blowing agents are additionally used.

A common formulation for thermal insulation PU rigid foam is shown in Table 7.8.

Another important characteristic required for thermal insulation foams is combustibility. The behavior of polyurethane foams in fire is not good. Improved flame resistance is achieved by the incorporation of either reactive or additive-type fire retardants. Chlorinated or phosphorus-containing polyols may be used as reactive hydroxy compounds. Phosphorus promotes

Table 7.8 A common formulation for thermal insulation PU rigid foam

	Parts by weight
Polyether polyol	100
Glycerol	10
PMDI	stoichiometric + 5%
CFC-11	50
Triethylenediamine	0.5
Silicone block copolymer	1.0

char formation which serves as a protective barrier. On the other hand, halogens are free radical scavengers that suppress flame propagation.

An even more significant improvement was introduced by the isocyanurate rigid foams which are based on isocyanate cyclotrimerization presented in Section 7.2.1 [61]. Cyclotrimerization results in highly crosslinked foams that are too brittle. Trimerization was coupled with urethane formation resulting in polyurethane–isocyanurate rigid foams and composite materials. Isocyanate trimerization was found to be a slower process than either urethane or urea formation and thus it takes place at the end of the polyaddition process with the remaining isocyanate groups in the presence of specific catalysts.

Some properties of polyurethane–polyisocyanurate foams are given in Table 7.9.

Polyurethane–polyisocyanurate foams have good resistance to burning and flame spread and very good hydrolytic stability [61,62].

Besides trimerization, isocyanates can react with each other in the presence of suitable catalysts to form carbodiimide structures (Section 7.2.1). The simultaneous evolution of carbon dioxide gives rise to foamed products. As carbodiimide groups further react with free isocyanate groups, the resulting foams are crosslinked rigid materials with densities of 0.016–0.020 g/cm^3 and high flame resistance, and good thermal and acoustical insulation characteristics. Among their applications are underfloor sound insulation, pipe insulation and thermal insulation between cavity walls [62].

Table 7.9 Some properties of polyurethane–polyisocyanurate foams

Property	*Value*
Density	0.024–0.096 g/cm^3
Initial *K*-factor	0.019 W/m · K
Equilibrium *K*-factor	0.024 W/m · K
Service temperature	−185 to 150°C (compared to −40 to 95°C for rigid PU foam)
Compression strength	0.14–0.28 MPa
Shear strength	0.10–0.24 MPa

Rigid polyurethane foams can be produced by a large number of methods either in plant or by portable equipment and pour-in-place methods. One-shot, prepolymer and quasi prepolymer techniques can all be used. Besides continuous slabbing (pouring into large blocks onto a moving conveyor system), other processing alternatives include pouring in place, spraying, frothing and laminating. Foam density is largely a function of the concentration of the blowing agents.

The rigid urethane foams have proved to be far superior to any other polymeric foam in thermal insulation properties over a wide range of temperatures. They are extensively used in home and building insulation for walls, roofs and structural components [60], for refrigerated trucks, trailers and railway cars, for refrigerators and freezers (household, commercial and industrial), for insulation of storage tanks and pipes, and for insulation in automobiles, aircraft and ships. Another application is for flotation in barge compartments, small boats and in the oil industry.

Less expanded, high density structural and decorative foams, the so called 'simulated wood' (different wood grains may be captured perfectly on the foam surface) are used for doors, windows and other accessories, and in furniture manufacture [60,63].

The recent abundant patent literature demonstrates the effort to find various ways for the reutilization of PU waste materials [5]. Glycolysis is the general chemical method for the processing of large amounts of waste. On the other hand, the physical method consists of the production of reutilizable PU powder. For example, the foam originating from disused refrigerators is already recuperated as powders.

7.4.3 Polyurethanes by the reaction injection molding process

A development of one-shot injection molding gave rise to the independent process known as 'reaction injection molding' (RIM) which has become a high productivity manufacturing process. The RIM technique is applied to produce parts made from PU that range from high density, high modulus elastomers to lower density structural foam parts [64–67].

The RIM process consists of a precise and individual metering of all the components, which must be liquid, and their thorough and extremely rapid mixing; the highly reactive liquid mixture is poured into molds where the reactions take place, and the polymeric material is formed and shaped into the final products.

The considerations made before on raw materials and additives, as well as on isocyanate reactions, all apply to RIM components. RIM polyols are di- and trifunctional polypropylene oxide capped with ethylene oxide, MW = 4000–6000, as well as amine terminated polyethers, MW = 2000–5000. The resulted soft segments of urethane or urea-type respectively, impart good low temperature impact, elongation, tear strength and abrasion resistance.

Glycol and amine chain extenders are both used. The most widely used chain extender is 2,4/2,6-diethyltoluene diamine, 80/20 isomer mixture (DETDA) that forms urea hard segments contributing characteristics of stiffness and thermal stability [68]. Full polyurea RIM formulations are so reactive that the reactions occur almost instantaneously (0.8–1.2 s) in the RIM mix head and the use of any catalyst can be avoided. For a better control, the reaction can be slowed down by blending low levels of amine terminated polyethers with polyether polyols. When the use of catalysts is necessary, their selection is even more critical in order to balance the speed of each individual reaction in the highly reactive system. Typical catalysts are organometallics such as dibutyltin dilaurate and tertiary amines, especially DABCO.

A comparison of the relative reactivities of amine and hydroxyl end groups toward the isocynate group is presented in Table 7.10. The isocyanate component is a liquid carbodiimide modified MDI or other liquid glycol modified MDI derivative.

Individual components are continuously recirculated from their storage tanks through the delivery lines, mix head and return lines to ensure their temperature consistency within a ±1°C limit.

The mixing head is the critical point of the system. It is a 0.3–4 cm^3 chamber where the components are continuously introduced and mixed. Mixing is realized by high pressure 'impingement', i.e. a counter-current introduction of component streams at 10–20 MPa and their highly efficient turbulent mixing. The simple configuration of the impingement mixing head avoids the need of cleaning the head and delivery lines between shots (self-cleaning system). The resultant polymerizing liquid mixture, which is highly reactive, is then injected into molds that are part of the RIM processing system. Reaction rates and injection times are adjusted to achieve a gelation time at the point of complete fill of the mold, which is in the range of 1–2 s.

The reaction between –OH and –NCO groups to form urethane linkages is highly exothermic (approximately 104 kJ/mol) and it takes place in the molds. The molds are provided with good thermal conductivity

Table 7.10 Relative reactivities of hydroxyl and amine groups toward isocyanates [2]

Reactive group	*Relative reactivity*
Secondary hydroxyl	1.0
Primary hydroxyl	5–10
Primary aromatic amine	10–100
Secondary aliphatic amine	2000–3000
Primary aliphatic amine	10 000

Table 7.11 Typical conditions for reaction injection molding [69]. (Reproduced by permission of Taylor & Francis, Inc. from *Principles of Polymer Systems* by F. Rodriguez, 2nd edn, 1982, pp. 354. All rights reserved)

Formulation	*Parts by weight*
Resin stream:	
Polyols	150
Dibutyltin dilaurate (catalyst)	0.045
Amine (cocatalyst)	0.6
Isocyanate stream:	
Polyfunctional isocyanate	96
Blowing agent (Freon type)	3
Methylene chloride	1
Processing parameter	*Condition*
Resin temperature	54°C
Isocyanate temperature	21°C
Mold temperature	57 ± 3°C
Resin pressure	120 atm
Isocyanate pressure	100 atm
Throughput	1.1 kg/s
Demolding time	1–2 min
Postcure	1 hour at 120°C

(metallic molds) and are designed with temperature control. Table 7.11 illustrates a typical formulation and the operation conditions for a RIM process [69].

Further productivity improvements have resulted in the introduction of internal mold release (IMR) based on zinc stearate solubilized in the polymerizing mixture by the presence of the amine-type compounds (amine polyether and amine chain extender) [68]. When the amine compatibilizers are rapidly consumed on contact with the isocyanate, zinc stearate precipitates as a uniform dispersion in the reaction mass. The IMR minimizes the time spent for mold preparation.

Milled glass up to 40–50%, treated with aminosilane adhesion promoter, fibrous calcium silicate mineral or treated mica are used as fillers in reinforced RIM (RRIM) to provide a high flexural modulus and low thermal expansion coefficient, especially in thin parts. Some properties of unfilled and glass filled PU/urea RIM polymer are presented in Table 7.12.

When RIM is used to produce microcellular or foamed parts, the reaction mixture is injected into the mold without filling it completely. Filling the mold is completed by the expansion of the material, which can vary from 5–10% expansion for microcellular elastomers to 60–75% expansion for structural foam systems. Surfactants such as

Table 7.12 Some properties of PU/urea RIM for automotive applications [68]

Property	*Unfilled*	*15.8% Glass filled*
Density (g/cm^3)	1.02	1.07
Hard segment content (%)	41.1	30.9
Flexural modulus (MPa)		
−29°C	849	618/392*
22°C	405	360/194*
70°C	304	260/137*
Tensile strength (MPa)	26.7	15.2
Elongation (%)	195	125/180*
Coefficient of thermal expansion $\times 10^{-6}$/°C (from 66°C to 121°C)	170	51/145*

* parallel/perpendicular

polyoxyalkylene polysiloxane copolymers are used to facilitate the formation of a fine cell structure.

Low boiling halogenated alkanes such as CFC-11 or the environmentally safer CFC-123 are used as blowing agents. They are vaporized by the reaction heat in the interior of the mold but are condensed in the surface layer by contact with the cooled mold. An expanded foam core is formed in the interior of the part that compresses the exterior layer of material to the mold walls, resulting in a continuous skin-like surface. These self-skinning foams produced by RIM technology are also called integral skin or structural foams. The densities of RIM foamed products may range from 0.2 g/cm^3 to 1.2 g/cm^3 for microcellular elastomers.

The same variations as practiced in the manufacture of cast elastomers and foams are encountered in RIM processes, namely:

- true one-shot system: all components are individually and simultaneously introduced into the mixer;
- two-component system: isocyanate constitutes component A, and all other ingredients are premixed and constitute component B;
- quasi prepolymer system: the isocyanate is reacted with a part of polyol to give the quasi prepolymer (component A); the remaining polyol is premixed with all other ingredients and constitutes component B.

PU, polyurea, PU/urea and PU/isocyanurate elastomers and foams produced by RIM and RRIM processes have found numerous applications in the automotive industry for the production of exterior large size components such as front and rear bumpers, instrument housings, head rests, steering wheels, pneumatic tires for tractors and other parts. The shoe industry utilizes RIM microcellular elastomers or flexible and semiflexible foams with or without an integral skin to produce shoe soles. Rigid foams are used in the furniture and building industries.

7.5 POLYURETHANE COATINGS, ADHESIVES AND SEALANTS

Polyurethanes are largely used for coatings, adhesives and sealants owing to their outstanding properties such as abrasion resistance, toughness, flexibility, tear strength, adhesive properties and solvent resistance.

The chemistry of PU coatings, adhesives and sealants is similar to that encountered with other urethane materials. TDI, MDI or PMDI are the isocyanates used for non-light exposed applications. Aliphatic and cycloaliphatic isocyanates such as isophorone, hydrogenated MDI, biuret or isocyanurate derivatives of HDI and a number of newer isocyanates are chosen in those applications where weatherability and appearance are important. Polyethers, polyesters, acrylic polyols, glycols and amines are used in prepolymers and for chain extension. For some applications, the formulation is based on thermoplastic PU. Besides the PU resin, catalysts, fillers, pigments, plasticizers and other additives and sometimes solvents compose the formulation.

To meet pollution and safety regulations, an increasing importance is given to PU water dispersions for coatings and adhesives, thus avoiding the utilization of organic solvents [26].

Available systems are classified as one-component and two-component [70,71]. The one-component system actually consists of several components mixed and stored together and can be cured by oxygen (air), moisture or heat. They are based on PU with high MW but which are still soluble. The two-component systems are stored as two separate components that contain ingredients terminated in active groups and they are mixed just before the application (short pot life) and then cured. Two-component systems are used in more demanding applications for physical characteristics.

7.5.1 Coatings

The US consumption of PU coatings in 1990 was 111 000 tons [2]. They can be formulated into a large number of types of coatings; the early attempts to classify them have been outgrown [70].

Several one-component systems are available on the market. Uralkyds are urethane modified alkyd resins cured by atmospheric oxygen and used as wood finishes and enamels. Moisture cured coatings based on NCO terminated prepolymers are limited to thin films and in places with relative humidities of 40–75%. Major markets are for concrete and wood seamless floors [60]. Coatings based on blocked isocyanates can be cured at room temperature (phenol blocked and amine coreactants) or at 190°C (alcohol blocked and hydroxylic coreactants). These coatings are also available as water dispersions.

The most important types of PU coatings are based on two-component systems: an isocyanate prepolymer and a polyol. They need meter-mix equipment. Clear topcoats have recently entered the automotive market. PU coatings are largely used for aircraft, heavy machinery, buses, trucks and maintenance coatings.

Coatings based on high MW thermoplastic PU dissolved in a solvent are used on textiles and leather (equally as water dispersions) and for magnetic media coatings used for audio, video and computer tapes and computer floppy disks [2]. They are cured by the evaporation of the solvent. Other PU coatings are formulated for radiation cure (flooring), vapor cure (foundry molds) or as urethane powder coats (appliances and metal furniture).

The water dispersions of PU elastomers are of growing importance. They are non-toxic, non-flammable and do not pollute air or waste water [26]. They are based on PU ionomers which have hydrophilic ionic groups incorporated into the polymer. These ionic groups function as an ionic emulsifier making unnecessary the use of external emulsifiers. Due to the absence of low MW emulsifiers, these systems show extremely good film forming properties even in the case of especially hard PUs such as air drying lacquers.

PU ionomers are prepared to form anionic or cationic dispersions. Anionic ionomers have sodium sulfonate (a) or sodium carboxylate (b) ionic groups attached to side branches. They are prepared by the chain extension of NCO terminated prepolymers with special diamines that contain the respective ionic groups:

(a) $$\underset{\text{Prepolymer}}{\text{OCN-R-NCO}} + \text{H}_2\text{N-R'-}\underset{\substack{| \\ \text{R''SO}_3^{\ominus}\text{Na}^{\oplus}}}{\text{NH}} \longrightarrow [\text{-R-NH-CO-NH-R'-}\underset{\substack{| \\ \text{R''SO}_3^{\ominus}\text{Na}^{\oplus}}}{\text{N}}\text{-CO-NH-}]_n$$

(b) $$\underset{\text{Prepolymer}}{\text{OCN-R-NCO}} + \text{H}_2\text{N-R'-}\underset{\substack{| \\ \text{R''COO}^{\ominus}\text{Na}^{\oplus}}}{\text{NH}} \longrightarrow [\text{-R-NH-CO-NH-R'-}\underset{\substack{| \\ \text{R''COO}^{\ominus}\text{Na}^{\oplus}}}{\text{N}}\text{-CO-NH-}]_n$$

These ionomers can be readily melt dispersed into water (without the use of any solvent) and subsequently crosslinked by the polycondensation of the urea groups from the polymer chain with formaldehyde.

Cationic PU ionomers are prepared from NCO terminated prepolymers chain extended with alkyl substituted ethanolamines such as

N-methyldiethanolamine, $CH_3{-}N(CH_2CH_2OH)_2$:

$OCN{-}R{-}NCO + CH_3N(CH_2CH_2OH)_2 \longrightarrow$

Prepolymer

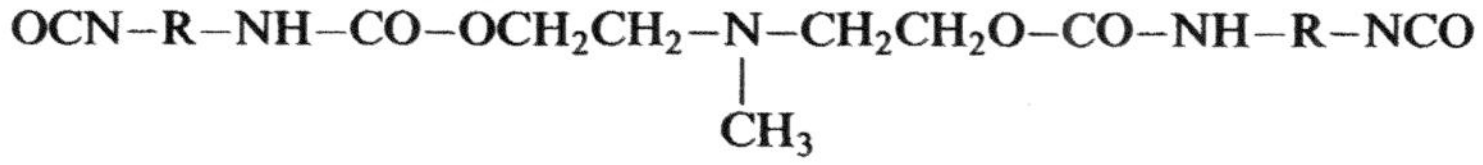

Significantly more light stable PUs are obtained from aminodiols which contain the tertiary nitrogen atom in a side branch such as 2-alkyl-2-dialkylaminomethyl-1,3-propanediol:

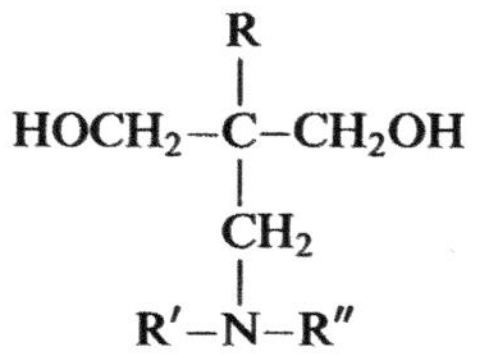

Emulsification occurs when these extended prepolymers are added to 3% aqueous acetic acid under high speed mixing. The subsequent curing of the latex takes place by the reaction of terminal free isocyanate groups either with water or with water soluble diamines [11].

Another way to prepare cationic dispersions from the above mentioned extended prepolymers is by quaternization of tertiary nitrogen atoms. They can react readily with formaldehyde in the presence of water to form crosslinked products.

Ionomers with a high percentage of ionic groups are excellent dispersants, superior to the usual surface active agents for the emulsification of non-ionic PU. The hydrophilic ionomer component forms the surface layer of the dispersed particles.

7.5.2 Adhesives

PU adhesives are complex mixtures containing PU resin, fillers, pigments and plasticizers. The USA's consumption of PU adhesives in 1990 was 88 000 tons [2]. PUs manifest very good adhesion to materials as varied as rubber, other plastics, synthetic fibers, wood, glass, cements and metals. Adhesive properties may be attributed to the polar nature of the compounds used in their formulation. Furthermore, the isocyanates present in PU compositions may react with any active hydrogen present in the adherent or with the thin films of water often present on the surfaces of materials such as ceramics, glass and metals. These reactions permit the attainment of high bond strength [71]. PU adhesives are very useful when elastomeric bonds are required on flexible substrates such as rubber or plastic.

The best of PU adhesives are superior in performance to virtually any other adhesive type at cryogenic temperatures. They maintain their shear strength and toughness at temperatures far below those that cause serious embrittlement of adhesives designed for use at ambient temperatures and above [60]. Polyurethanes have the drawback of poor chemical and thermal stability and loss of bond strength at temperatures above 80–100°C.

The oldest types were based on triphenylmethane-4,4′,4″-triisocyanate or NCO terminated prepolymers as solvent-borne or 100% solids, cured by air moisture. They are used today in rebonded foam, tire cords and furniture. Conventional hot melt adhesives are based on high MW TPU in solution and are used in the shoe industry. After the evaporation of the solvent, heat and pressure are applied to melt the polymer and press the parts for adhesion. Reactive hot melt adhesives also contain some free isocyanate groups that will moisture-cure and contribute to some crosslinking [72].

Two-component adhesives result in more flexible and softer elastomers. One of the components is an NCO terminated prepolymer and the other is a polyol or polyamine. They are extensively used in the automotive industry, to laminate panels for construction, foam to textiles, or plastic films together for food packaging. Hybrid adhesives have been prepared from PU–polyepoxy or PU–acrylics.

Aqueous PU latex adhesives are growing in importance in order to avoid the use of large amounts of volatile organic solvents in open systems. PU latices can be prepared in several ways. In one instance, a thermoplastic PU made in solution is emulsified in water, then the solvent is removed. Or an isocyanate terminated prepolymer is blocked with an appropriate isocyanate blocking agent and subsequently emulsified in water. Chain extension and crosslinking are accomplished by water or diamines present in water, and they take place within each particle of the dispersion. Eventually particles are considered to form a macromolecule. Adhesive latices can also be prepared from PU ionomers in a similar way to that presented in Section 7.5.1.

7.5.3 Sealants

Sealants are materials introduced into joints in order to prevent the transmission of water vapor and air, and to absorb the expansion and contraction of the joint. They contain 30–50% PU resins, the remainder being fillers, pigments, plasticizers and other additives. They can be formulated as one-component moisture cured systems or as two-component fast curing systems. PU sealants exhibit low modulus, low hardness (10–50 Shore A), excellent adhesion, 100% elasticity (high joint movement capability), high abrasion resistance, low temperature flexibility and low moisture transmission.

The USA's consumption of PU sealants was 30 000 tons in 1990 with a growing demand in modern high rise construction for insulated glass windows, for sealing windshields and clay sewer pipes, for highway and airport sealants and different marine applications [71].

REFERENCES

1. Ulrich, H. (1987) *Proceedings of the Polyurethane World Congress*, Aachen, FRG, Society of the Plastics Ind., New York, p. 254.
2. Gum, W.F., Riese, W. and Ulrich, H. (eds) (1992) *Reaction Polymers. Chemistry, Technology, Applications, Markets*, Hanser Publishers, Munich.
3. Richter, R. and Ulrich, H. (1977) *The Chemistry of Cyanates and their Thio Derivatives, Part 2, Syntheses and Preparative Applications of Isocyanates*, J. Wiley, London, pp. 619–818.
4. Twitchett, H.J. (1974) *Chem. Soc. Reviews*, **3**(2), 209.
5. Uhlig, K. (1993) *Kunststoffe* (Engl.), **10**, 51.
6. Thomas, M.R. (1982) *Preprints of papers by the Division of Organic Coatings and Plastic Chemistry*, 183rd ACS National Meeting, Las Vegas, p. 506.
7. Entelis, S.G. and Nesterov, O.V. (1966) *Uspehi Khimii*, **35**, 2178.
8. Yen, Y.C. and Tsao, T.S. (1982) *Polyols for Polyurethanes*, Process Economics Report 45A, Stanford Research Institute, Menlo Park, CA.
9. Gagnon, S.D. (1986) in *Encyclopedia of Polymer Science and Engineering*, Vol. 6, Wiley Interscience, New York, p. 273.
10. Clinton, N. and Matlock, P. (1986) in *Encyclopedia of Polymer Science and Engineering*, Vol. 6, Wiley-Interscience, New York, p. 225.
11. Oertel, G. (1985) *Polyurethane Handbook*, Hanser Publications, Munich.
12. Friedly, H.R. (1992) in *Reaction Polymers* (eds W.F. Gum, W. Riese and H. Ulrich), Hanser Publishers, Munich, p. 66.
13. Hawkins, W.L. (1972) *Polymer Stabilization*, Wiley-Interscience, New York.
14. Inoue, S. and Aida, T. (1984) in *Ring Opening Polymerization* (eds K.J. Ivin and T. Saegusa), Vol. 1, Elsevier Applied Science, London, p. 4185.
15. Dreyfuss, P. and Dreyfuss, M.P. (1982) in *Kirk–Othmer Encyclopedia of Chemical Technology*, 3rd edn, Vol. 18, J. Wiley & Sons, New York, p. 645.
16. Dreyfuss, P. (1982) *Poly(tetrahydrofuran)*, Gordon and Breach, New York.
17. Mark, J.E., Erman, B. and Eirich, F.R. (1994) *Science and Technology of Rubber*, 2nd edn, Academic Press, New York, p. 58.
18. Plowman, K.R. (1992) in *Reaction Polymers* (eds W.F. Gum, W. Riese and H. Ulrich), Hanser Publishers, Munich, p. 86.
19. Habermann, C.E. and Hartwell, G.E. (1992) in *Reaction Polymers* (eds W.F. Gum, W. Riese and H. Ulrich), Hanser Publishers, Munich, p. 96.
20. Kirland, C. (1986) *Plast. Technol.*, March, p. 83.
21. Hassel, A. (1987) *Plast. Technol.*, December, p. 81.
22. Friedly, H.R. (1992) in *Reaction Polymers* (eds W.F. Gum, W. Riese and H. Ulrich), Hanser Publishers, Munich, p. 80.
23. Kennedy, J.P. (1989) in *Handbook of Polymer Science and Technology* (ed. N.P. Cheremisinoff), Vol. 2, Marcel Dekker, New York, p. 599.
24. Zachariasiewicz, M. (1987) *Adhesives and Sealants Council Fall Conference*, p. 115.
25. Chang, V.S.C. and Kennedy, J.P. (1982) *Polym. Bull.*, **8**, 69.
26. Hepburn, C. (1992) *Polyurethane Elastomers*, 2nd edn, Elsevier Applied Science, London.
27. Lee, F.T.H. (1989) in *Handbook of Polymer Science and Technology* (ed. N. Cheremisinoff), Vol. 2, Marcel Dekker, New York, p. 291.

28. Stuber, F.A. (1992) in *Reaction Polymers* (eds W.F. Gum, W. Riese and H. Ulrich), Hanser Publishers, Munich, p. 114.
29. Cornell, M.C. (1992) in *Reaction Polymers* (eds W.F. Gum, W. Riese and H. Ulrich), Hanser Publishers, Munich, p. 486.
30. Flynn, K.G. and Nenortas, D.R. (1963) *J. Org. Chem.*, **28**, 3527.
31. Ricci, R.L., Heijden, B., Andrew, G.D. and Swaintek, D.J. (1985) *J. Elastomers and Plastics*, **17**, 249.
32. Hoestettler, F. and Cox, E.F. (1960) *Ind. Eng. Chem.*, **52**, 609.
33. Britain, J.W. and Gemeinhardt, P.G. (1960) *J. Appl. Polym. Sci.*, **4**, 208.
34. Ulrich, H. (1992) in *Reaction Polymers* (eds W.F. Gum, W. Riese and H. Ulrich), Hanser Publishers, Munich, p. 247.
35. Alberino, L.M. (1992) in *Reaction Polymers* (eds W.F. Gum, W. Riese and H. Ulrich), Hanser Publishers, Munich, p. 269.
36. Engh, M. (1992) in *Reaction Polymers* (eds W.F. Gum, W. Riese and H. Ulrich), Hanser Publishers, Munich, p. 642.
37. Rader, C.P. (1993) *Kunststoffe*, **83**(10), 36.
38. Orchon, S. (1990) in *Handbook of Plastic Materials and Technology* (ed. I. Rubin), J. Wiley & Sons, New York, p. 501 and 511.
39. Van Bogart, J.W.C., Lilaonitkul, A. and Cooper, S.L. (1979) in *Multiphase Polymers* (eds S.L. Cooper and G.M. Estes), American Chemical Society, Washington.
40. Seefried, C.G., Koleske, J.V. and Critchfield, F.E. (1975) *J. Appl. Polym. Sci.*, **19**, 2503 and 3185.
41. Farrissey, W.J. (1992) in *Reaction Polymers* (eds W.F. Gum, W. Riese and H. Ulrich), Hanser Publishers, Munich, p. 260.
42. Goyert, W. and Hespe, H. (1978) *Kunststoffe*, **68**, 819.
43. Sung, C.S.P., Hu, C.B. and Wu, C.S. (1980) *Macromolecules*, **13**, 111.
44. Kwei, T.K. (1982) *J. Appl. Polym. Sci.*, **27**, 2891.
45. Guise, G.B. and Smith, G.C. (1980) *J. Appl. Polym. Sci.*, **25**, 149.
46. Bonk, H.W., Drzal, R.S., Georgacopoulos, G.N. and Shah, T.M. (1985) *Proceedings of the SPE*, New York, 43rd Annual Technical Conference, Plastics, p. 1300.
47. Georgacopoulos, C.N. (1987) *Rubber World*, **196**(6), 15.
48. Bonk, H.W. (1992) in *Reaction Polymers* (eds W.F. Gum, W. Riese and H. Ulrich), Hanser Publishers, Munich, p. 279.
49. Huber, L.M. (1992) in *Reaction Polymers* (eds W.F. Gum, W. Riese and H. Ulrich), Hanser Publishers, Munich, p. 297.
50. Woods, G. (1982) *Flexible Urethane Foams*, Applied Science Publishers, London.
51. Malwitz, N. and Kresta, J.E. (1987) *Proceedings of the Polyurethane World Congress*, SPI, New York, p. 826.
52. Herrington, R.M. and Turner, R.B. (1992) in *Reaction Polymers* (eds W.F. Gum, W. Riese and H. Ulrich), Hanser Publishers, Munich, p. 306.
53. Thoen, J.A. (1992) in *Reaction Polymers* (eds W.F. Gum, W. Riese and H. Ulrich), Hanser Publishers, Munich, p. 288.
54. Kennedy, E., Deguiseppi, D., Knis, S. and Tichy, J.J.(1992) in *Reaction Polymers* (eds W.F. Gum, W. Riese and H. Ulrich), Hanser Publishers, Munich, p. 537.
55. Canaday, J.S. and Skowronski, M.J. (1985) *J. Cell. Plast.*, **21**, 338.
56. Bodnar, T.W. (1992) in *Reaction Polymers* (eds W.F. Gum, W. Riese and H. Ulrich), Hanser Publishers, Munich, p. 359.

57. Glicksman, L.R. and Torpey, M.R. (1987) *Proceedings of the Polyurethane World Congress*, SPI, New York, p. 80.
58. Brasseur, G. and Hitchman, M.H. (1988) *Science*, **240**, 634.
59. Bhattacharjee, D. (1992) in *Reaction Polymers* (eds W.F. Gum, W. Riese and H. Ulrich), Hanser Publishers, Munich, p. 367.
60. Feldman, D. (1989) *Polymeric Building Materials*, Elsevier Applied Science, London.
61. Briggs, P.J. (1985) *Cell. Polym.*, **4**, 261.
62. Blaga, A. (1974) *Canadian Building Digest*, NRC Canada, p. 168.
63. Brydson, J.A. (1995) *Plastics Materials*, 6th edn, Butterworths, London.
64. Sweeney, F.M. (1990) in *Handbook of Plastic Materials and Technology* (ed. I. Rubin), J. Wiley & Sons, New York, p. 1273.
65. Lee, L.J. (1980) *Rubber Chem. Technol.*, **53**, 542.
66. Frisch, K.C. (1980) *Rubber Chem. Technol.*, **53**, 126.
67. Cornell, M.C. (1987) *Proceedings of the Polyurethane World Congress*, SPI, New York, p. 198.
68. Cornell, M.C. (1992) in *Reaction Polymers* (eds W.F. Gum, W. Riese and H. Ulrich), Hanser Publishers, Munich, p. 486.
69. Rodriguez, F. (1982) *Principles of Polymer Systems*, 2nd edn, Hemisphere Publishing Corp., New York.
70. Hickner, R.A. (1992) in *Reaction Polymers* (eds W.F. Gum, W. Riese and H. Ulrich), Hanser Publishers, Munich, p. 664.
71. Cranley, P.E. (1992) in *Reaction Polymers* (eds W.F. Gum, W. Riese and H. Ulrich), Hanser Publishers, Munich, p. 692.
72. Markevka, V., Mackeis, S.J. and Bunnelte, W.L. (1988) US Patent 4,775,719, H.B. Fuller.

8 Epoxy polymers

8.1 INTRODUCTION

These polymers may be defined as those materials in which chain extension crosslinking occur through reactions of epoxy groups. Epoxy polymers were first synthesized by P. Schlack, the investor of nylon 6, in 1934. The curing reactions were discovered by P. Castan and polymer products were introduced on the market by Ciba towards the end of World War II [1].

The growth of the world production of epoxy polymers in shown in Figure 8.1. More than 220 000 tons of epoxies are now used annually in the USA; worldwide capacity is 400 000 tons [2]. The steady increase in consumption is due to the rapid development of several fields of application. At present, this includes mainly adhesives, tooling, composite applications for conventional epoxy systems and for special resins used in the electronic industry.

The most important epoxy polymers available on the market (monoglycidyl compounds excepted) comprise the following [3]:

1. aromatic di- and polyglycidyl compounds of bis-phenol A (including advancement products), bis-phenol F (isomers mixture), phenol novolacs, cresol novolacs, tris(*p*-hydroxyphenyl)methane, tetrakis(*p*-hydroxyphenyl) ethane (glyoxal derivative);
2. aromatic bis-methylglycidyl ether of bis-phenol A;
3. aliphatic diglycidyl ethers of 1,4-butanediol, 1,6-hexanediol, neopentyl glycol, polypropylene glycol 425;
4. *N*-glycidyl compounds of aniline, 4,4′-diaminodiphenylmethane, isocyanuric acid, 1,3-bis(3-hydantoinyl)-2-hydroxypropane, *p*-aminophenol (*N*,*N*,*O*-triglycidyl compound);
5. diglycidyl esters of tetrahydrophthalic acid, hexahydrophthalic acid;
6. cycloaliphatic epoxies: 3,4-epoxycyclohexylmethyl, 3,4-epoxycyclohexanecarboxylate, 3-(3,4-epoxycyclohexyl)-7,8-epoxy-2,4-dioxaspiro[5,5]-undecane, bis(3,4-epoxycyclohexylmethyl)adipate.

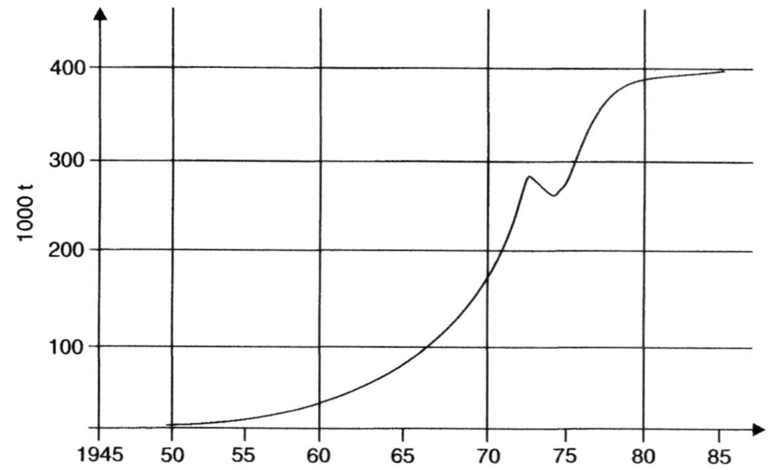

Figure 8.1 World production of epoxy resins [3].

Although a number of different types are now commercially available, the major share of the market is still held by systems based upon the reaction products of epichlorohydrin and 2,2-bis(4′-hydroxyphenyl) propane (bisphenol A):

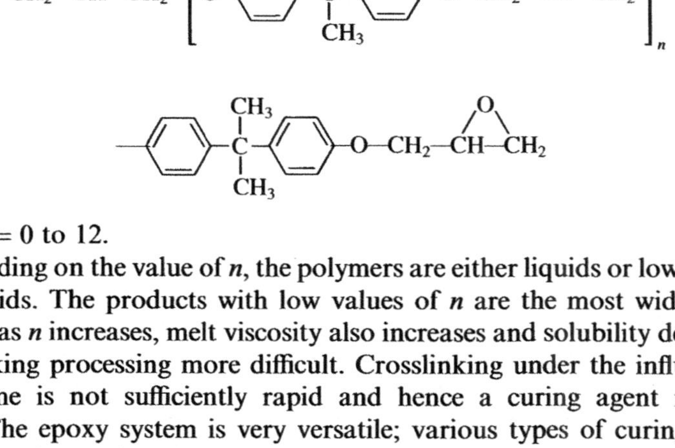

where $n = 0$ to 12.

Depending on the value of n, the polymers are either liquids or low melting point solids. The products with low values of n are the most widely used because, as n increases, melt viscosity also increases and solubility decreases, thus making processing more difficult. Crosslinking under the influence of heat alone is not sufficiently rapid and hence a curing agent must be added. The epoxy system is very versatile; various types of curing agents may be used, creating the possibility of tailoring, to some extent, the final network structure to meet specific requirements. Homopolymerization of the prepolymer is possible in the presence of a boron trifluoride complex.

Most other curing agents actually copolymerize with the epoxy (or hydroxyl) groups. The compounds used to the greatest extent are acid anhydrides and polyfunctional amines [4].

The term curing is used to describe the process by which one or more kinds of reactants (i.e. an epoxy prepolymer and a crosslinking agent) are transformed from low MW compounds to a highly crosslinked network. The network is composed of segments involving only the epoxide or both the epoxide and the curing agent.

The type of curing agent has an important effect on the structure and properties of the final epoxy polymer. Anhydrides provide desirable chemical stability (except to alkalis) and have good electrical insulating properties; aliphatic amines can be effective at ambient temperature and result in fast curing. A high degree of crosslinking leads to very compact structures and provides greater thermal stability. Most of the hardeners in common use belong to the following groups: tertiary amines, polyfunctional amines, acid anhydrides and polysulfides. The curing of epoxy polymers with such crosslinking agents, its features and kinetics have been reported in many papers [5–21].

T.F. Mika [22] has categorized the hardeners for epoxy polymers into two general classes:

- active hydrogen compounds which cure by polyaddition reaction;
- anionic or cationic catalysts.

The following points should be considered when selecting a crosslinking agent for a particular application: curing conditions (i.e. temperature), compatibility prepolymer, curing agent, volatility of the hardener, properties of cured epoxy and cost.

Depending on the choice of crosslinking agent, epoxies can be made to cure slowly (24 h) or quickly (1 min) at room temperature and/or high temperature, above T_g or even at 150°C.

8.2 EPOXY POLYMER SYNTHESIS

Epoxy polymer is commercially produced through the Taffy process or advancement process.

The bis-phenol that plays the largest role in epoxy polymer synthesis is 2,2-bis(4-hydroxyphenyl)propane generally referred to as dian in Europe, whereas the common designation in the USA is bis-phenol A (BPA). Its preparation involves treating acetone with an excess (7.5 mol) of phenol and anhydrous hydrogen chloride, recovering a crystalline molecular compound of phenol and BPA, and then recycling the filtrate containing the by-products. The mechanism is believed to involve electrophilic attack of a protonated acetone molecule on the aromatic ring followed by acid catalyzed elimination of water to give a cation:

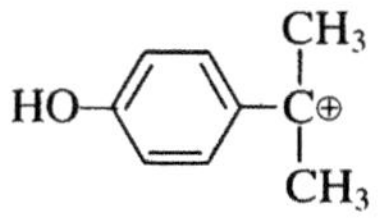

This process is probably involved in the formation of not only BPA, but also of some of the by-products of the reaction. Although the reaction of acetone with phenol is the most important preparative method of BPA and the only one known to have been operated on a commercial scale, several others have been reported in the technical literature.

Epichlorohydrin is produced by chlorination of propylene to allyl chloride which is converted to glycerol dichlorohydrin by reaction with hypochlorous acid followed by dehydrochlorination to epichlorohydrin.

$$CH_2{=}CHCH_3 \longrightarrow CH_2{=}CHCH_2Cl \longrightarrow ClCH_2CH(OH)CH_2Cl \longrightarrow$$

$$\overset{\;\;O}{\overbrace{CH_2-CH}}-CH_2Cl$$

Epichlorohydrin is useful in the formation of glycidyl compounds by reaction with a wide variety of substrates containing reactive hydrogen atoms.

Most commercial epoxy prepolymers are in fact low MW oligomers and form relatively tough products when crosslinked with an appropriate curing agent (hardener).

Commercial epoxy polymers, about 95% of which are diglycidyl ethers of BPA (DGEBA) are mixed oligomers characterized by their epoxide equivalent and have a number average MW of up to about 500 and are viscous liquids. Above this MW, the polymers are low melting point solids [23]:

8.2.1 Taffy process

In this procedure, BPA is reacted at 90–95°C with a controlled excess (for 1 mol BPA, 1.1–2.8 mol epicholorohydrin) of epichlorohydrin to give oligomers with glycidil ether groups

$$-O-CH_2-\underset{\;\;O}{\underbrace{CH-CH_2}}$$

on both ends in the presence of a concentrated (20–50%) solution of NaOH and an inert solvent (i.e. toluene).

The flow chart of the Taffy process is presented in Figure 8.2. The basic operations are copolyaddition of BPA with epichlorohydrin in the presence of alkali, washing, prepolymer (oligomer) dehydration, filtration, solvent distillation and packaging.

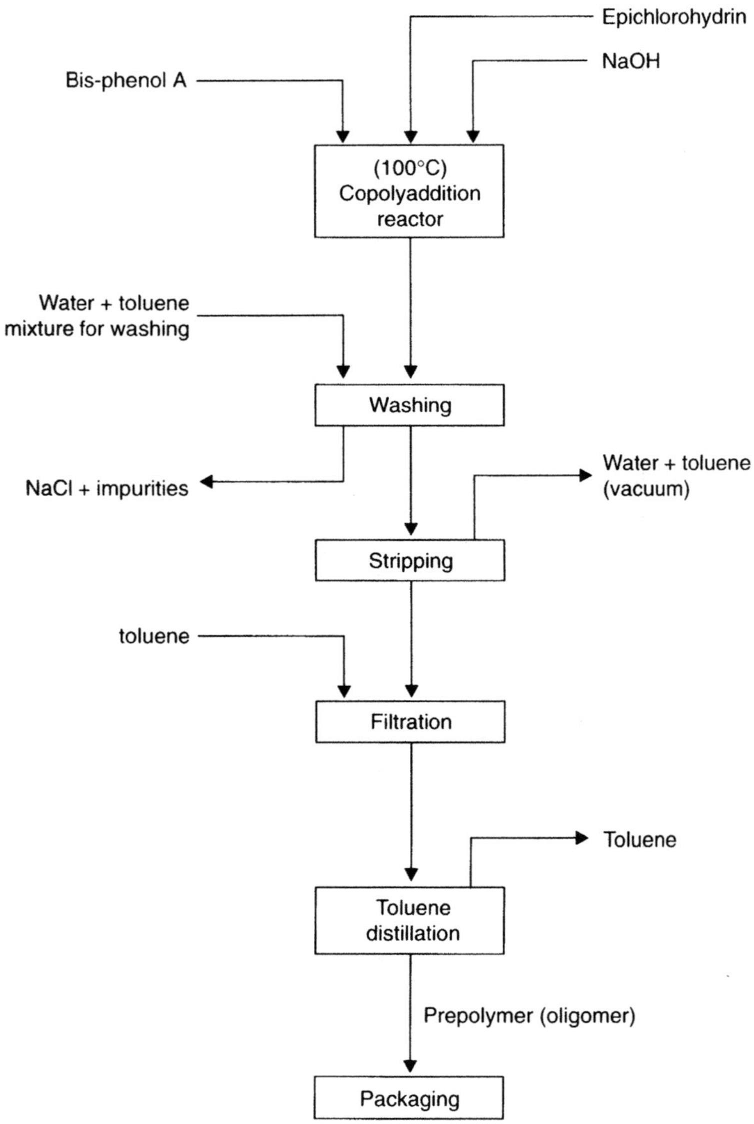

Figure 8.2 Taffy process for epoxy prepolymer synthesis.

The copolyaddition takes place in a heterogeneous medium. NaOH reacts with the HCl formed and gives NaCl; the system must have a neutral pH. The white putty-like product formed, called 'Taffy', is an emulsion of molten polymer in brine. Depending on the initial concentrations of the reactants the MW of the oligomer may reach 3000. A water–toluene mixture is used for removing NaCl and toluene soluble impurities. The washed product is stripped of solvent and water under vacuum at about 150°C. An even higher MW oligomer can be made by treating this polymer in the second reactor with additional BPA in the presence of suitable catalysts. After drying, the oligomer is sent to packaging. It is recommended to wash the product twice. During the second wash, CO_2 is introduced to neutralize the alkalis [24,25].

Using NMR spectroscopy, H. Batzer and S.A. Zaahir [26] have found that about 0.09–0.6 branches are formed for every oligomer of MW 1500–4000.

8.2.2 Advancement process

In this process, BPA is reacted with a prepolymer, usually the DGEBA, along with some higher oligomers in the presence of suitable catalysts, at a higher temperature (170–180°C) and pressure. The prepolymer and BPA mixture is stirred under these conditions for 4–12 hours. By adding alkaline activators such as pyridine, dimethylaniline or sodium acetate, the reaction rate increases significantly. The oligomers necessary for such a process are obtained in two different stages. In the first stage, the DGEBA synthesis is performed in the presence of high alkaline activators such as secondary and tertiary amines, sodium acetate and carbonate. Epoxy polymers can also be obtained by starting with fractions of biatomic phenols.

8.2.3 Curing

In the second, or curing, stage epoxy polymers are considered structoterminal or structopendant depending on whether crosslinking occurs through the epoxy end groups or the hydroxyl groups. The prepolymer is structoterminal when polyamines are used for crosslinking; in this case, curing involves the base catalyzed opening of the epoxide groups.

Hardening occurs primarily through the hydroxyl groups when an anhydride is used as the crosslinking agent; in this situation, epoxy prepolymer is considered as structopendant. Phthalic anhydride is often used as the curing agent, although other anhydrides, including maleic and pyromellitic may be used in specialty applications [2,27].

Most of the curing agents in common use can be classified in the following groups:

- tertiary amines;
- polyfunctional amines;
- anhydrides;
- polymerization catalysts.

Among these hardeners, amines are the most versatile both at room temperature and at elevated curing temperature. The curing mechanisms with amines and the structures of the amine cured epoxy polymers have been thoroughly studied and the systems of epoxy resins with amine hardeners are used extensively in many industrial applications.

The curing mechanism of a resin with a primary amine is as follows:

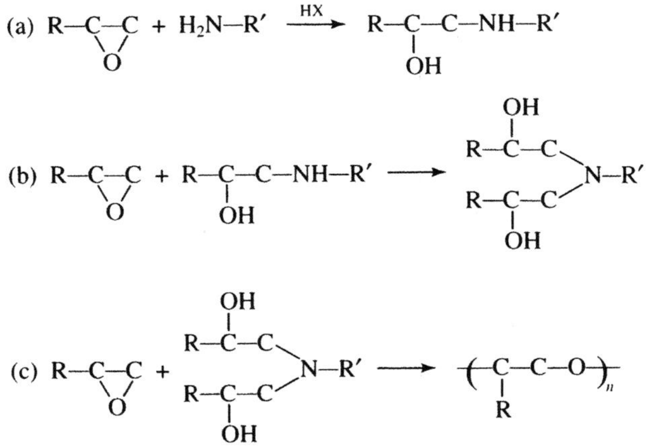

As shown in the above reaction scheme, the epoxy groups are successively opened by amine active hydrogens [Reactions (a) and (b)]. In this reaction, the presence of impurities comprising active hydrogen compounds (HX), such as water and alcohols, is required. The alcohols produced by Reactions (b) and (c) accelerate the curing reaction [19].

Tertiary amines used as hardeners for epoxy prepolymers include:

- triethylamine (TEA);
- benzyldimethylamine (BDA);
- dimethylaminoethylphenol (DMAMP);
- tri(dimethylaminomethyl)phenol (TDMAMP);
- tri-2-ethylhexoate salt of TDMAMP.

A more complete list can be found in the technical literature [22].

Tertiary amines are commonly referred to as catalytic curing agents since they induce the direct linkage of epoxy groups to one another. The reaction mechanism is believed to be as follows:

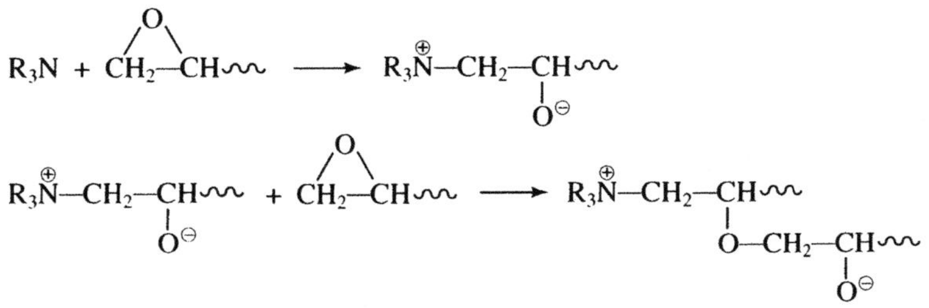

Since the reaction can occur at both ends of the diglycidyl ether molecule, a crosslinked structure will be built up. The overall reaction may, however, be more complicated because the epoxy group also reacts with hydroxyl groups, particularly when catalyzed [28].

There is a large variation of such curing agents on the market because of the necessity to improve specific properties or applications, such as pot life, flexibility, fast curing, lower toxicity, lower exoterm in large castings, etc.

Improved polyamines are achieved via mixtures of amines, tertiary monoamine plus a polyamine or other such ring-nitrogen derivatives as *N*-(2-hydroxyalkyl)aziridine; imino bis(aziridinylphosphine) oxides; aziridines in combination with polyamines or polyamines from aziridines; monotertiary diprimary triamines, such as $RN(CH_2CH_2CH_2NH_2)_2$, where R is 12 to 22 carbon atoms; and alkane triamines. Polyamine–epoxide reactions can be accelerated with phenol/C_1–C_7 aldehyde resins; aromatic orthosilicic esters; thiocyano compounds such as tetrahydrothiadiazine; quaternary ammonium salts of strong acids; α-metallometal salt of a carboxylic acid, such as α-sodio-sodium propionate and BF_3-amine and BF_3-phenol complexes [22].

Details of this group of hardeners and of the structures and dynamic mechanical properties of epoxy prepolymers cured with them are given by T. Kamon and H. Furukawa [19], K. Ausek [21] and others.

The mechanism of curing with cyclic anhydrides is less understood than that with polyamines. In the presence of tertiary amines, the reaction was found to proceed in the absence of proton donors, most probably via a zwitterion [21]. The simplified mechanism is:

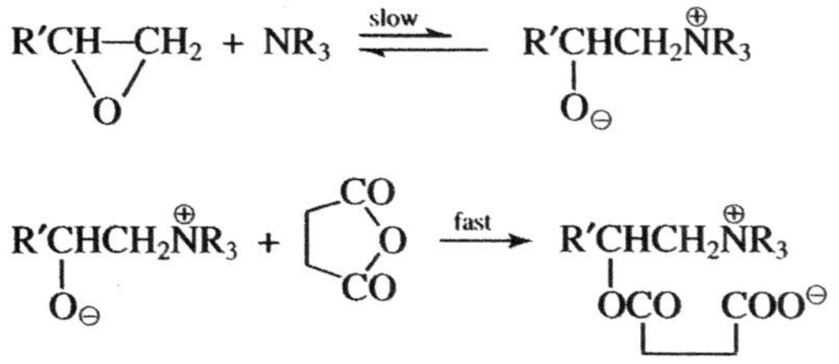

Propagation

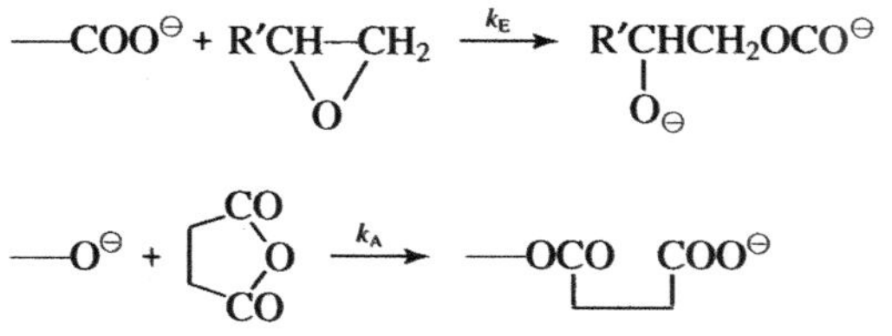

The tertiary amine is thus irreversibly bound in the structure, and further chain growth occurs by an anionic mechanism.

Cures with Lewis acid- and base-type curing agents proceed by homopolymerization of an epoxy group that is initiated by both Lewis acids and bases.

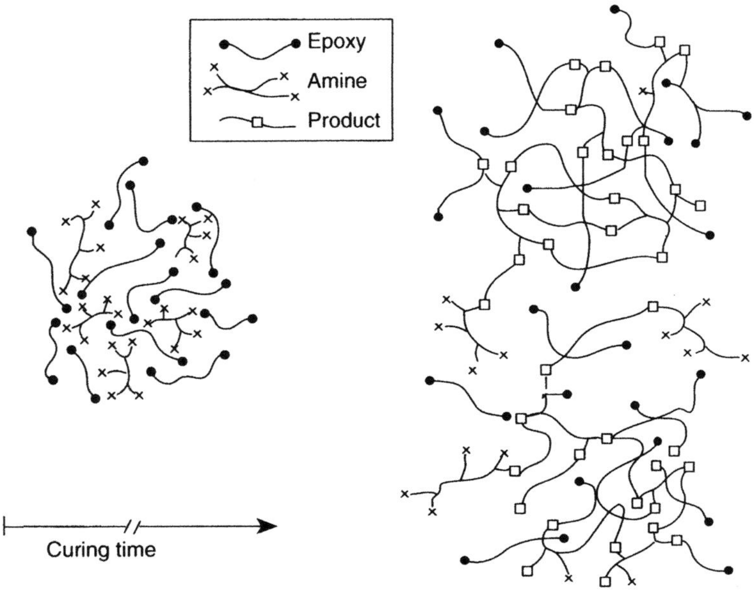

Figure 8.3 Schematic presentation of the initial formation of nodules in the resin curing agent mixture [30]. (Reprinted by permission of the publishers Butterworth–Heinemann Ltd from *Polymer*, **21**, 903, 1981.)

Studies on their morphology show that epoxies form networks with heterogeneous crosslink density distributions. These are due to synthesis conditions such as temperature, nature of the organic solvent, and composition of mixing of the initial components.

Some other factors have also been suggested, such as reactivity difference of the functional groups, unreacted functional groups, phase separation and intramolecular cyclization.

The high crosslink density regions have been described as agglomerates of colloidal particles (or flocculus) in an interstitial fluid with a lower MW [29].

Based on data of dynamic mechanical analysis (DMA) measurements and transmission electron microscopy, J. Mijovic and L. Tsay [30] believe that the formation of nodules (agglomerates) commences at random spots in the prepolymer–hardener mixture (Figure 8.3). These authors highlight the importance of the crosslinking agent concentration and consider that higher hardener concentrations lead to more excessive intramolecular reactions at the expense of the intermolecular matrix; this phenomenon effects a decrease of the Young's modulus. A simplified schematic presentation of a relative variation in the noduli of modules and matrix as a function of hardener concentration is shown in Figure 8.4.

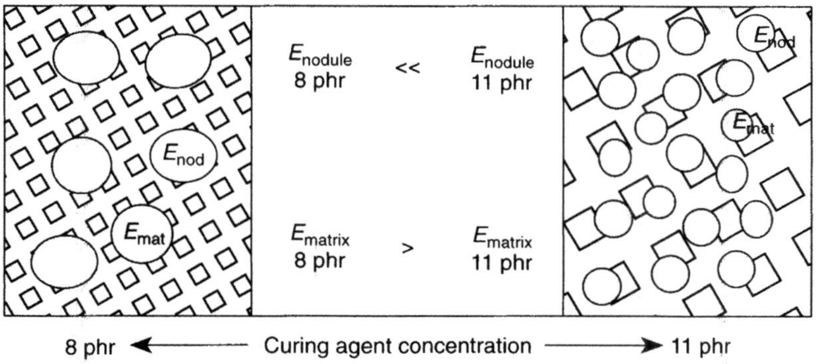

Figure 8.4 Schematic presentation of the variations in moduli of nodules and matrix as a function of curing agent concentration [30]. (Reprinted by permission of the publishers Butterworth–Heinemann Ltd from *Polymer*, **21**, 903, 1981.)

8.2.4 Other types of epoxies

There are many other types of epoxies and some of the more commonly used ones are glycidyl ethers of phenol–hydrocarbon novolacs, glycidyl ethers of aliphatic polyols and glycidyl esters.

Hydrocarbon epoxy novolacs (HEN) are members of a family of epoxy resins derived from the alkylation products of hydrocarbons and phenols. HENs are attractive candidates as matrices in electrical printed wiring boards and electronic encapsulation formulations [31].

Glycidyl ethers of aliphatic polyols are generally made by a two-step reaction. In the first step, the aliphatic hydroxy compound is reacted with epichlorohydrin to produce chlorohydrin ether which is then dehydrohalogenated via a base to yield the glycidyl ether. Glycidyl ethers, such as polypropylene glycol

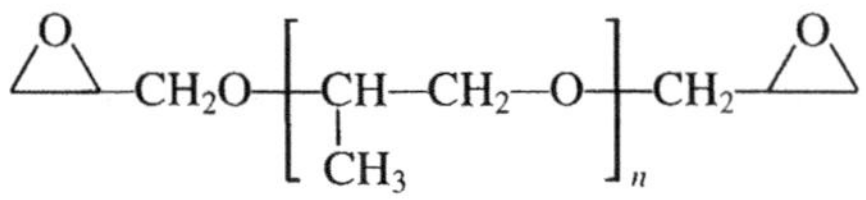

are used as plasticizers to increase the elongation and impact resistance of BPA based epoxies [32].

Glycidyl esters are a sub-class of epoxy resins derived from any aromatic and/or aliphatic carboxylic acid and contain the following functional group:

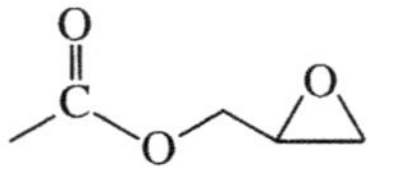

The chemical reactions produced when a glycidyl ester reacts with an amine, amide-amine, anhydride, acid or thiourea are thought to be similar to those with glycidyl ethers based on BPA. They are used in coatings and for electrical potting/encapsulation [33].

8.2.5 Additives

To provide certain characteristics, diluents, flexibilizers, fortifiers, fillers or other modifiers are frequently added to epoxy polymers.

1. Diluents can be used in applications with composites (i.e. impregnation of fiber mats) where the viscosity can be reduced by adding non-reactive organic solvents such as xylene or toluene at around 10% by weight. Sometimes reactive diluents like low MW mono- or di-epoxide compounds or tri-phenylphosphite are preferred because they become bound into the system and do not cause problems through volatilization [34].
2. Flexibilizers provide the slight flexibility and reduced brittleness needed for impact resistance. They achieve this by decreasing the degree of cross-linking. This may be realized by adding a higher MW prepolymer, by increasing aliphatic amine concentration or by decreasing acid anhydride concentration [35]. In many products low MW polymers are incorporated into the network to provide flexible links in the network. They are poly-amides (which are also curing agents), polyanhydrides and polysulfides. Whilst flexibilizers impart toughness, it is at the expense of stiffness strength and heat resistance [34]. Other flexibilizers are dimerized and trimerized fatty acids, epoxidized oils, aliphatic monoamines, asphaltic compounds and glycols. Of these, the trimerized fatty acids are outstanding in producing cured flexible polymers having good all around strength [35].
3. Fortifiers and their effect on the curing and thermal stability of epoxy polymers were studied by R.A. Patel *et al.* [36,37]. Epoxy fortifiers were prepared by condensing phenyl glycidyl ether (PGE) with 4-hydroxy-acetanilide (HA) and acetoacetanilide (AA) to give PGEHA and PFEAA. A condensation reaction of vinyl cyclohexane dioxide (VCD) with HA and AA produces the fortifiers VCDHA and VCDAA. When *p*-cresyl glycidyl ether (CGE) and DGEEG (diluent) were condensed with HA, CGEHA and DGEEGHA, respectively, were obtained. The condensation products from the immediately foregoing processes were solid or semisolid in nature and were used as such for curing purposes. A commercial grade resin DGEBA of epoxy equivalent $190.0\,\text{g}\cdot\text{eq}^{-1}$ was used. Phthalic anhydride was purified by crystallization from acetic anhydride (m.p. 131°C) where the catalyst used was triethyl amine, a laboratory grade reagent. These authors found that the use of fortifiers to the epoxy-anhydride system in the presence of a diluent lowers the necessary cure temperature; thermal stability is also affected.

Table 8.1 Effect of fillers and epoxy resin properties [38]. (Reprinted with permission of Elsevier Science Publishers, from *Polymers in Microelectronics* by D.S. Soane and Z. Martynenko, 1989)

Filler or extender	*Property improved*
Silica; calcium carbonate	Cost
Glass microballoons	Density
Fibrous glass, chopped nylon	Impact strength Tensile strength Flexural strength
Aluminum powder, calcium carbonate	Machinability
Carbon black asbestos	Heat resistance
Silica; metal powders	Dimensional stability Thermal conductivity
Lithium aluminum silicate	Thermal expansion coefficient
Inorganic fillers	Fire retardance
Asbestos; powdered coal	Moisture resistance
Mica; silica; asbestos	Electrical insulation
Silver or aluminum powder; carbon; graphite	Electrical conductivity
Graphite; mica; molybdenum disulfide	Lubricity

4. Fillers are the most common additives. Their primary role is to reduce shrinkage and cost. Since they do not shrink themselves and may occupy a significant volume of the total epoxy polymer system, overall shrinkage is reduced. In addition, fillers reduce the coefficient of thermal expansion, add thermal stability and increase thermal conductivity. Some common fillers are aluminum oxide, aluminum powder, iron powder, copper powder, silica, talc and calcium carbonate [35]. The influence of a filler, be it organic or inorganic (metallic or non-metallic), is generally related to the amount incorporated rather than to type. Mechanical properties can be affected by the nature and amount of filler. The improved properties are shown in Table 8.1.
5. Other additives which can be introduced into the epoxy systems, depending on the applications, are flame retardants, thixotropic agents, pigments, silver powder for the increase of electrical conductivity, etc.

8.3 PROPERTIES OF CURED EPOXIES

The properties of cured epoxy polymers depend on:

- the type of epoxy prepolymer, its MW (viscosity);
- the type of hardener and its amount;
- the degree of crosslinking;
- the nature and amount of additives.

A proper selection of prepolymer, hardener and additives allows the properties of the cured epoxies to be tailored to achieve specific performance characteristics. This versatility has been a major factor in the steady growth rate of epoxy polymers over the years.

Besides this versatility feature, properly cured epoxies have other attributes [39]:

- excellent chemical resistance, particularly to alkaline environments;
- outstanding adhesion to a variety of substrates;
- very high tensile, compressive and flexural strengths;
- low shrinkage on cure;
- dimensional stability;
- excellent electrical insulation properties and retention thereof on aging or exposure to difficult environments;
- remarkable resistance to corrosion;
- a high degree of resistance to physical abuse;
- ability to cure over a wide range of temperature;
- superior fatigue strength.

Toughness can be improved by blending with thermoplastics [40]. Table 8.2 shows the general range of properties of different epoxy polymers.

The mechanical properties of cured epoxies are not very different from those of unsaturated polyesters. An important difference is that much higher heat distortion temperatures can be obtained and some formulations (e.g. those based on epoxidized novolacs) may reach even 300°C. Other important differences are that the rigid epoxies have lower coefficients of thermal expansion. Shrinkage during curing is less (2–5%) and water absorption is lower. Their electrical properties are somewhat better than those of polyesters for insulation purposes and chemical resistance is also superior, as shown in Table 8.2. Cured epoxies are resistant to acids and

Table 8.2 Properties of cured epoxies

Glass transition temperature (°C)	120–190
Maximum continuous service temperature (°C)	130
Specific gravity	1.11–1.14
Thermal conductivity (W/m · K)	0.88
Coefficient of linear thermal expansion ($1/K \times 10^{-6}$)	11–35
Tensile modulus (GPa)	2.1–5.5
Tensile strength (MPa)	28–90
Compressive strength (MPa)	241
Water absorption (vol%)	0.05–0.2
Stability to acids and bases	Excellent
Stability to solvents	Variable (less stable to ketones and halogenated hydrocarbons)

bases, although acid cured products are less resistant than the amine cured ones. The drawbacks of cured epoxy polymers are their higher cost and difficulty of processing [34].

8.4 APPLICATIONS

Due to their specific performance characteristics these polymers have found a broad range of applications. They are used mainly as adhesives, sealants, paints, coatings and composites. Their existence has greatly stimulated the development of high performance structural adhesives and composites, for application in microelectronics, aerospace transportation and construction engineering. Epoxies are the principal polymers for high performance composites and account for about 80% of the polymers used in this field. The principal reinforcement is carbon fiber, with aramid fibers being used where toughness is required.

Some of the more important applications are adhesives and coatings. The high performance of epoxy adhesives is due to the high polarity of the macromolecules and to the fact that the presence of aliphatic hydroxyl and ether groups create electromagnetic bonding attraction between the macromolecules and the surfaces being bonded. Also, the oxirane ring forms chemical bonds with most surfaces, particularly those having active hydrogens, such as metals [35]. Polyepoxy adhesives are most commonly used as two-component liquids or pastes and cure at room or elevated temperature. Curing is done with a polyamide or polyamine or in the quick setting type with an amine catalyzed polymercaptan. Other epoxy adhesives are supplied as supported tape, that is a glass fabric tape impregnated with adhesive or as non-supported tape. These novel adhesives are 'one pack' systems which contain, beside the prepolymer, a latent curing agent such as dicyandiamide and BF_3. These epoxy adhesives are used in the aircraft industry to replace many of the mechanical fasteners once used and also for automotive adhesives. They have a pot life of more than one year at room temperature and have increased the use of epoxies for specialty adhesives and sealants. The 'one-pack' system provides fast cure where heated, for instance five minutes at 100°C.

A significant increase in the use of epoxy adhesives in the civil engineering industry has taken place over the last few years. This has been mainly due to two factors.

1. The increasing realization that the bond strengths of epoxy adhesives are considerably greater than the cohesive strength of concrete, which is the most common load bearing construction material used in civil engineering. The tensile strength of good quality concrete varies between 1.75 and 5 MN/m^2 (253.7–725 psi) according to the mix used, whereas the

tensile strength of epoxy adhesives exceeds 5 MN/m^2 (725 psi) and is often much greater, in the range of 56 MN/m^2 (8120 psi).

2. The ability to make significant savings in time by using an adhesive. The rate of development of mechanical strength of epoxy adhesives is therefore important in this context, and it can vary widely depending upon the epoxy system used, the ambient temperature and other factors. However, it is certainly faster than the rate at which concrete develops comparable proportions of its final strength.

The uses of epoxy adhesives and grouts can be classified broadly into:

- remedial work such as strengthening and repair of existing structure (concrete crack repair, bonding concrete to concrete, bonding reinforcements);
- new work, where the use of adhesive was envisaged at the design stage [41].

Various modifiers, bituminous or synthetic polymers, have been used to improve the properties of polyepoxides. Furthermore, different thermoplastics and thermosetting polymers, including elastomers, have also been incorporated into polyepoxides to modify their properties [42]. A polyamide soluble in an ethanol–water mixture is used in epoxy–nylon film adhesives to obtain high peel strength as well as good heat resistance. A thermoplastic polyurethane modified epoxy resin has been developed which is reported to give better peel strength at cryogenic temperatures than that obtained with epoxypolyamide.

Poly(vinyl acetal)s show good compatibility with epoxy resins and improve the peel strength of adhesives. Among the thermosetting resins, phenolics have long been blended to obtain heat resistant adhesive. Epoxy–nitrile rubber blends yield high peel strength adhesives [43].

With carboxyl terminated butadiene–acrylonitrile copolymer, the thermal characteristics of an epoxy have also been modified [44]. Also, the adhesion of epoxies can be improved and the cost decreased by blending with certain amounts of kraft lignin [45–50].

In recent years, a number of halogenated epoxy polymers with substantial hydrophobic characteristics have been developed. Among them are fluoroepoxy resins which show great promise as adhesives; however, a number of significant problems, such as toughness, remain to be overcome.

For coatings, the most common commercially available is the 'two-package' type coating; this type is used today as heavy duty industrial and marine maintenance products, tank linings, floor toppings, etc. [51].

The crosslinking agents reacting with epoxide and hydroxyl groups result in highly chemical and solvent-resistant films because all the bonds are relatively stable C–C, C–O (ether) and C–N (amine) bonds. Many of the more common epoxy resin coatings and their end uses are summarized in Figure 8.5.

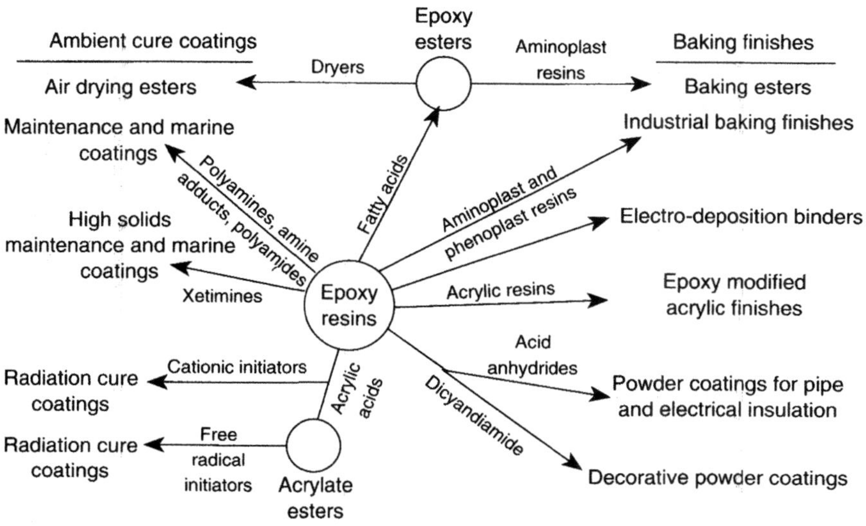

Figure 8.5 Epoxy resins in surface coating applications [39]. (Reprinted by permission of American Chemical Society from *Applied Polymer Science*, 2nd edn, by R.W. Tess and G.E. Poehlein (eds), 1985.)

High solid content and solventless epoxy coatings have been used for marine, maintenance and tank lining applications for many years. However, compared to conventional solvent based epoxy coatings, they have relatively short pot lives. Because of the versatility of the reactants, new information on formulations, based on diglycidyl ethers of BPA has been reported. UV curable cycloaliphatic epoxies, high performance proprietary epoxies, one-component systems, high solids, two-pack systems, high solids, epoxy–polyamide coatings, and epoxidized jatropha seed oil films have been described. The reaction products of higher oligomer and epichlorohydrin have been used as coatings [25].

The applications in electrical and electronic fields are adhesives, composites, printed circuit boards, coatings, solder maskants, fiber reinforced laminates, encapsulation of electrical and electronic equipment and components, fluidized bed coating for electronic devices, corrosion protective coating for metal surfaces, spray and dip coating for printed wiring board assemblies, connector and transformer molded cases [38,53]. The advantages of using epoxy polymers in this field are: no volatiles released during curing; very good adhesion; and it is suitable for all thermosetting processing techniques. Among the disadvantages are: poor oxidation and moisture stability; limited thermal stability (225°C); and specialty grades are expensive [38].

Epoxy polymers are used as matrices in fiber reinforced composites, providing high strength to weight ratios, and good thermal and electrical properties. The fabrication of fiber composites with epoxy resins is dominated by those involving continuous fibers: lamination in autoclaves, pultrusion, filament winding and hot compression molding [54]. Filament wound epoxy composites are used for rocket motor casings, pressure vessels and tanks. Reinforced epoxies use a variety of fibers such as glass, boron, graphite, Kevlar and cellulose for the preparation of laminates and fiber reinforced structures. Glass fiber reinforced epoxy pipes are used in the oil, gas, mining and chemical industries [28]. Treatment of carbon fibers with polyamides has an important role in the reinforcing of epoxies [55].

About 70 000 tons of epoxies are used annually as composites in the USA, much of it by the electronics and transportation industries. While epoxy has a low order of toxicity and is not a carcinogen or mutagen, the amine curing agents may cause dermatitis if handled without protective gloves or protective clothing [2].

The properties of typical fiber glass filled and unfilled epoxy resins are shown in Table 8.3.

Its good adhesion to concrete makes epoxy polymer a good material for concrete repair. Epoxies are used in concrete composites and in grout mixtures for flooring. Decorative flooring and exposed aggregate systems make use of epoxies because of their low curing shrinkage, and the good bonding of glass, marble and quartz chip by epoxy matrix.

Due to their end groups, epoxy polymers can be grafted and some grafted epoxies can be used for coatings. Thermosetting acrylic polymers have been synthesized by reaction of an epoxy prepolymer with an acrylic to form a graft copolymer capable of being crosslinked. For certain coatings epoxy prepolymers are reacted in the same way with drying fatty acids [56].

Table 8.3 Properties of typical unfilled and filled epoxy polymer [2]. (Reprinted with permission of ASM International from *Reinforced Plastics, Properties and Applications*, by R.B. Seymour, 1991)

Property	*Unfilled epoxy*	*Fiber glass filled epoxy*
Heat deflection temperature at 1.82 MPa (°C)	140	150
Maximum resistance to continuous heat (°C)	120	135
Coefficient of linear expansion (10^{-5}cm/cm · °C)	2.5	2.0
Tensile strength (MPa)	52	83
Elongation (%)	5	4
Flexural strength (MPa)	124	103
Compressive strength (MPa)	70	100
Notched Izod impact strength (J/m)	11	25
Hardness (Rockwell)	M90	M105
Specific gravity	1.0	2.8

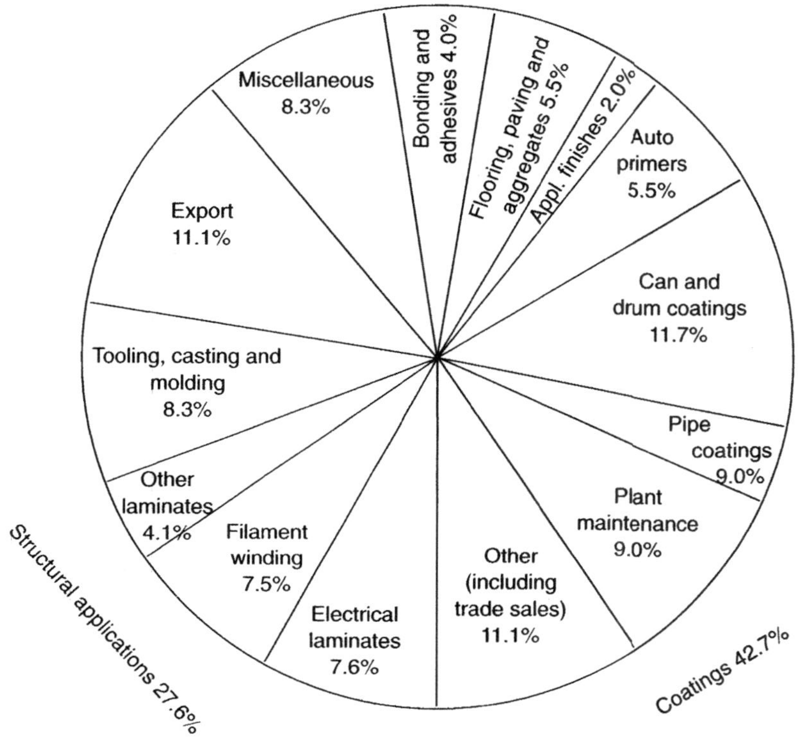

Figure 8.6 Epoxy consumption by end use [57]. (Reprinted from *Epoxy Resins – Chemistry and Technology*, 2nd edn by C.A. May, 1988. Courtesy of Marcel Dekker Inc.)

A general view of epoxy polymers applications is given by C.A. May (see Figure 8.6) [57].

REFERENCES

1. Ulrich, H. (1982) *Introduction to Industrial Polymers,* Hanser Publ., Munich.
2. Seymour, R.B. (1991) *Reinforced Plastics,* ASM International, USA.
3. Lohse, F. (1987) *Makromol. Chem. Macromol. Symp.*, **7**, 1–16.
4. Critchley, J.P., Knight, G.J. and Wright W.W. (1983) *Heat Resistant Polymers,* Plenum Press, New York.
5. Spacek, V., Pouchly, J. and Biros, J. (1987) *Eur. Polym. J.*, **23**(5), 377.
6. Fedtke, M. (1987) *Makromol. Chem. Makromol. Symp.* **7**, 153.
7. Bair, H.E. (1985) *Polymer Reprints*, **26**(1), 10.
8. Schiraldi, A. and Baldini, P. (1983) *J. Thermal Anal.*, **28**, 295.
9. Cizmeciogly, M. and Gupta, A. (1982) *Sampe Quarterly*, April.
10. Mijovic, J. (1986) *J. Appl. Polym. Sci.*, **31**, 1177.
11. Schiraldi, A. (1985) *Thermochimica Acta*, **96**, 283.

12. Renner, A., Moser, R., Bellus, M. *et al.* (1985) *J. Polym. Sci., Polym. Chem. Ed.*, **23**, 2341.
13. Barton, J.M. (1980) *Polymer*, **21**, 603.
14. Dickie, R.A., Labana, S.S. and Bauer, R.S. (1988) *Cross-linked Polymers*, American Chemical Society, Washington.
15. Mijovic, J. and Wang, H.T. (1989) *J. Appl. Polym. Sci.*, **37**(9), 2661.
16. Pang, K.P. and Gilham, J.K. (1989) *J. Appl. Polym. Sci.*, **37**, 1969.
17. Duffy, J.V. and Lee, G.F. (1988) *J. Appl. Polym. Sci.*, **35**, 1367.
18. Jozavi, H. and Sancaktar, E. (1988) *J. Adhesion*, **25**, 185.
19. Kamon, T. and Furukawa, H. (1986) *Advances in Polym. Sci.*, **80**, 173.
20. Varma, I.K. and Satyabhama, P.V. (1986) *Journal of Composite Materials*, **20**, 410.
21. Dusek, K. (1984) in *Advances in Chemistry Series 208. Rubber Modified Thermosets Resins* (eds C.K. Riew and J.K. Gilham), American Chemical Society, Washington, p. 3.
22. Mika, T.F. (1973) in *Epoxy Resins* (eds C.A. May and Y. Tanaka), Marcel Dekker, New York, p. 239.
23. Alger, M.S.M. (1983) *Polymer Science Dictionary*, Elsevier Applied Science, London.
24. Feldman, D. (1974) *Technology of Macromolecular Compounds*, Ed. Technica, Bucharest.
25. Gupta, S.K. and Kumar, A. (1987) *Reaction Engineering of Step Growth Polymerization*, Plenum Press, New York.
26. Batzer, H. and Zahir, S.A. (1975) *J. Appl. Polym. Sci.*, **19**, 601.
27. Odian, G. (1981) *Principles of Polymerization*, 2nd edn, J. Wiley, New York.
28. Chanda, M. and Roy, S.K. (1987) *Plastics Technology Handbook*, Marcel Dekker, New York.
29. Morgan, R.J. (1985) in *Epoxy Resins and Composites I, Advances in Polymer Science*, Springer-Verlag, Berlin, p. 1.
30. Mijovic, J. and Tsay, L. (1981) *Polymer*, **22**, 903.
31. Bogan, G.W. (1992) in *Reactive Polymers* (eds W.F. Gum, W. Riese and H. Ulrich), Hanser Publ., Munich, p. 130.
32. Bertram, J.L. (1992) in *Reactive Polymers* (eds W.F. Gum, W. Riese and H. Ulrich), Hanser Publ., Munich, p. 134.
33. Detloff, M.L. and Gold, M.L. (1992) in *Reactive Polymers* (eds W.F. Gum, W. Riese and H. Ulrich), Hanser Publ., Munich, p. 140.
34. Dyson, R.W. (1990) in *Engineering Polymers* (ed. R.W. Dyson), Blackie, Glasgow.
35. Nielsen, P.O. (1982) *Adhesives Age*, April, 42.
36. Patel, R.A. and Patel, R.G. (1988) *Die Angewandte, Makrom. Chem.*, **161**, 89.
37. Patel, R.H., Patel, R.S. and Patel, R.G. (1989) *Thermochimica Acta*, **141**, 77.
38. Soane, D.S. and Martynenko, Z. (1989) *Polymers in Microelectronics*, Elsevier, Amsterdam, p. 257.
39. Bauer, R.S. (1985) in *Applied Polymer Science*, 2nd edn (eds R.W. Tess and G.W. Poehlein), ACS Symposium Series, 285, American Chemical Society, Washington, p. 931.
40. Kinloch, A.J., Yuen, M.L. and Jenkins, S.D. (1994) *J. Mater. Sci.*, **29**(14), 3781.
41. Potter, N.G. (1976) *Use of Epoxy Resins*, Chemical Publishing Co., New York.
42. Yorkgitis, E.M., Eiss Jr, N.S., Tran, C., Wilkes, G.L. and McGrath, J.E. (1985) in *Epoxy Resins and Composites* (ed. I.K. Dusek), Springer-Verlag, Berlin, p. 80.

43. Updergaff, J.H. and Suen, T.J. (1977) in *Polymerization Processes* (eds C.E. Schildknecht and J. Skiest), J. Wiley & Sons, New York, p. 497.
44. Walker, J.M., Richardson, W.E. and Smith, C.H. (1976) *Mod. Plast.*, **62**, May.
45. Feldman, D. and Banu, D. (1988) *J. Polym. Sci.*, **26**(4), 973.
46. Feldman, D. and Khoury, M. (1988) *J. Adhes. Sci. Technol.*, **2**(2), 107.
47. Feldman, D., Banu, D. and Khoury, M. (1989) *J. Appl. Polym. Sci.*, **37**, 877.
48. Feldman, D., Banu, D., Nathanson, A. and Wang, J. (1991) *J. Appl. Polym. Sci.*, **42**, 1537.
49. Feldman, D., Banu, D., Luchian, C. and Wang, J. (1991) *J. Appl. Polym. Sci.*, **42**, 1307.
50. Wang, J. and Feldman, D. (1991) *J. Adhes. Sci. Technol.*, **5**(7), 565.
51. Shaw, S.J. and Tod, D.A. (1994) *Materials World*, **2**(10), 523.
52. Seymour, R.B. (1988) *J. Coat. Technol.*, **60**(759), 57.
53. Breslow, A.J. (1973) in *Epoxy Resins Chemistry and Technology* (eds C.A. May and Y. Tanaka), Marcel Dekker, New York, p. 485.
54. Jones, F.R. (1993) in *Chemistry and Technology of Epoxy Resins* (ed. B. Ellis), Blackie, London.
55. Skourlis, T.S., Duvis, T. and Papaspyrides, C.B. (1993) *Compos. Sci. Technol.*, **48**(1–4), 119.
56. Stevens, M.P. (1990) *Polymer Chemistry*, Oxford University Press, New York, p. 379.
57. May, C.A. (1988) *Epoxy Resins, Chemistry and Technology*, 2nd edn, Marcel Dekker, New York, p. 7.

9 Phenolics (PFs)

9.1 INTRODUCTION

Phenol–formaldehyde polymers known as poly(phenol–formaldehyde), phenoplasts or phenolics are produced by polycondensation of phenol or a mixture of phenols with an aldehyde, in most cases, formaldehyde [1].

The acid catalyzed polycondensation of phenol with formaldehyde was reported in 1872 by Baeyer but it was not until February 1909 that L. Baekelend traced the history of research on phenol–formaldehyde polymers and early attempts to obtain a useful product [2]. The latter had patented this polycondensation process in 1907. The commercial development of the phenol–formaldehyde products is considered to be the beginning of the truly synthetic polymers era, the beginning of plastic industry, although cellulose nitrate had been known and in use for some time.

Increasingly continuously, the production of polymers obtained from the above mentioned components has reached 1.3 million metric tons in 1985 in the USA [3]. In 1988 global demand for phenolics reached 2.7 million metric tons. The USA accounted for 51% followed by Western Europe (14%), Japan (13%), Eastern Europe (7%), the rest of Asia (7%), Canada (4%), Latin America (3%) and Mexico (1%) [4]. In 1994 the output of PF in the USA was 1.454 million tons compared to 1.395 million tons in 1993, which represents an annual increase of 4.2% [5].

The synthesis of PF is essentially a polycondensation between phenol or a mixture of cresols and formaldehyde, and an acid or base catalyst can be used. These differences lead to PFs with different microstructure and properties. Baekelend defined the differences between the two catalyst systems and underlined the effects of the mole ratio of the reactants.

The most important process, known as the Hock process, for the synthesis of phenol is the cumene process; besides phenol it also produces acetone. Phenol is the most important raw material for PF. Small amounts of other unsubstituted phenols, such as resorcinol, are used. Of the substituted phenols, the most important are the following alkylated phenols: cresols, *p-tert* butylphenol, *p-tert* amylphenol and *p-tert* octylphenol. Bis-phenol A, generally used for epoxies, is growing in importance as a raw material in PF.

Though many aldehydes react with phenol and its derivatives, only formaldehyde, the most reactive aldehyde, and with the lowest MW is of major commercial importance. Furfural is also used, forming phenolics that readily oxidize resulting in dark brown to black products. Formaldehyde is used as formalin, which is its solution in water, or as solid paraform, which consists of a mixture of polymethylene glycol esters having a degree of polymerization of 6–100 [6].

Formaldehyde is produced by dehydrogenation of methanol, over either an iron oxide/molybdenum oxide catalyst or over silver catalyst.

9.2 REACTION MECHANISMS

Acid catalysis (pH 1–4) of a mixture with an excess of phenol leads to soluble novolacs; base catalysis (pH 7–11) with an excess of formaldehyde, to soluble resoles (A-stage prepolymers), then to resitols (B-stage polymers) and finally to crosslinked resites (C-stage polymers) [7]. The rate of reaction between phenol and formaldehyde in the strongly acidic conditions (pH 1–4) is proportional to $[H^+]$, but above pH 5 it is proportional to $[HO^-]$, indicating a change in reaction mechanism. Polycondensation under strongly acidic conditions leads to linear products (novolacs or novolaks) of MW up to 2000. A stable A-stage novolac is produced by using a molar ratio of phenol to formaldehyde of 1.25. The reaction product after the removal of water by vacuum distillation yields a solid which is then pulverized. In a subsequent phase this linear polymer is transformed into a thermoset polymer with additional formaldhyde supplied by hexamethylenetetramine (HMTA).

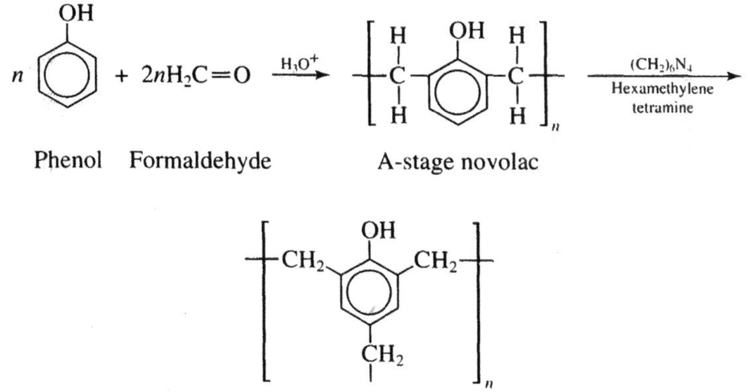

C-stage novolac

High *ortho–ortho* novolacs are produced under weakly acidic conditions by the use of specific catalysts, usually metal ion carboxylates [8].

Under basic conditions (pH 7–11), polycondensation produces resoles which contain methylene and benzylic ether linkages. The process is catalyzed by hydrates of sodium, potassium, lithium or magnesium. An increase in basicity usually leads to a lower *ortho–para* ratio; thus, in going from pH 8.7 to pH 13.0 the *ortho–para* ratio decreases from 1.1 to 0.38 [8].

The resole polymer is cured by heating and further polycondensation occurs by reaction of the methylol groups. This is a disadvantage of the system in that it contains within itself the means of crosslinking, and precautions must be taken during synthesis that gelation does not occur. The crosslinked network structure is represented as [9]:

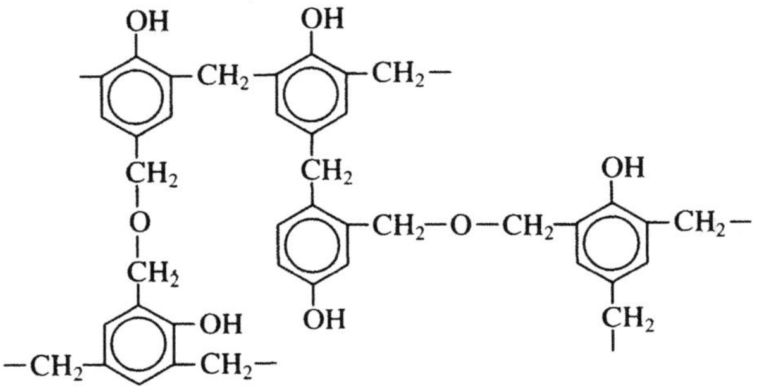

Heat curing, by far the most important crosslinking process for resoles, is conducted at temperatures up to 200°C.

9.3 NOVOLAC PRODUCTION

This prepolymer can be produced in a batch or continuous process, starting with a molar ratio of phenol to formaldehyde in the range of 1:(0.75–0.85). The continuous technology uses two stages in two stirred reactors at temperatures of 120–180°C, the second stage being necessary to complete the polycondensation started in the first one. From the second reactor, the reacted mixture is conveyed into a flash drum acting also as a vapor/liquid separator. The flashed vapor is condensed, collected in a receiver and sent to purification; the liquid phase separating on the bottom of the flash drum settles in two layers. The upper, aqueous layer, containing small amounts of phenol is withdrawn and sent to purification; the lower layer which contains the novolac is sent to a vacuum evaporator for the separation of water which contains small amounts of phenol [10]. The flow chart of the continuous polycondensation is presented in Figure 9.1.

Yields of conventional novolacs of high MW are 105–110% based on the amount of phenol taken in the process and a maximum of 5 wt% phenol in the final prepolymer. Yields of high *ortho*-polymers are 90–100%. The latter have excellent rheological properties and excellent cure speed.

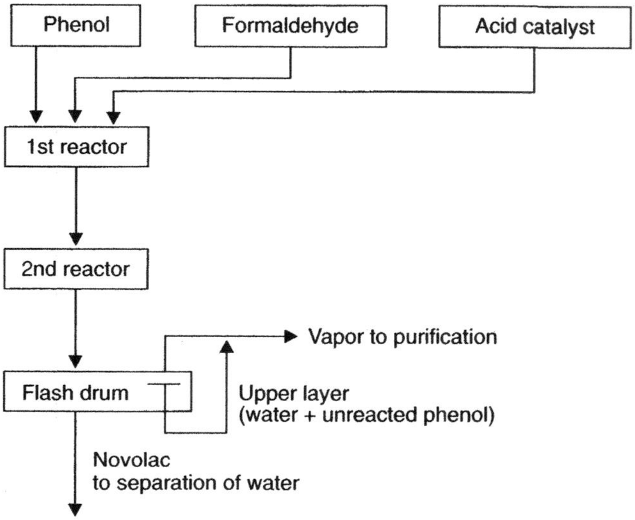

Figure 9.1 Production of novolac.

For novolacs obtained at pH lower than 4, sulfonic acid, hydrochloric acid, sulfuric acid or other acids are used as catalysts. For the so called neutral novolacs obtained at pH greater than 4, acetates of zinc, lead or magnesium are preferred.

9.4 RESOLE PRODUCTION

The polycondensation with alkalis is more versatile and related to the main parameters, such as molar ratios, catalysts and the reaction conditions applied, than the process in acidic medium. Resoles are generally made at a 1.25–1.4 formaldehyde to phenol molar ratio. Besides the hydroxides, liquid ammonia or gaseous ammonia, tertiary amine, etc. can be used as catalysts. Polycondensation takes place in a reactor equipped with an agitator to prevent local gelation and to circulate the viscous fluids. For polycondensation, molten phenol is charged to the reactor followed by the catalyst. Formalin is added at a rate to maintain the temperature of the mixture at 60–70°C. Reaction cycles are short, *c.* 45 minutes at 100°C or 3 hours at 65°C; the products are recovered as 40–50% aqueous solutions. Resole dehydration must be completed rapidly, preferably at temperatures lower than 100°C.

Low molecular weight, water soluble resoles are then cooled and neutralized. To advance the process, partial condensation is accomplished in a second step, and water is removed by vacuum dehydration [10,11].

The flow chart of the batch technology for the production of resoles is presented in Figure 9.2. The same equipment can also be used for novolacs.

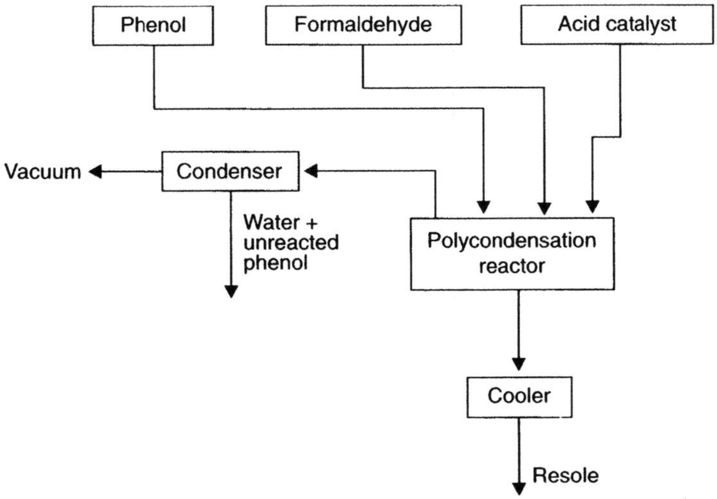

Figure 9.2 Production of resole.

Exact temperature and time control are required. Adequate vacuum, at least 0.5×10^{-6} Pa (50 mbar), and a cooling water temperature as low as possible, are necessary to maintain a low polymer temperature (maximum 60°C) and a relatively brief distillation period. The duration of the distillation depends on the desired polymer content [10]. To obtain the resitol (the B stage) a post reaction is necessary. Further reaction in excess of formaldehyde increases the MW leading to a highly crosslinked microstructure, known as resite [12].

The physical state of phenolics is dependent on their MW and the amount of reacted aldehyde. After starting the polycondensation, viscous reactive liquids soluble in alcohol and polar solvents, and difficult to isolate in an anhydrous form, are obtained. A post polycondensation decreases the solubility and increases the softening point until an infusible, insoluble tridimensional gel is obtained [6].

9.5 CURING MECHANISM

The mechanism of the curing process is complex. Novolacs are thermoplastics and can be cured by the addition of crosslinking agents such as paraformaldehyde or HMTA. For a novolac with a phenol/formaldehyde ratio of 1:0.8 about 10–15% HMTA is required.

Heat curing of resoles is obtained at 130–200°C. In the first stage, methylene and ether bridges are produced. At temperatures above 210°C the ether linkage is converted to a methylene bridge and oxidation occurs [8]. Novolac has more ether linkages than resole [13].

The exact number and positions of crosslinkage in various grades of phenol–formaldehyde polymers are not known, but it is possible to postulate a very complex highly idealized three-dimensional structure if the ratio of the reactants is correct.

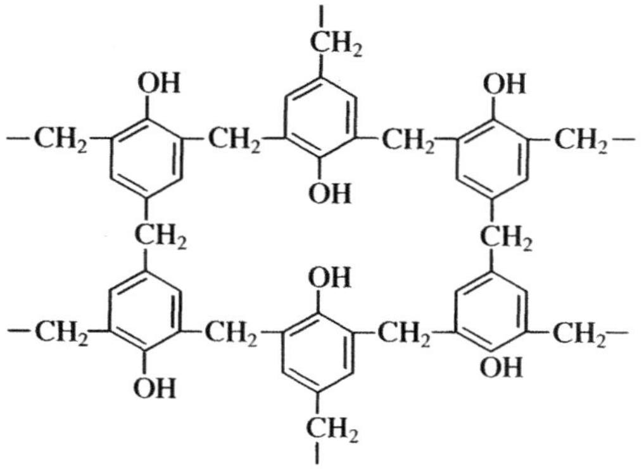

This is a planar representation of a very complex tridimensional structure; there is considerably more irregularity than is implied in this representation.

These polymers, when classified according to the nature of the polycondensation occurring during their production, are of two fundamental types, one- or two-stage phenolics.

1. One-stage phenolics: in these, all the necessary reactants required to produce a thermosetting polymer are charged in the polycondensation reactor in the proper proportions and react together in the presence of an alkaline catalyst. The phenolic as discharged is thermosetting or heat reactive and requires only further heating to complete polycondensation to an infusible, insoluble polymer.
2. Two-stage phenolics: only part of the necessary aldehyde is added in the reactor and polycondensation takes place in the presence of an acid catalyst. The product is a fusible thermoplastic able to react in the secondary stage with additional formaldehyde and to form a thermosetting polymer. This additional formaldehyde is furnished by HMTA. The latter, under the influence of heat, breaks into formaldehyde and ammonia which serves as a catalyst for this second stage.

Both one- and two-stage phenolics are used separately or in combination in commercial molding materials [14].

The resoles are sometimes referred to as one-stage polymers (resins), since crosslinked products can be made from the initial reaction mixture only by adjusting the pH. The resole process is also known as the one-stage process. On the other hand, the resoles are sometimes referred to as two-stage polymers because, in this case, it is necessary to add some agent to enable formation of tridimensional products.

A significant characteristic of novolacs is that they represent completed reactions and as such have no ability to continue increasing their MW and there is no danger of gelation (crosslinking) during their production [15].

Phenolic prepolymers can be crosslinked with epoxies instead of HMTA. Such epoxidized novolacs are generally used as powders. The epoxy modified phenolics, both novolac and resole, are readily obtained by the reaction of epichlorohydrine with a PF of significantly high MW.

Epoxy modified PF exhibit high strength, good adhesion, excellent dielectric properties and improved oxidation resistance [10]. In general, phenolic polymers can also be classified between heat reactive and non-heat reactive and between polymers made by using unsubstituted or substituted phenols [6].

9.6 PROPERTIES

PFs have very low manufacturing costs. However, the processing is difficult; the elimination of water during crosslinking requires conditions (such as applying pressure) to prevent formation of voids due to water vapor. The high crosslinking density of phenolics leads to very low mobilities of chain segments between crosslinking points and thus to an excellent dimensional stability (heat distortion temperature 112–125°C). Certain poor mechanical properties such as low notched impact strength (1.5 J/mm) are probably caused by void formation during hardening [16,17].

Because they are multifunctional, unsubstituted and heat reactive, phenolics, made of phenol, cresols and xylenols, can be crosslinked to form films. They are soluble in alcohols, ketones and esters, insoluble in aromatic and aliphatic hydrocarbons, and they are compatible with polar polymers.

Since they are brittle, the novolacs which are unsubstituted and non-heat reactive are non-film formers, permanently fusible and soluble in alcohols, ketones and esters. Heating a novolac under pressure in the presence of a component able to release formaldehyde will result in a crosslinked, C-staged phenolic, insoluble and infusible.

The third class of substituent and heat reactive phenolics are made by using *para*-substituted phenols where the substituted is *t*-butyl, *t*-octyl and phenyl. They are brittle solids and non-film formers. The presence of substituents makes them less polar and hence they are soluble in ketones, esters and aromatic hydrocarbons. The substituted and non-heat reactive phenolics are made with the same substituted phenols as those of third class phenolics. They are brittle, permanently fusible and soluble solids, and non-film formers. Similar to their non-heat reactive analogs, they have broad solubility and compatibility [6].

In general, the maximum continuous service temperature for phenolics is 110°C. The first step during thermo-oxidative degradation in air is the

Table 9.1 Some physical and mechanical properties of phenolics

Property	*Value*
Specific gravity	1.4
Specific heat	1.3 kJ/kg · K
Thermal conductivity	0.35 W/m · K
Tensile strength	25 MPa (3600 psi)
Compressive strength	69–207 MPa (10 000–30 000 psi)

oxidation of a methylene group to a hydroperoxide followed by decomposition of the latter to dihydroxy benzophenone and benzhydrol structures. The benzophenone structure will then oxidize further. At temperatures between 250 and 400°C, water, carbon dioxide and formaldehyde are the major degradation products [9]. The values of a few physical and mechanical properties [17] are given in Table 9.1.

Because of the exothermicity of the curing phase and the accompanying shrinkage, it is necessary for some applications to incorporate inert fillers; they reduce cost and may provide additional benefits such as improving impact resistance [15]. Each phenolic is designed to fill certain types of applications at the lowest overall cost. No phenolic has been found that will fill several types of special requirements. Fillers and resins play an important role in the special properties of the product available [18].

9.7 APPLICATIONS

The main applications of phenolics are as molding powders, foundry resins, abrasive binders, coatings, foams, thermal insulation binders, laminates, wood derivatives and ion exchangers.

Phenolics and their molding compounds are of considerable interest for use in electrical components, in electrical and structural elements of space vehicles, electronics, for handling radioactive materials, and as protective coatings in nuclear power plants. In these applications high thermal resistance is required in conjunction with high energy radiation resistance [10,19].

9.7.1 Molding powders

Inert fillers are necessary to produce useful moldings and not only to reduce the cost. Commonly used are wood flour, mica, clay, glass fibers, asbestos, cotton fabric and chopped fabric [20]. The latter improves the impact strength, the asbestos can be used to improve heat and chemical resistance, and the mica powder is used for superior electrical insulation. Many other special fillers are used to impart certain properties of phenolic molding compositions. The ratio of filler to polymer also has an important effect on properties. Normally, the amount of filler is between 30 and 60% by weight of the molding powder. Sometimes the filler also acts as a coloring agent. Carbon black imparts a black color and is used to produce a non-blending grade for materials that might come in contact with alcohol or other solvents.

When novolac is the starting phenolic, other additives are necessary besides the filler, such as hardener, accelerator, lubricant, pigment and in some cases a plasticizer [21]. HMTA is used almost universally as the cross-linking agent.

Some of the uses of molding powders are for general purpose powders, high impact strength compounds, heat-stable compounds and compounds for electrical insulation. The common means of producing different items from phenolics are compression molding, transfer molding and injection molding.

Mouldings are glossy, opaque and dark colored but have the tremendous advantage of being inherently flame retardant (especially those with mineral fillers), low smoke materials which do not require halogen or phosphorus based products. They are stiff, hard, have good electrical insulation properties, and good temperature, water, solvent and chemical resistance. These main characteristics led to their acceptance in diverse market areas such as automotive, transportation, electronics, aeronautics, aerospace, electrical appliances and business equipment [20,22,23].

9.7.2 Foundry polymers

Both novolac and resole-type PFs are used in the bonding of sand for the manufacture of shell molds and cores for the casting of metals. PFs constitute 6–8% of the total composition [4].

9.7.3 Abrasive binders

Phenolics are used to manufacture grinding wheels and coated abrasives. A grinding wheel has three main components: the abrasive grain, the bonding agent and the filler. Wheels made of PF are much less sensitive to impact, push and side pressure than ceramic wheels. The polymeric bond is stronger, more resilient and thermally shock resistant than the vitreous bond. In the formulation for the manufacturing of cut-off wheels the amount of PF (liquid and powder) is in the range of 14–20%, pbw. PF are also used for abrasive paper, abrasive tissues and vulcanized fiber abrasives [10,21].

9.7.4 Coatings

Since PFs result in brittle coatings due to their structural rigidity after curing, efforts have been expended in flexibilizing the rigid polymer without adversely affecting the adhesion to metals, galvanic protection, chemical and abrasion resistance, and high temperature characteristics. Hence PF coatings are used in combination with more flexible, hydrophobic polymers such as epoxy, natural resins, poly(vinyl butyral), etc. The most important application areas for PF coatings are automotive primers, coatings for metallic containers, anticorrosive marine paints and printing inks [10].

Phenolic varnishes offer many excellent coating properties. As they have good flow characteristics, they brush well. They have good adhesion to wood, due to the polarity of hydroxyl groups. The cured films have fair resistance to dilute acids, alkalis, alcohols and aliphatic hydrocarbon solvents [6]. Protective coatings are used in nuclear power plants against corrosion, for maintenance and decontamination [10].

9.7.5 PF foams

Early techniques for phenolic foam manufacturing were based on the expansion of novolac–HMTA mixtures. Now, these foams' manufacturing processes are completely dominated by the acid catalyzed process of a resole-type PF with added blowing agent and surfactant [10].

PF foam has good chemical and thermal resistance, high resistance to water transmission and water uptake, good dimensional stability, high strength to weight ratio and less flammability than most polymer foams. However, due to its high open cell content it has relatively low thermal resistance. Thermal insulation efficiency can be improved by the application of a skin of suitable material [14]. Favorable foam characteristics can be obtained in a two-step process whereby *o*-cresol is prereacted with formaldehyde at an alkaline pH followed by the addition of an equimolar amount of phenol and formaldehyde [10].

PF foam technology has progressed in recent years so that slabstock and laminated boards can now be produced using continuous processes. PF composites cannot match the versatility of foam/facing combinations developed by PU and polyisocyanurate manufacturers; however, they do offer the construction industry a useful combination of physical and fire properties [24]. The thermal conductivity of phenolics is 0.35 W/m · K [17]. PF foam achieves the highest classification as the result of fire tests and produces smoke obscuration of less than 5%, compared with 50–90% for most commercial grades of PS or PU foam [14].

9.7.6 PF laminates

Phenolic laminates made with papers, cotton fiber, asbestos and glass fiber were among the first to achieve commercial interest and are still important. For these products, partly cured phenolics in the form of alcoholic solutions (or in special cases water) are used to impregnate the continuous reinforcement phase. Under certain conditions of heat and pressure a coherent tough sheet is produced. Common applications for thermal materials include high voltage insulating structures, gear wheels, water lubricated bearings, and the core of decorative table top laminates. In the latter case a melamine–formaldehyde polymer is used for impregnating the top decorative layer.

PF laminates are of value not only because of good electrical insulation, but also because of their good strength, high rigidity and processing. Sheet, tubular and molded laminates are used [21].

9.7.7 PF adhesives

Phenolic adhesives were originally developed for bonding wood members at room temperature where higher temperatures could not be obtained. They have since been used in the bag molding industry and in the manufacture of laminated wood beams and other types of wooden articles where higher temperatures had previously been employed. Good results have also been obtained in bonding other materials such as plastics, glass and ceramic wares. In the field of metal bonding and particularly the bonding of metal to other materials, PF adhesives produce excellent bonds if the metallic surface has been adequately prepared with a special adhesive primer [14].

Phenolic adhesives are used in the manufacture of plywood, particle board, wafer board and bonded wood products such as decorative beams of large proportions. Novolacs or resoles are used for such applications; they can be modified with resorcinol, crude phenolic products or filled with low cost cellulosics [4]. Unlike laminating PF, the adhesive must not saturate the wood veneer or chips, but must remain in the glue line of the surface of the ply or chips. The adhesives are designed to have high viscosity, so that they will remain within the glue line. Water present in the adhesive does not need to be removed before curing since it can diffuse through the wood.

9.7.8 PF ion exchangers

The first completely synthetic ion exchange polymers were prepared by B.A. Adams and E.L. Holmes in 1935. The stage of obtaining the crosslinked PF is followed by the introduction of the ion exchange functional group, namely sulfonic acid ($-SO_3H$) by reaction with hot sulfuric acid.

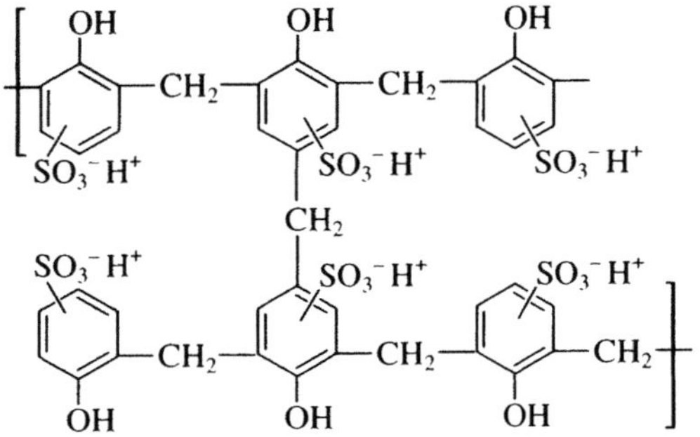

Phenol–formaldehyde sulfonic acid cation resin

Often it is possible to incorporate the functional, or inorganic group during the polycondensation by using an appropriately substituted benzene such as 3-hydroxybenzenesulfonic acid (metaphenolsulfonic acid).

PF cation exchangers are obtained by producing a matrix carrying weakly basic secondary amine groups [25].

Cured PF, as the matrix material for functional groups with exchange characteristics, has the advantage of high resistance to the usual regeneration agents (dilute acids or bases), solvents and chemicals. PF ion exchangers have good resistance to breakage, hydrophilic behavior and difunctional characteristics due to phenolic hydroxyl and sulfonyl acid or amine groups [10].

9.7.9 Other PF polymers

Resorcinol is a very reactive phenol able to react with formaldehyde at room temperature. Since resorcinol formaldehyde polymers are expensive, only resorcinol and *m*-aminophenol modified PF are presently being used. The temperature resistance of the resorcinol adhesives is *c.* 200°C. The adhesion to various materials is very good such that they can also be used to bond leather, rubber, plastics, ceramic, etc. [10].

Although non-modified, inorganically filled PFs are already considered temperature resistant, their thermo-oxidative resistance can be further improved by chemical modification such as:

- etherification or esterification of the phenolic hydroxyl group;
- complex formation with polyvalent elements (Ca, Mg, Cd, etc.);
- replacement of the methylene linking group by heteroatoms (O, S, N, Si, P, etc.) [8,10].

REFERENCES

1. Campbell, J.M. (1994) *Introduction to Synthetic Polymers*, Oxford Science Publications, Oxford.
2. Moravetz, H. (1985) *Polymers. The Origins and Growth of a Science*, J. Wiley & Sons, New York, p. 58.
3. Seymour, R.B. (1989) in *Pioneers in Polymer Science* (ed. R.B. Seymour), Kluwer Academic Publishers, Dordrecht, pp. 81–92.
4. Nelson, D.L. (1992) in *Reaction Polymers* (eds W.F. Gum, W. Riese and H. Ulrich), Hanser Publishers, Munich.
5. Kirshner, E.M. (1995) *Chem. Eng. News*, **73**(15), 21.
6. Fry, J.S., Merriam, C.N. and Boyd, W.H. (1985) in *Applied Polymer Science*, 2nd edn (eds R.W. Tess and G.W. Poehlein), ACS Symposium Series 285, American Chemical Society, Washington.
7. Seymour, R.B. and Carraher Jr, C.E. (1992) *Polymer Chemistry*, 3rd edn, Marcel Dekker, New York.

8. Robins, J. (1986) in *Structural Adhesives, Chemistry and Technology* (ed. S.R. Hartshorn), Plenum Press, New York.
9. Crithley, J.P., Knight, G.J. and Wright, W.W. (1983) *Heat Resistant Polymers*, Plenum Press, New York, p. 23.
10. Knop, A. and Pilato, L.A. (1985) *Phenolic Resins*, Springer-Verlag, Berlin.
11. Kirk–Othmer (1985) *Concise Encyclopedia of Chemical Technology*, J. Wiley & Sons, New York.
12. Gupta, S.K. and Kumar, A. (1987) *Reaction Engineering of Step Growth Polymerization*, Plenum Press, New York.
13. Dunn Jr, L.B. (1994) in *Polymers from Agricultural Coproducts* (eds M.L. Fishman, R.B. Friedman and S.J. Huang), ACS Symposium Series 575, American Chemical Society, Washington.
14. Feldman, D. (1989) *Polymeric Building Materials*, Elsevier Applied Science, London.
15. Chanda, M. and Roy, S.K. (1987) *Plastics Technology Handbook*, Marcel Dekker, New York.
16. Elias, H.G. (1993) *An Introduction to Plastics*, VCH, Weinheim.
17. Daniels, C.A. (1989) *Polymers, Structures and Properties*, Technomic Publishing Co., Lancaster, p. 79.
18. Du Bois, J.H. and John, F.W. (1981) *Plastics*, 6th edn, Van Nostrand Reinhold Co., New York.
19. Seymour, R.B. and Carraher, C.E. (1990) *Giant Molecules*, J. Wiley & Sons, New York.
20. Wright, R.E. (1991) *Moulded Thermosets*, Hanser Publishers, Munich, pp. 25–33.
21. Brydson, J.A. (1989) *Plastics Materials*, 5th edn, Butterworths, London, pp. 605–621.
22. Whelan, T. and Goff, J. (1990) *Molding of Thermosetting Plastics*, Van Nostrand Reinhold, New York.
23. Wright, R.E. (1995) *Injection/Transfer Molding of Thermosetting Plastics*, Hanser Publishers, Munich.
24. Anon. (1981) *Mod. Plast. Int.*, Oct., 48–50.
25. Harland, C.E. (1994) *Ion Exchange, Theory and Practice*, 2nd edn, Royal Society of Chemistry, Letchworth, pp. 22–25.

Aminoplasts 10

10.1 INTRODUCTION

The successful use of poly(phenol–formaldehyde) catalyzed the development of other polycondensation products based on formaldehyde with urea and melamine, known as amino resins, aminoplasts or aminoplastics. Such polymers were first described by Tollens in 1884 and were patented by John in 1918 [1].

The term aminoplast is normally applied to polycondensation products of formaldehyde and functional amides and amidines such as urea (carbamide) and melamine (2,4,6-triamino-1,3,5-triazine) compounds.

It is of special interest that the very first commercial product based on amino polymers should be a complex formulation such as a molding compound. The reason that the more complicated products came first is that the amino molding compound was unique for its time. In contrast to the phenolics which offer only a limited range of dark colors, the aminoplasts could be supplied in pastel, translucent colors, all the brilliant colors of the spectrum. The USA's output of urea based aminoplasts was 856 000 tons in 1994, representing an 8.3% annual growth compared to 1993; for those based on melamine an output of 131 000 tons was recorded in 1994 and an annual growth of 5.6% compared to 1993 [2].

Formaldehyde addition to amino groups results in methylol compounds of urea:

$$HO-CH_2-NH-CO-NH_2$$
$$HO-CH_2-NH-CO-NH-CH_2OH$$

and melamine:

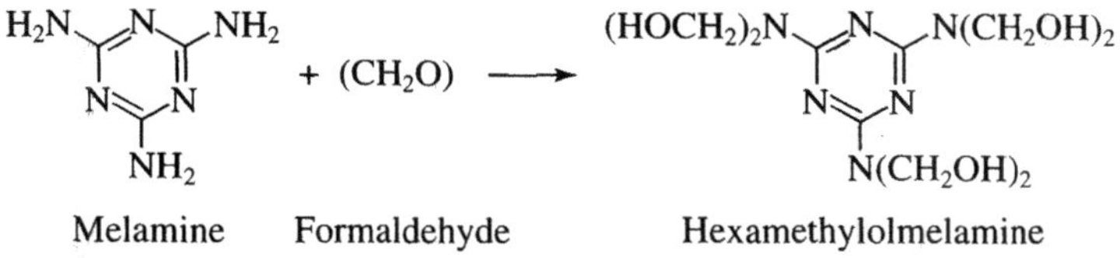

Melamine Formaldehyde Hexamethylolmelamine

Secondary amine groups (–NH–) either add additional formaldehyde or cure by reaction with this aldehyde and elimination of water to branched and finally crosslinked structures.

Urea is obtained at high pressure and temperature from ammonia and carbon dioxide and melamine by trimerization of cyanamide.

Benzoguanamine and acetoguanamine may be used in place of melamine for greater solubility in organic solvents and greater chemical stability.

10.2 UREA–FORMALDEHYDE PRODUCTION

The precise control of the course, speed and extent of the reaction is essential for successful production. The main parameters are the mole ratio of the initial components, the type of catalyst, duration and temperature. UF polymers are usually made by a batch process. Formalin is the most convenient to handle and it is the lowest in cost [3].

The flow chart (Figure 10.1) shows the manufacturing of the UF syrup normally used for molding compounds or other aminoplast products. This polycondensation takes place usually in two stages. In the first one, the

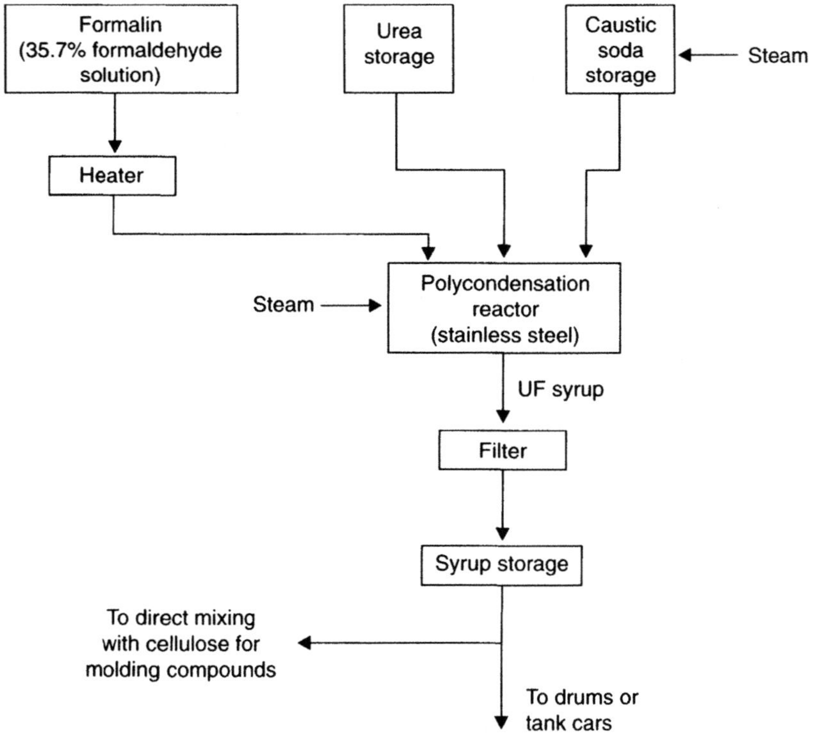

Figure 10.1 Production of urea–formaldehyde syrup.

components in a mole ratio of 1:1.3 to 1:1.5 urea to formaldehyde are reacted at a pH of 8 leading to monomethylol and dimethylol urea. In the second stage, this mixture, which contains unreacted urea and aldehyde, is subjected to acid conditions at higher temperature; the solution increases in viscosity and sets to an insoluble and irreversible gel; with further heating, this gel is transformed, with an excess of formaldehyde or with HMTA, to a hard, colorless, transparent, insoluble and infusible product having a crosslinked microstructure:

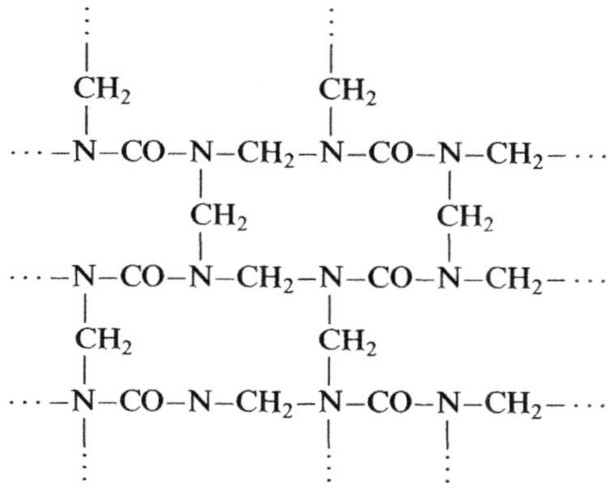

The basic chemistry of the UF polymers is the same as that for most applications, except foams must cure at room temperature within the first few minutes, as compared to adhesives or moldings which cure at a high temperature (150°C and above). In contrast to the above mentioned technology a butylated UF polymer for use in the formulation of fast curing baking enamels can be manufactured as follows. Urea, paraformaldehyde and butanol are charged in the mole ratio of 1.0 urea:2.12 formaldehyde:1.5 butanol. The alkaline pH is controlled by adding triethanolamine (about 1% of the weight of the urea); the mixture is heated until all the paraformaldehyde is dissolved. Phthalic anhydride is then added to produce a pH of 4.0 and the water is azeotroped off until the batch temperature reaches 117°C. The mixture is then cooled and diluted to the desired solid content [4].

10.3 MELAMINE–FORMALDEHYDE POLYMER PRODUCTION

The manufacturing of this polymer has been known since the mid 1930s. Under slightly alkaline conditions, melamine reacts with formaldehyde to give methylol derivatives with up to six methylol groups per molecule. When the mole ratio of formaldehyde to melamine is greater than 2:1, all the possible methylolmelamines have been shown to be produced in the reaction mixture. On further heating the methylolmelamines condense and a point is reached where the hydrophobic product separates but only on cooling.

The rate constants and the crosslinking strongly depend upon the pH, and formaldehyde concentration in the reaction mass. The process occurs over a wide range of pH. The crosslinking is at a minimum pH of about 10–10.5. Basic conditions promote further addition of formaldehyde until the functionality is satisfied. As opposed to this, acid conditions lead to a rapid polycondensation.

In the case of a pH above 9, there is a substantial loss of formaldehyde to form formic acid [5]. The final network structure may be represented as [6]:

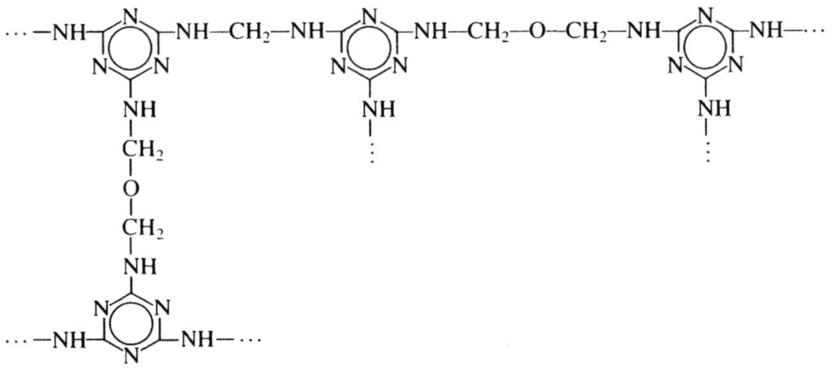

A melamine–formaldehyde polymer for use in molding compounds can be produced by using formalin (37% concentration) with melamine in the molar ratio of 2:1 formaldehyde to melamine. The reaction conditions should be neutral or slightly alkaline (pH 8–8.5), with a temperature of 60°C to 85°C. The process should be continued until some polymer is produced, polymer which can be detected through the turbidity of an aqueous sample. When proper conversion has been reached, the syrup obtained is pumped to a dough mixer where it is blended with α-cellulose in a ratio of 1 part of cellulose to 3 parts of polymer solids. The mass formed is dried to produce a hard, brittle popcorn-like intermediate. The molding compound is produced by applying heat and pressure to the mixture based on the cellulose–melamine intermediate, a stabilizer, a colorant, a mold-lubricant, etc.

Using a mole ratio of 6.5 to 10 formaldehyde to melamine and an initial pH of 9.0–9.5, in the presence of methanol, and after a few stages at different pHs and temperatures, a highly methylated melamine–formaldehyde textile treating product containing 60% solids can be obtained [4]. Such polymers can also be used for the treatment of paper and leather.

10.4 PROPERTIES

Urea–formaldehyde polymers are soluble in formaldehyde and concentrated sulfuric acid, and can be recovered from the latter unchanged. Studies done on the structure of cured polymer indicate the presence of crystalline domains and absence of water, respectively, both consistent with hydrogen bonded protein-like or polyamide-like close chain packing. The structure of UF polymers may thus resemble that of proteins more than that of PF. FTIR studies show that the true structure of UF polymers is a function of both feed ratio, pH and processing [7].

UF polymer has a specific gravity of 1.5, maximum continuous service temperature of 70°C, water absorption of 0.4–0.8% (of volume), it is self-extinguishing, and has good stability to acids, bases, solvents, oils and fats [8].

The melamine–formaldehyde prepolymers are soluble in water but can be made soluble in organic solvents by etherification with alcohols like 1-butanol. These aminoplasts are generally harder and more moisture resistant than UF polymers [7].

Melamine–formaldehyde polymer has a specific gravity of 1.5, maximum continuous service temperature of 80°C, water absorption of 0.1–0.6% (of volume), it is self-extinguishing, has good stability to solvents, oil and fats, but poor chemical resistance to acids and bases [8].

10.5 APPLICATIONS

UF polymers are used in applications similar to those of PF polymers, for molding powders, laminates, adhesives and foams. Melamine polymers are used in molding and laminating formulations, and like UF as alkyd enamel modifications and textile fiber finishes. Their largest use is in the manufacture of decorative dinnerware (tableware), table tops and coatings.

10.5.1 Molding compounds

The formulation for such products contains, besides UF, additives such as fiber, pigment, hardener (more commonly referred to as an accelerator), stabilizer, plasticizer and lubricant. As hardener (accelerator), an acid catalyst such as ammonium phenoxyacetate is used; in order to obtain sufficient rate of curing, usually about 0.2–2.0% is necessary. Stabilizer such as hexamine is added to improve the shelf life.

Commercial UF molding powders may differ in the following aspects: nature of the filler, ease of flow (depending on heating during drying and densifying), speed of cure, type of grind and presence of plasticizer.

UF moldings have a low cost, good electrical insulation properties and resistance to continuous heat up to 70°C.

Molding compounds made of melamine–formaldehyde powders are superior to those made of UF from the following points of view: lower water absorption, better resistance to staining, better maintained electrical characteristics in damp conditions and higher temperatures, better heat resistance and greater hardness [9,10].

Compared with UF polymers, melamine–formaldehyde polymers have a better color range, track resistance and scratch resistance, but are more expensive than similar, general purpose UF and PF products [11].

10.5.2 Adhesives

Almost 60% of UF products are used as adhesives for the particle board, plywood and furniture industries. UF polymers are used to produce the so called 'cold setting' adhesives and the various grades of powders and water solutions requiring heat for final bonding. Simple application methods have made these adhesives extremely important. The cold setting type has a catalyst incorporated in the powder, which does not become active until water is added. The use of a catalyst makes it possible to use this adhesive at a temperature of 24°C, but it should not be used at lower temperatures, as the final cure is too slow.

It is not uncommon for UF adhesives to contain a modifier, such as benzyl-alcohol or furfural, which is crosslinked and thus able to relieve stress imposed by shrinkage due to cure and loss of water. UF adhesives may also be mixed with PVAc latex [12].

UF adhesives have been shown to be emitters of formaldehyde, a noxious chemical and potential carcinogen [13–16].

Although superior properties are obtained through the use of melamine–formaldehyde adhesives, those used in the largest volume are UF and combinations of UF with melamine–formaldehyde polymers.

Melamine-like PF adhesives have excellent water resistance. UF adhesives are lower in cost, but somewhat more water sensitive. Often the best all-round balance between cost and performance is achieved with blends of these two kinds of adhesive.

10.5.3 UF foam

Manufacturing of UF foam implies, in addition to the principal polymer, the use of additives such as surfactants, catalyst, water, foaming agents, etc. The surfactant plays an important role in increasing the hydrophobicity of the foam with a view to better resisting the hydrolysis process. Generally these additives are aromatic sodium salts of sulfonic acids.

In cold settings the pH must be lower than in hot setting adhesives. It is usually desirable to have a pH of about 1–2 which may be adjusted with an acidic catalyst such as phosphoric acid. Great care must also be taken to keep the use of acid to a minimum because any excess will hydrolyze the finished foam. The hydrolysis is the most important cause of formaldehyde odor, from older, fully cured foam. Excess acid can cause an initially normal looking foam to crumble into powder after some time.

The water hardness influences foam shrinkage and also the effectiveness of the surfactant. Before preparing the final ready-to-foam and catalyst mixtures, the water hardness must be adjusted with a suitable softener.

In bulk insulating foams, air is used as the foaming gas, yielding 99% open cell structures. If CFC (chlorinated, fluorinated hydrocarbon) or other foaming agents are used, the cell structure can change, yielding predominantly closed cells.

In the application of UF foam, its compatibility with adjoining materials must be considered. UF foam can be easily peeled from PVC, PE, PS, polyesters and polyisobutylene and behaves as an inert material from the chemical point of view. It sticks to butyl rubber foils, but peels off asphalt papers used in roofing or foundation water proofing.

Other additives have been included to modify and tailor make urea–formaldehyde foam structures to yield specific characteristics, such as very low density ($11.2\,kg/m^3$, $0.7\,lb/ft^3$), flexibility, etc. More flexible structures have been made with urea–formaldehyde condensates with resorcinol. Urea–formaldehyde foams modified with alpha, omega-bis-epoxies are reported to be less friable and more impact resistant at low densities. Very fine textured low density foams were produced by using hexamine and melamine.

Because factory made foam slabs are easily damaged during transportation and occupy excessive and uneconomical shipping volume, UF foams are produced on site from portable foaming equipment. A good foam sets within the first minute, when it can be sliced if necessary. It hardens fully within 1 day as the residual water vaporizes from the cured foam. If the foam dries too quickly, it tends to warp, shrink or crack. Home insulating foam in closed spaces, on occasions, retains moisture for several years, especially in cold climates.

Urea–formaldehyde is the lightest of the insulating foams; its density is $10–14\,kg/m^3$ ($0.6–0.9\,lb/ft^3$). It has virtually no mechanical strength even in compression, is white in color and does not support combustion; it is very friable. Typical products have a closed cell content of about 80%. UF foams have a very low thermal conductivity with a K value of 0.022–0.029 $W/m \cdot K$ ($0.15–0.20$ Btu $in/ft^2 \cdot h^1 \cdot °F^1$) thus comparing very favorably with other insulating materials.

UF foam deteriorates continuously at a moderate to rapid rate; thus it has a short service life as compared with other building materials. The rate of deterioration depends on the conditions to which it is exposed. In residential applications, the factors most likely to accelerate degradation are high temperatures and high humidities [17,18]. Temperatures such as those encountered in a roof space on a summer day cause rapid deterioration. High humidity resulting from the flow of warm, humid air from the living spaces into insulated cavities also causes accelerated degradation. UF foams have higher water absorption and a high water vapor permeability.

Deterioration of the foams leads to a breakage of the cell walls and shrinkage of the insulation, decreasing the foam's ability to resist heat and air flow. It has a tendency to undergo considerable shrinkage upon curing and drying. Formaldehyde gas is released and may be carried into the living space by air infiltration and, at a slower rate, can diffuse through the wall materials. About 20% of the population will be affected by a concentration below 0.25 ppm. There is also a lot of evidence that even at 0.1 ppm a significant number of persons might react adversely to formaldehyde [19].

Formaldehyde has a pungent odor at high concentrations but a less definite odor at low concentrations. Most people can detect the presence of a gas concentration of 0.1 ppm; few can detect it at levels below 0.05 ppm. Based on available animal and human data, the EPA (Environmental Protection Agency) decided to classify formaldehyde as a probable human carcinogen [15].

REFERENCES

1. Seymour, R.B. (1989) in *Pioneers in Polymer Science* (ed. R.B. Seymour), Kluwer Academic Publishers, Dordrecht, p. 88.
2. Kirschner, E.M. (1995) *Chem. Eng. News*, **73**(15), 21.
3. Kirk–Othmer (1985) *Concise Encyclopedia of Polymer Science*, J. Wiley & Sons, New York, pp. 90–92.
4. Williams, L.L., Updergraff, J.H. and Petropoulos, J.C. (1995) in *Applied Polymer Science*, 2nd edn (eds R.W. Tess and G.W. Poehlein), ACS Symposium Series 285, American Chemical Society, Washington, pp. 1101–1115.
5. Gupta, S.K. and Kumar, A. (1987) *Reaction Engineering of Step Growth Polymerization*, Plenum Press, New York, p. 389.
6. Critchley, J.P., Knight, G.J. and Wright, W.W. (1983) *Heat Resistant Polymers*, Plenum Press, New York.
7. Stevens, M.P. (1990) *Polymer Chemistry*, 2nd edn, Oxford University Press, New York, pp. 481–484.
8. Daniels, C.A. (1989) *Polymers: Structure and Properties*, Technomic Publishing Co., Lancaster.
9. Brydson, J.A. (1989) *Plastics Materials*, 5th edn, Butterworths, London, pp. 627–651.
10. Seymour, R.B. and Carraher Jr, C.F. (1992) *Polymer Chemistry*, 3rd edn, Marcel Dekker, New York, pp. 231–234.
11. Chanda, M. and Roy, S.K. (1987) *Plastic Technology Handbook*, Marcel Dekker, New York, pp. 411–419.
12. Feldman, D. (1989) *Polymeric Building Materials*, Elsevier Applied Science, London.
13. Tabor, R.G. (1988) *Comments Toxicology*, **2**(3), 191–200.

14. Hanson, D.J. (1986) *Chem. Eng. News*, 22 September, 17–18.
15. Anon. (1987) *Chem. Eng. News*, 27 April, 18.
16. Myers, G.E. (1986) *Forest Products Journal*, **36**(6), 41–51.
17. Brown, S.K. (1990) *Polymer Degradation and Stability*, **27**, 121–143.
18. Myers, G.E. (1982) *Wood Science*, **15**(2), 127–138.
19. Hanson, D.J. (1982) *Chem. Eng. News*, 29 March, 34–37.

Index